Fundamentals of Linear Control

A Concise Approach

Taking a different approach from standard thousand-page reference-style control textbooks, *Fundamentals of Linear Control* provides a concise yet comprehensive introduction to the analysis and design of feedback control systems in fewer than 300 pages.

The text focuses on classical methods for dynamic linear systems in the frequency domain. The treatment is, however, modern and the reader is kept aware of contemporary tools and techniques, such as state-space methods and robust and nonlinear control.

Featuring fully worked design examples, richly illustrated chapters, and an extensive set of homework problems and examples spanning across the text for gradual challenge and perspective, this textbook is an excellent choice for senior-level courses in systems and control or as a complementary reference in introductory graduate-level courses. The text is designed to appeal to a broad audience of engineers and scientists interested in learning the main ideas behind feedback control theory.

Maurício C. de Oliveira is Adjunct Professor of Dynamic Systems and Control in the Department of Mechanical and Aerospace Engineering at the University of California, San Diego.

Fundamentals of Linear Control

A Concise Approach

MAURÍCIO C. DE OLIVEIRA
University of California, San Diego

CAMBRIDGE
UNIVERSITY PRESS

University Printing House, Cambridge CB2 8BS, United Kingdom

One Liberty Plaza, 20th Floor, New York, NY 10006, USA

477 Williamstown Road, Port Melbourne, VIC 3207, Australia

4843/24, 2nd Floor, Ansari Road, Daryaganj, Delhi - 110002, India

79 Anson Road, #06-04/06, Singapore 079906

Cambridge University Press is part of the University of Cambridge.

It furthers the University's mission by disseminating knowledge in the pursuit of education, learning, and research at the highest international levels of excellence.

www.cambridge.org
Information on this title: www.cambridge.org/9781107187528
DOI: 10.1017/9781316941409

© Maurício C. de Oliveira 2017

This publication is in copyright. Subject to statutory exception and to the provisions of relevant collective licensing agreements, no reproduction of any part may take place without the written permission of Cambridge University Press.

First published 2017

Printed in the United Kingdom by Clays, St Ives plc

A catalog record for this publication is available from the British Library.

Library of Congress Cataloging-in-Publication Data
Names: Oliveira, Mauricio C. de, author.
Title: Fundamentals of linear control : a concise approach / Maurbicio de Oliveira, University of California, San Diego.
Description: Cambridge, United Kingdom ; New York, NY, USA : Cambridge University Press, [2017] | Includes bibliographical references and index.
Identifiers: LCCN 2016052325 | ISBN 9781107187528
Subjects: LCSH: Linear control systems.
Classification: LCC TJ220 .O45   2017 | DDC 629.8/32 – dc23
LC record available at https://lccn.loc.gov/2016052325

ISBN 978-1-107-18752-8 Hardback

Cambridge University Press has no responsibility for the persistence or accuracy of URLs for external or third-party internet websites referred to in this publication, and does not guarantee that any content on such websites is, or will remain, accurate or appropriate.

To Beatriz and Victor

Contents

	Preface	*page* xi
	Overview	xiii
1	**Introduction**	1
	1.1 Models and Experiments	1
	1.2 Cautionary Note	5
	1.3 A Control Problem	6
	1.4 Solution without Feedback	6
	1.5 Solution with Feedback	7
	1.6 Sensitivity	10
	1.7 Disturbances	11
	Problems	13
2	**Dynamic Systems**	17
	2.1 Dynamic Models	18
	2.2 Block-Diagrams for Differential Equations	19
	2.3 Dynamic Response	20
	2.4 Experimental Dynamic Response	22
	2.5 Dynamic Feedback Control	24
	2.6 Nonlinear Models	27
	2.7 Disturbance Rejection	30
	2.8 Integral Action	33
	Problems	35
3	**Transfer-Function Models**	47
	3.1 The Laplace Transform	47
	3.2 Linearity, Causality, and Time-Invariance	52
	3.3 Differential Equations and Transfer-Functions	55
	3.4 Integration and Residues	57
	3.5 Rational Functions	63
	3.6 Stability	69
	3.7 Transient and Steady-State Response	71
	3.8 Frequency Response	73

Contents

3.9	Norms of Signals and Systems	75
	Problems	78

4 Feedback Analysis — 91

4.1	Tracking, Sensitivity, and Integral Control	93
4.2	Stability and Transient Response	97
4.3	Integrator Wind-up	102
4.4	Feedback with Disturbances	105
4.5	Input-Disturbance Rejection	106
4.6	Measurement Noise	111
4.7	Pole–Zero Cancellations and Stability	111
	Problems	117

5 State-Space Models and Linearization — 126

5.1	Realization of Dynamic Systems	126
5.2	State-Space Models	133
5.3	Minimal State-Space Realizations	139
5.4	Nonlinear Systems and Linearization	141
5.5	Simple Pendulum	143
5.6	Pendulum in a Cart	146
5.7	Car Steering	148
5.8	Linear Control of Nonlinear Systems	150
	Problems	154

6 Controller Design — 165

6.1	Second-Order Systems	165
6.2	Derivative Action	172
6.3	Proportional–Integral–Derivative Control	174
6.4	Root-Locus	175
6.5	Control of the Simple Pendulum – Part I	183
	Problems	192

7 Frequency Domain — 201

7.1	Bode Plots	201
7.2	Non-Minimum-Phase Systems	213
7.3	Polar Plots	216
7.4	The Argument Principle	218
7.5	Stability in the Frequency Domain	223
7.6	Nyquist Stability Criterion	227
7.7	Stability Margins	235

		7.8 Control of the Simple Pendulum – Part II	238
		Problems	246
8	**Performance and Robustness**		255
	8.1	Closed-Loop Stability and Performance	255
	8.2	Robustness	261
	8.3	Small Gain	262
	8.4	Control of the Simple Pendulum – Part III	266
	8.5	Circle Criterion	271
	8.6	Feedforward Control and Filtering	278
		Problems	284
	References		293
	Index		295

Preface

The book you have in your hands grew out of a set of lecture notes scribbled down for MAE 143B, the senior-level undergraduate *Linear Control* class offered by the Department of Mechanical and Aerospace Engineering at the University of California, San Diego.

The focus of the book is on classical methods for analysis and design of feedback systems that take advantage of the powerful and insightful representation of dynamic linear systems in the frequency domain. The required mathematics is introduced or revisited as needed. In this way the text is made mostly self-contained, with accessory work shifted occasionally to homework problems.

Key concepts such as tracking, disturbance rejection, stability, and robustness are introduced early on and revisited throughout the text as the mathematical tools become more sophisticated. Examples illustrate graphical design methods based on the root-locus, Bode, and Nyquist diagrams. Whenever possible, without straying too much from the classical narrative, the reader is made aware of contemporary tools and techniques such as state-space methods, robust control, and nonlinear systems theory.

With so much to cover in the way of insightful engineering *and* relevant mathematics, I tried to steer clear of the curse of the engineering systems and control textbook: becoming a treatise with 1000 pages. The depth of the content exposed in fewer than 300 pages is the result of a compromise between my utopian goal of *at most* 100 pages on the one hand and the usefulness of the work as a reference and, I hope, inspirational textbook on the other. Let me know if you think I failed to deliver on this promise.

I shall be forever indebted to the many students, teaching assistants, and colleagues whose exposure to earlier versions of this work helped shape what I am finally not afraid of calling the *first* edition. Special thanks are due to Professor Reinaldo Palhares, who diligently read the original text and delighted me with an abundance of helpful comments.

I would like to thank Sara Torenson from the UCSD Bookstore, who patiently worked with me to make sure earlier versions were available as readers for UCSD students, and Steven Elliot from Cambridge University Press for his support in getting this work to a larger audience.

Maurício de Oliveira
San Diego, California

Overview

This book is designed to be used in a quarter- or semester-long senior-level undergraduate linear control systems class. Readers are assumed to have had some exposure to differential equations and complex numbers (good references are [BD12] and [BC14]), and to have some familiarity with the engineering notion of signals and systems (a standard reference is [Lat04]). It is also assumed that the reader has access to a high-level software program, such as MATLAB, to perform calculations in many of the homework problems. In order to keep the focus on the content, examples in the book do not discuss MATLAB syntax or features. Instead, we provide supplementary MATLAB files which can produce all calculations and figures appearing in the book. These files can be downloaded from http://www.cambridge.org/deOliveira.

Chapters 1 and 2 provide a quick overview of the basic concepts in control, such as feedback, tracking, dynamics, disturbance rejection, integral action, etc. Math is kept at a very basic level and the topics are introduced with the help of familiar examples, such as a simplistic model of a car and a toilet bowl.

Chapter 3 formalizes the concept of a transfer-function for dynamic linear system models. Its first part is a review of the Laplace transform and its application to linear ordinary differential equations. The second part introduces systems concepts such as stability, transient and steady-state response, and the frequency response method. Some topics, e.g. complex integration, the calculus of residues, and norms of signals and systems, are covered in more depth than is usually found in typical introductory courses, and can be safely skipped at first read.

Equipped with the concept of a transfer-function, Chapter 4 formalizes fundamental concepts in feedback analysis, such as tracking, sensitivity, asymptotic and internal stability, disturbance rejection, measurement noise, etc. Homework problems in this chapter expose readers to these concepts and anticipate the more sophisticated analytic methods to be introduced in the following chapters.

Chapter 5 takes a slight detour from classic methods to introduce the reader to state-space models. The focus is on practical questions, such as realization of dynamic systems and controllers, linearization of nonlinear systems, and basic issues that arise when using linear controllers with nonlinear systems. It is from this vantage point that slightly more complex dynamic systems models are introduced, such as a simple pendulum and a pendulum in a cart, as a well as a simplified model of a steering car. The simple pendulum model is used in subsequent chapters as the main illustrative example.

Table I.1 Homework problems classified by theme per chapter

Problem theme	Ch. 1	Ch. 2	Ch. 3	Ch. 4	Ch. 5	Ch. 6	Ch. 7	Ch. 8
DC motor		2.41–2.46	3.95–3.99	4.29–4.37	5.30–5.33	6.30–6.33	7.29–7.31	8.41–8.43
Elevator		2.18–2.26	3.62–3.70	4.23–4.28	5.19–5.22	6.14–6.15	7.15–7.16	8.37
Free-fall		2.4–2.7	3.53		5.8–5.13			
Inclined plane	1.10							
Insulin homeostasis					5.49–5.52	6.38–6.40	7.36–7.37	
Mass–spring–damper		2.27–2.33	3.71–3.79		5.23–5.27	6.16–6.18	7.17–7.19	8.38–8.40
Modulator			3.50					
One-eighth-car model						6.19–6.23	7.20–7.23	
One-quarter-car model						6.24–6.29	7.24–7.28	
OpAmp circuit		2.38–2.40	3.90–3.94		5.28–5.29			
Orbiting satellite					5.41–5.45	6.37	7.35	8.47
Pendulum in a cart					5.6			
Population dynamics					5.46–5.48			
RC circuit		2.34–2.35	3.80–3.84		5.39–5.40			
Rigid body								
RLC circuit		2.36–2.37	3.85–3.89					
Rotating machine		2.10–2.17	3.54–3.61	4.20–4.22	5.15–5.18	6.11–6.13	7.12–7.14	8.35–8.36
Sample-and-hold			3.51–3.52					
Simple pendulum						6.9–6.10		
Sky-diver		2.8–2.9			5.14			
Smith predictor				4.11–4.14				
Steering car					5.7			
Water heater		2.49–2.56		4.38–4.43	5.36–5.38	6.34–6.36	7.32–7.34	8.44–8.46
Water tank			3.100–3.104		5.34–5.35			

Chapter 6 takes the reader back to the classic path with an emphasis on control design. Having flirted with second-order systems many times before in the book, the chapter starts by taking a closer look at the time-response of second-order systems and associated performance metrics, followed by a brief discussion on derivative action and the popular proportional–integral–derivative control. It then introduces the root-locus method and applies it to the design of a controller with integral action to the simple pendulum model introduced in the previous chapter.

Chapter 7 brings a counterpoint to the mostly time-domain point of view of Chapter 6 by focusing on frequency-domain methods for control design. After introducing Bode and polar plots, the central issue of closed-loop stability is addressed with the help of the Nyquist stability criterion. The same controller design problem for the simple pendulum is revisited, this time using frequency-domain tools.

An introductory discussion on performance and robustness is the subject of the final chapter, Chapter 8. Topics include Bode's sensitivity integral, robustness analysis using small gain and the circle criterion, and feedforward control and filtering. Application of some of these more advanced tools is illustrated by certifying the performance of the controllers designed for the simple pendulum beyond the guarantees offered by local linearization.

In a typical quarter schedule, with 20 or 30 lectures, the lightweight Chapters 1 and 2 can be covered rather quickly, serving both as a way to review background material and as a means to motivate the reader for the more demanding content to come. Instructors can choose to spend more or less time on Chapter 3 depending on the prior level of comfort with transfer-functions and frequency response and the desired depth of coverage.

Homework problems at the end of Chapters 1 through 3 introduce a variety of examples from various engineering disciplines that will appear again in the following chapters and can be used as effective tools to review background material.

Chapters 4 through 7 constitute the core material of the book. Chapters 5 and 7, especially, offer many opportunities for instructors to select additional topics for coverage in class or relegate to reading, such as discussions on nonlinear analysis and control, a detailed presentation of the argument principle, and more unorthodox topics such as non-minimum-phase systems and stability analysis of systems with delays.

The more advanced material in Chapter 8 can be covered, time permitting, or may be left just for the more interested reader without compromising a typical undergraduate curriculum.

This book contains a total of almost 400 homework problems that appear at the end of each chapter, with many problems spanning across chapters. Table I.1 on page xiv provides an overview of select problems grouped by their motivating theme. Instructors may choose to follow a few of these problems throughout the class. As mentioned previously, many of the problems require students to use MATLAB or a similar computer program. The supplementary MATLAB files provided with this book are a great resource for readers who need to develop their programming skills to tackle these problems.

1 Introduction

In controls we make use of the abstract concept of a *system*: we identify a phenomenon or a process, the *system*, and two classes of *signals*, which we label as *inputs* and *outputs*. A signal is something that can be measured or quantified. In this book we use real numbers to quantify signals. The classification of a particular signal as an input means that it can be identified as the *cause* of a particular system behavior, whereas an output signal is seen as the *product* or *consequence* of the behavior. Of course the classification of a phenomenon as a system and the labeling of input and output signals is an abstract construction. A mathematical description of a system and its signals is what constitutes a *model*. The entire abstract construction, and not only the equations that we will later associate with particular signals and systems, is the model.

We often represent the relationship between a system and its input and output signals in the form of a *block-diagram*, such as the ones in Fig. 1.1 through Fig. 1.3. The diagram in Fig. 1.1 indicates that a system, G, produces an output signal, y, in the presence of the input signal, u. Block-diagrams will be used to represent the interconnection of systems and even algorithms. For example, Fig. 1.2 depicts the components and signals in a familiar controlled system, a water heater; the block-diagram in Fig. 1.3 depicts an algorithm for converting temperature in degrees Fahrenheit to degrees Celsius, in which the output of the circle in Fig. 1.3 is the algebraic sum of the incoming signals with signs as indicated near the incoming arrows.

1.1 Models and Experiments

Systems, signals, and models are often associated with concrete or abstract experiments. A model reflects a particular setup in which the outputs appear *correlated* with a prescribed set of inputs. For example, we might attempt to model a car by performing the following experiment: on an unobstructed and level road, we depress the accelerator pedal and let the car travel in a straight line.[1] We keep the pedal excursion constant and let the car reach constant velocity. We record the amount the pedal has been depressed and the car's terminal velocity. The results of this experiment, repeated multiple times with different amounts of pedal excursion, might look like the data shown in Fig. 1.4. In this experiment the signals are

---

[1] This may bring to memory a bad joke about physicists and spherical cows...

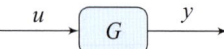

Figure 1.1 System represented as a block-diagram; u is the input signal; y is the output signal; y and u are related through $y = G(u)$ or simply $y = Gu$.

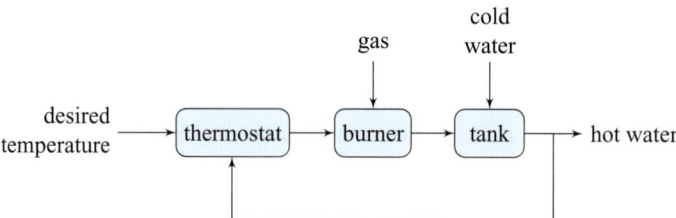

Figure 1.2 Block-diagram of a controlled system: a gas water heater; the blocks thermostat, burner, and tank, represent components or sub-systems; the arrows represent the *flow* of input and output signals.

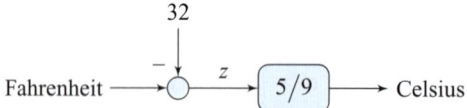

Figure 1.3 Block-diagram of an algorithm to convert temperatures in Fahrenheit to Celsius: Celsius = 5/9(Fahrenheit − 32); the output of the circle block is the algebraic sum of the incoming signals with the indicated sign, i.e. z = Fahrenheit − 32.

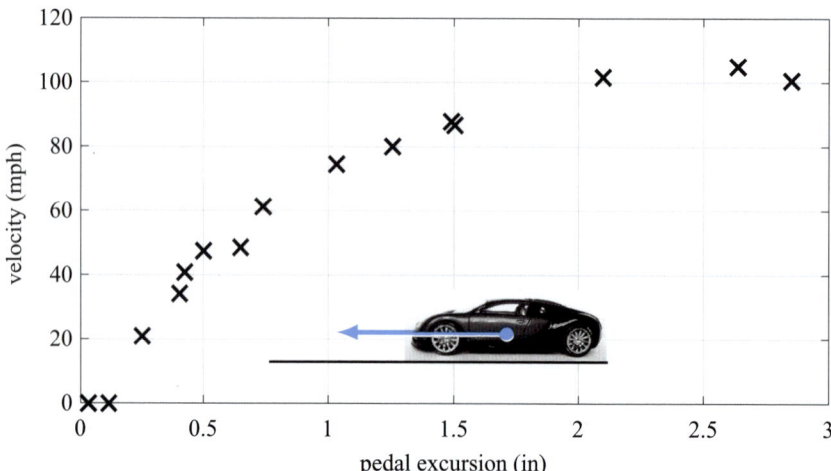

Figure 1.4 Experimental determination of the effect of pressing the gas pedal on the car's terminal velocity; the pedal excursion is the input signal, u, and the car's terminal velocity is the output signal, y.

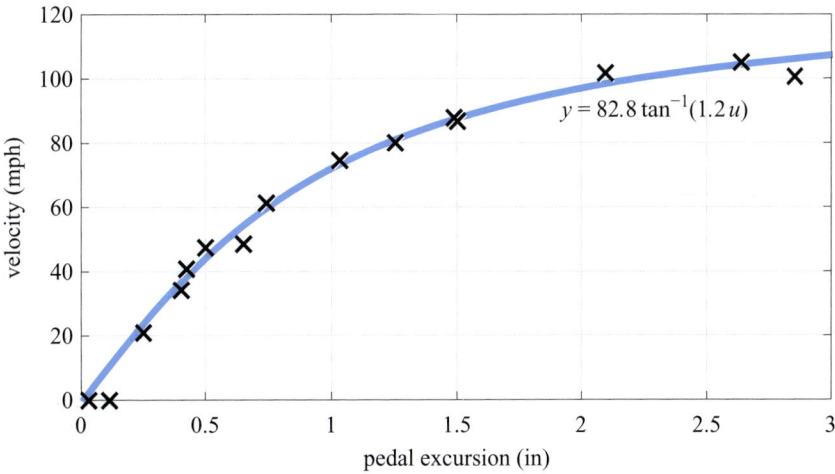

Figure 1.5 Fitting the curve $y = \alpha \tan^{-1}(\beta u)$ to the data from Fig. 1.4.

input: pedal excursion, in cm, inches, etc.;
output: terminal velocity of the car, in m/s, mph, etc.

The **system** is the car *and* the particular conditions of the experiment. The data captures the fact that the car does not move at all for small pedal excursions and that the terminal velocity *saturates* as the pedal reaches the end of its excursion range.

From Fig. 1.4, one might try to *fit* a particular mathematical function to the experimental data[2] in hope of obtaining a *mathematical model*. In doing so, one invariably loses something in the name of a simpler description. Such trade-offs are commonplace in science, and it should be no different in the analysis and design of control systems. Figure 1.5 shows the result of fitting a curve of the form

$$y = \alpha \tan^{-1}(\beta u),$$

where u is the input, pedal excursion in inches, and y is the output, terminal velocity in mph. The parameters $\alpha = 82.8$ and $\beta = 1.2$ shown in Fig. 1.5 were obtained from a standard least-squares fit. See also P1.11.

The choice of the above particular function involving the arc-tangent might seem somewhat arbitrary. When possible, one should select candidate functions from first principles derived from physics or other scientific reasoning, but this does not seem to be easy to do in the case of the experiment we described. Detailed physical modeling of the vehicle would involve knowledge and further modeling of the components of the vehicle, not to mention the many uncertainties brought in by the environment, such as wind, road conditions, temperature, etc. Instead, we make an "educated choice" based on certain physical aspects of the experiment that we believe the model should capture. In this case, from our daily experience with vehicles, we expect that the terminal velocity

---

[2] All data used to produce the figures in this book is available for download from the website http://www.cambridge.org/deOliveira.

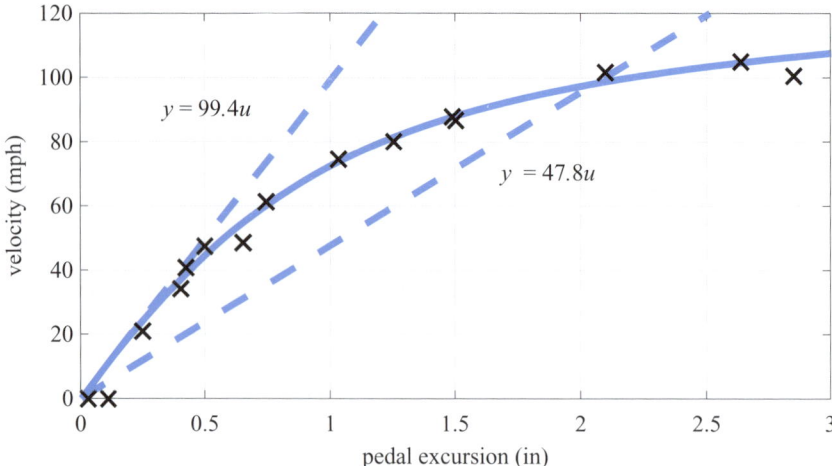

Figure 1.6 Linear mathematical models of the form $y = \gamma u$ for the data in Fig. 1.4 (dashed); the model with $\gamma = 47.8$ was obtained by a least-squares fit; the model with $\gamma = 99.4$ was obtained after linearization of the nonlinear model (solid) obtained in Fig. 1.5; see P1.12 and P1.11.

will eventually *saturate*, either as one reaches full throttle or as a result of limitations on the maximum power that can be delivered by the vehicle's powertrain. We also expect that the function be *monotone*, that is, the more you press the pedal, the larger the terminal velocity will be. Our previous exposure to the properties of the arc-tangent function *and* engineering intuition about the expected outcome of the experiment allowed us to successfully select this function as a suitable candidate for a model.

Other families of functions might suit the data in Fig. 1.5. For example, we could have used *polynomials*, perhaps constrained to pass through the origin and ensure monotonicity. One of the most useful classes of mathematical models one can consider is that of *linear models*, which are, of course, first-order polynomials. One might be tempted to equate linear with simple. Whether or not this might be true in some cases, simplicity is far from a sin. More often than not, the loss of some feature neglected by a linear model is offset by the availability of a much broader set of analytic tools. It is better to *know* when you are wrong than to *believe* you are right. As the title suggests, this book is mostly concerned with linear models. Speaking of linear models, one might propose describing the data in Fig. 1.4 by a linear mathematical model of the form

$$y = \gamma u. \tag{1.1}$$

Figure 1.6 shows two such models (dashed lines). The curve with slope coefficient $\gamma = 47.8$ was obtained by performing a least-squares fit to all data points (see P1.11). The curve with coefficient $\gamma = 99.4$ is a first-order approximation of the nonlinear model calculated in Fig. 1.5 (see P1.12). Clearly, each model has its limitations in describing the experiment. Moreover, one model might be better suited to describe certain aspects of the experiment than the other. Responsibility rests with the engineer or the scientist to select the model, or perhaps set of models, that better fits the problem in hand, a task that at times may resemble an art more than a science.

1.2 Cautionary Note

It goes without saying that the mathematical models described in Section 1.1 do not purport to capture every detail of the experiment, not to mention reality. Good models are the ones that capture *essential* aspects that we *perceive* or can experimentally validate as real, for example how the terminal velocity of a car responds to the acceleration pedal in the given experimental conditions. A model does not even need to be *correct* to be useful: for centuries humans used[3] a model in which the sun revolves around the earth to predict and control their days! What is important is that models provide a way to express *relevant* aspects of reality using mathematics. When mathematical models are used in control design, it is therefore with the understanding that the model is bound to capture only a subset of features of the actual phenomenon they represent. At no time should one be fooled into *believing* in a model. The curious reader will appreciate [Fey86] and the amusingly provocative [Tal07].

With this caveat in mind, it is useful to think of an idealized *true* or *nominal model*, just as is done in physics, against which a particular setup can be *mathematically* evaluated. This nominal model might even be different than the model used by a particular control algorithm, for instance, having more details or being more complex or more accurate. Of course *physical* evaluation of a control system with respect to the underlying natural phenomenon is possible only by means of experimentation which should also include the physical realization of the controller in the form of computer hardware and software, electric circuits, and other necessary mechanical devices. We will discuss in Chapter 5 how certain physical devices can be used to implement the dynamic controllers you will learn to design in this book.

The models discussed so far have been *static*, meaning that the relationship between inputs and outputs is *instantaneous* and is independent of the past history of the system or their signals. Yet the main objective of this book is to work with *dynamic* models, in which the relationship between present inputs and outputs may depend on the present and past history[4] of the signals.

With the goal of introducing the main ideas behind feedback control in a simpler setup, we will continue to work with static models for the remainder of this chapter. In the case of static models, a mathematical *function* or a set of *algebraic equations* will be used to represent such relationships, as done in the models discussed just above in Section 1.1.

Dynamic models will be considered starting in Chapter 2. In this book, signals will be continuous functions of time, and dynamic models will be formulated with the help of *ordinary differential equations*. As one might expect, experimental procedures that can estimate the parameters of dynamic systems need to be much more sophisticated than the ones discussed so far. A simple experimental procedure will be briefly discussed in Section 2.4, but the interested reader is encouraged to consult one of the many excellent works on this subject, e.g. [Lju99].

---

[3] Apparently 1 in 4 Americans and 1 in 3 Europeans still go by that model [Gro14].
[4] What about the future?

1.3 A Control Problem

Consider the following problem:

Under the experimental conditions described in Section 1.1 and given a target terminal velocity, $\bar{y}$, is it possible to design a system, the controller, that is able to command the accelerator pedal of a car, the input, u, to produce a terminal velocity, the output, y, equal to the target velocity?

An *automatic* system that can solve this problem is found in many modern cars, with the name *cruise controller*. Of course, another *system* that is capable of solving the same problem is a human driver.[5] In this book we are mostly interested in solutions that can be implemented as an *automatic control*, that is, which can be performed by some combination of mechanical, electric, hydraulic, or pneumatic systems running without human intervention, often being programmed in a digital computer or some other logical circuit or calculator.

Problems such as this are referred to in the control literature as *tracking* problems: the controller should make the system, a car, *follow* or *track* a given target output, the desired terminal velocity. In the next sections we will discuss two possible approaches to the cruise control problem.

1.4 Solution without Feedback

The role of the controller in tracking is to compute the input signal u which produces the desired output signal y. One might therefore attempt to solve a tracking problem using a system (controller) of the form

$$u = K(\bar{y}).$$

This controller can use only the reference signal, the target output $\bar{y}$, and is said to be in *open-loop*,[6] as the controller output signal, u, is not a function of the system output signal, y.

With the intent of analyzing the proposed solution using mathematical models, assume that the car can be represented by a *nominal model*, say G, that relates the input u (pedal excursion) to the output y (terminal velocity) through the mathematical function

$$y = G(u).$$

The connection of the controller with this idealized model is depicted in the block-diagram in Fig. 1.7. Here the function G can be obtained after fitting experimental data as done in Figs. 1.5 and 1.6, or borrowed from physics or engineering science principles.

---

[5] After some 16 years of *learning*.
[6] As opposed to *closed-loop*, which will be discussed in Section 1.5.

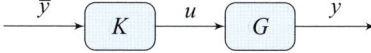

Figure 1.7 Open-loop control: the controller, K, is a function of the reference input, $\bar{y}$, but not a function of the system output, y.

The block-diagram in Fig. 1.7 represents the following relationships:
$$y = G(u), \qquad u = K(\bar{y}),$$
that can be combined to obtain
$$y = G(K(\bar{y})).$$
If G is *invertible* and K is chosen to be the inverse of G, that is $K = G^{-1}$, then
$$y = G(G^{-1}(\bar{y})) = \bar{y}.$$
Matching the controller, K, with the nominal model, G, is paramount: if $K \neq G^{-1}$ then $y \neq \bar{y}$.

When both the nominal model G and the controller K are linear,
$$y = Gu, \qquad u = K\bar{y}, \qquad y = GK\bar{y},$$
from which $\bar{y} = y$ only if the product of the *constants* K and G is equal to one:
$$KG = 1 \quad \implies \quad K = G^{-1}, \qquad u = G^{-1}\bar{y}.$$
Because the control law relies on knowledge of the nominal model G to achieve its goal, any imperfection in the model or in the implementation of the controller will lead to less than perfect tracking.

1.5 Solution with Feedback

The controller in the open-loop solution considered in Section 1.4 is allowed to make use only of the target output, $\bar{y}$. When a measurement, even if imprecise, of the system output is available, one may benefit from allowing the controller to make use of the measurement signal, y. In the case of the car cruise control, the terminal velocity, y, can be measured by an on-board speedometer. Of course the target velocity, $\bar{y}$, is set by the driver.

Controllers that make use of output signals to compute the control inputs are called *feedback controllers*. In its most general form, a feedback controller has the functional form
$$u = K(\bar{y}, y).$$
In practice, most feedback controllers work by first creating an *error signal*, $\bar{y} - y$, which is then used by the controller:
$$u = K(e), \qquad e = \bar{y} - y. \tag{1.2}$$

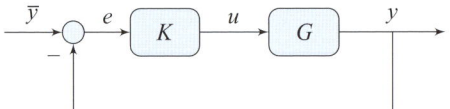

Figure 1.8 Closed-loop feedback control: the controller, K, is a function of the reference input, $\bar{y}$, and the system output, y, by way of the error signal, $e = \bar{y} - y$.

This scheme is depicted in the block-diagram in Fig. 1.8. One should question whether it is possible to implement a physical system that replicates the block-diagram in Fig. 1.8. In this diagram, the measurement, y, that takes part in the computation of the control, u, in the controller block, K, is the same as that which comes out of the system, G. In other words, the signals flow in this diagram is *instantaneous*. Even though we are not yet properly equipped to address this question, we anticipate that it will be possible to construct and analyze *implementable* or *realizable* versions of the feedback diagram in Fig. 1.8 by taking into account dynamic phenomena, which we will start discussing in the next chapter.

At this point, we are content to say that if the computation implied by feedback is performed *fast enough*, then the scheme *should* work. We analyze the proposed feedback solution only in the case of static linear models, that is, when both the controller, K, and the system to be controlled, G, are linear. Feedback controllers of the form (1.2), which are *linear* and *static*, are known by the name *proportional controllers*, or *P* controllers for short. In the *closed-loop* diagram of Fig. 1.8, we can think of the signal $\bar{y}$, the target velocity, as an input, and of the signal y, the terminal velocity, as an output. A mathematical description of the relationship between the input signal, $\bar{y}$, and output signal, y, assuming linear models, can be computed from the diagram:

$$y = Gu, \qquad u = Ke, \qquad e = \bar{y} - y.$$

After eliminating the signals e and u we obtain

$$y = GKe = GK(\bar{y} - y) \qquad \Longrightarrow \qquad (1 + GK)y = GK\bar{y}.$$

When $GK \neq -1$,

$$y = H\bar{y}, \qquad H = \frac{GK}{1 + GK}.$$

A mathematical relationship governing a particular pair of inputs and outputs is called a *transfer-function*. The function H calculated above is known as a *closed-loop transfer-function*.

Ironically, a first conclusion from the closed-loop analysis is that it is not possible to achieve exact tracking of the target velocity since H cannot be equal to one for any finite value of the constants G and K, not even when $K = G^{-1}$, which was the open-loop solution. However, it is not so hard to make H get close to one: just make K large! More precisely, make the product GK large. How large it needs to be depends on the particular system G. However, a welcome side-effect of the closed-loop solution is that the controller gain, K, does not depend directly on the value of the system model, G.

Table 1.1 Closed-loop transfer-function, H, for various values of K and G

	K				
G	0.02	0.05	0.5	1	3
47.8	0.4888	0.7050	0.9598	0.9795	0.9931
73.3	0.5945	0.7856	0.9734	0.9865	0.9955
99.4	0.6653	0.8325	0.9803	0.9900	0.9967

As the calculations in Table 1.1 reveal, the closed-loop transfer-function, H, remains within 1% of 1 for values K greater than or equal to 3 for *any* value of G lying between the two crude linear models estimated earlier in Fig. 1.6.

In other words, feedback control does not seem to rely on exact knowledge of the system model in order to achieve good tracking performance. This is a major feature of feedback control, and one of the reasons why we may get away with using incomplete and not extremely accurate mathematical models for feedback design. One might find this strange, especially to scientists and engineers trained to look for accuracy and fidelity in their models of the world, a line of thought that might lead one to believe that better accuracy *requires* the use of complex models. For example, the complexity required for accurately modeling the interaction of an aircraft with its surrounding air may be phenomenal. Yet, as the Wright brothers and other flight pioneers demonstrated, it is possible to design and implement effective feedback control of aircraft without relying explicitly on such complex models.

This remarkable feature remains for the most part true even if nonlinear[7] models are considered, although the computation of the transfer-function, H, becomes more complicated.[8] Figure 1.9 shows a plot of the ratio $y/\bar{y}$ for various choices of gain, K, when a linear controller is in feedback with the static nonlinear model, G, fitted in Fig. 1.5. The trends are virtually the same as those obtained using linear models. Note also that the values of the ratio of the terminal velocity by the target velocity are close to the values of H calculated for the linear model with gain $G = 99.4$ which was obtained through "linearization" of the nonlinear model, especially at low velocities.

Insight on the reasons why feedback control can achieve tracking without relying on precise models is obtained if we look at the control, the signal u, that is effectively computed by the closed-loop solution. Following steps similar to the ones used in the derivation of the closed-loop transfer-function, we calculate

$$u = Ke = K(\bar{y} - y) = K(1 - H)\bar{y} = \frac{K}{1 + GK}\bar{y} = \frac{1}{K^{-1} + G}\bar{y}.$$

Note that $\lim_{K \to \infty} u = G^{-1}\bar{y}$, which is exactly the same control as that computed in open-loop (see Section 1.4). This time, however, it is the feedback loop that *computes* the function G^{-1} based on the error signal, $e = \bar{y} - y$. Indeed, u is simply equal to

---
[7] Many but not all nonlinear models.
[8] It requires solving the nonlinear algebraic equation $y = G(K(\bar{y} - y))$ for y. The dynamic version of this problem is significantly more complex.

10 Introduction

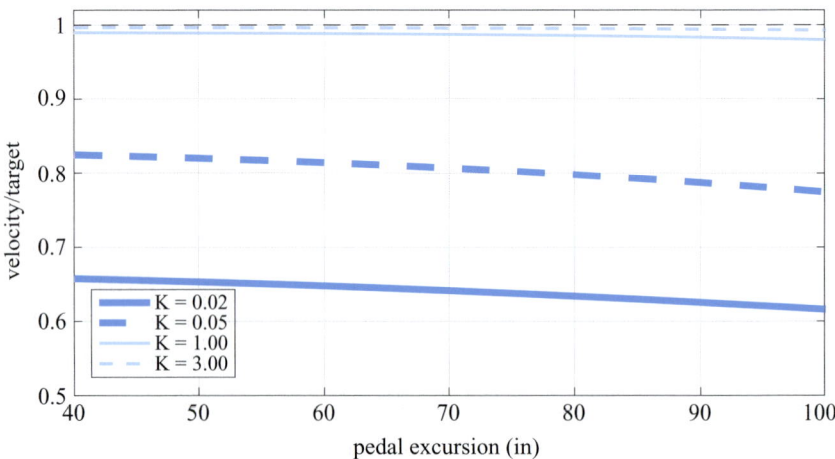

Figure 1.9 Effect of the gain K on the ability of the terminal velocity, y, to track a given target velocity, $\bar{y}$, when the linear feedback control, $u = K(y - \bar{y})$, is in closed-loop (Fig. 1.8) with the nonlinear model, $y = G(u) = 82.8\tan^{-1}(1.2u)$ from Fig. 1.5.

$K(\bar{y} - y)$, which, when K is made large, converges to $G^{-1}\bar{y}$ by virtue of feedback, no matter what the value of G is. A natural question is what are the side-effects of raising the control gain in order to improve the tracking performance? We will come back to this question at many points in this book as we learn more about dynamic systems and feedback.

1.6 Sensitivity

In previous sections, we made statements regarding how insensitive the closed-loop feedback solution was with respect to changes in the system model when compared with the open-loop solution. We can quantify this statement in the case of static linear models.

As seen before, in both open- and closed-loop solutions to the tracking control problem, the output y is related to the target output $\bar{y}$ through

$$y = H(G)\bar{y}.$$

The notation $H(G)$ indicates that the transfer-function, H, depends on the system model, G. In the open-loop solution $H(G) = GK$ and in the closed-loop solution $H(G) = GK(1 + GK)^{-1}$.

Now consider that G assumes values in the neighborhood of a certain nominal model $\bar{G}$ and that $H(\bar{G}) \neq 0$. Assume that those changes in G affect H in a continuous and differentiable way so that[9]

$$H(G) = H(\bar{G}) + H'(\bar{G})(\Delta G) + O\left(\Delta G^2\right), \qquad \Delta G = G - \bar{G},$$

---

[9] The notation $O(x^n)$ indicates a polynomial in x that has only terms with degree greater than or equal to n.

which is the Taylor series expansion of H as a function of G about $\bar{G}$. Discarding terms of order greater than or equal to two, we can write

$$H(\bar{G}) - H(G) \approx H'(\bar{G})(\bar{G} - G).$$

After dividing by $H(\bar{G})$ we obtain an expression for how changes in G affect the transfer-function $H(G)$:

$$\frac{H(\bar{G}) - H(G)}{H(\bar{G})} \approx S(\bar{G})\frac{\bar{G} - G}{\bar{G}}, \qquad S(G) = \frac{G}{H(G)}H'(G).$$

The function S is called the *sensitivity* function.

Using this formula we compute the sensitivity of the open-loop solution. In the case of linear models,

$$H(G) = GK, \qquad \Longrightarrow \qquad S(G) = \frac{G}{GK}K = 1.$$

This can be interpreted as follows: in open-loop, a relative change in the system model, G, produces a relative change in the output, y, of the same order.

In closed-loop, after some calculations (see P1.13),

$$H(G) = \frac{GK}{1 + GK}, \qquad \Longrightarrow \qquad S(G) = \frac{1}{1 + GK}. \qquad (1.3)$$

By making K large we not only improve the tracking performance but also reduce the sensitivity S. Note that $S + H = 1$, hence $S = 1 - H$, so that the values of S can be easily calculated from Table 1.1 in the case of the car cruise control. For this reason, H is known as the *complementary sensitivity* function.

In the closed-loop diagram of Fig. 1.8, the transfer-function from the reference input, $\bar{y}$, to the tracking error, e, is

$$e = \bar{y} - y = (1 - H)\bar{y} = S\bar{y},$$

which is precisely the sensitivity transfer-function that we have calculated based on model variations. The smaller the sensitivity, S, the better the controller tracks the reference input, $\bar{y}$. Perfect tracking would be achieved if we could make $S = 0$. This is a win–win *coincidence*: the closer controllers can track references, the less sensitive the closed-loop will be to variations in the model.

1.7 Disturbances

Another way of accounting for variability in models is to introduce additional *disturbance signals*. Consider, for example, the block-diagram in Fig. 1.10, in which the disturbance signal, w, adds to the input signal, u, that is,

$$y = G(u + w).$$

It is the scientist or engineer who must distinguish disturbance signals from regular input signals. Disturbances are usually nuisances that might be present during the operation of the system but are hard to model at the design phase, as well as other phenomena

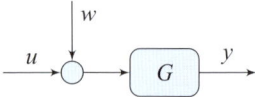

Figure 1.10 System with input disturbance w.

affecting the system that are not completely understood. For example, in Section 2.7, we will use a disturbance signal entering the block-diagram in Fig. 1.10 as w to model a road slope in the car cruise control problem. Because feedback control can be very effective in handling disturbances, delegating difficult aspects of a problem to disturbances is key to simplifying the control design process. Indeed, excerpts from the 1903 Wright brothers' patent for a *flying machine*, shown in Fig. 1.11, hint that this way of thinking might have played a central role in the conquest of flight.

It is easy to incorporate disturbances into the basic open- and closed-loop schemes of Figs. 1.7 and 1.8, which we do in Figs. 1.12 and 1.13. In both cases, one can write the output, y, in terms of the reference input, $\bar{y}$, and the disturbance, w. Better yet, we can write the transfer-function from the inputs, $\bar{y}$ and w, to the tracking error, $e = \bar{y} - y$. In open-loop we calculate with Fig. 1.12 that

$$e = \bar{y} - y = \bar{y} - G(K\bar{y} + w) = (1 - GK)\bar{y} - Gw.$$

Substituting the proposed open-loop solution, $K = G^{-1}$, we obtain

$$e = -Gw,$$

which means that open-loop control is very effective at tracking but has no capability to *reject* the disturbance w, as one could have anticipated from the block-diagram in Fig. 1.12. Open-loop controllers will perform poorly in the presence of disturbances. This is similar to the conclusion obtained in Section 1.6 that showed open-loop controllers to be sensitive to changes in the system model.

UNITED STATES PATENT OFFICE.

ORVILLE WRIGHT AND WILBUR WRIGHT, OF DAYTON, OHIO.

FLYING-MACHINE.

No. 821,393. Specification of Letters Patent. Patented May 22, 1906.

Application filed March 23, 1903. Serial No. 149,220.

lar manner. In either case owing to the varying conditions to be met there are numerous disturbing forces which tend to shift the machine from the position which it should occupy to obtain the desired results. It is the chief object of our invention to provide means for remedying this difficulty, and we will now proceed to describe the construction by means of which these results are accomplished.

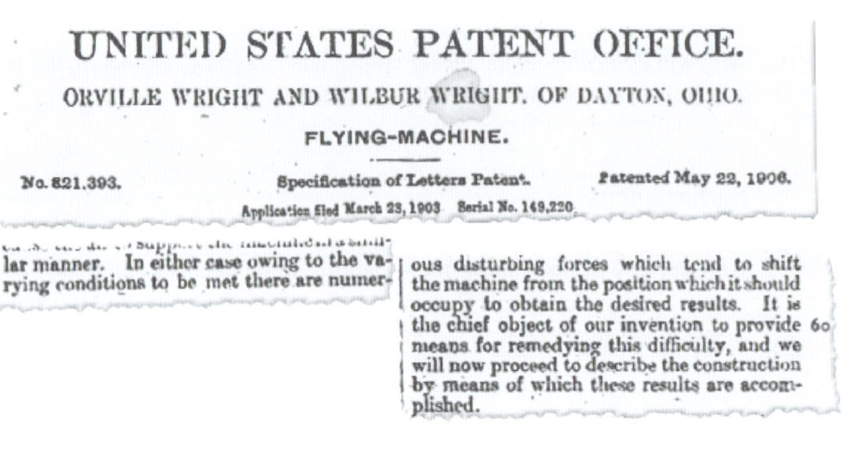

Figure 1.11 The Wright brothers' 1903 patent [WW06].

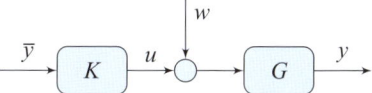

Figure 1.12 Open-loop configuration with input disturbance w.

In closed-loop, Fig. 1.13, we calculate that

$$e = \bar{y} - y = \bar{y} - G(Ke + w) \quad \Longrightarrow \quad (1 + GK)e = \bar{y} - Gw$$

and the tracking error is

$$e = \frac{1}{1 + GK}\bar{y} - \frac{G}{1 + GK}w. \tag{1.4}$$

The control gain, K, shows up in both transfer-functions from the inputs, w and $\bar{y}$, to the tracking error, e. High control gains reduce both terms at the same time. That is, the closed-loop solution achieves good tracking *and* rejects the disturbance. This is a most welcome feature and often the main reason for using feedback in control systems. By another coincidence, the coefficient of the first term in (1.4) is the same as the sensitivity function, $S(G)$, calculated in Section 1.6.

Problems

1.1 For each block-diagram in Fig. 1.14 identify *inputs*, *outputs*, and other relevant *signals*, and also describe what physical quantities the signals could represent. Determine whether the system is in *closed-loop* or *open-loop* based on the presence or absence of *feedback*. Is the relationship between the inputs and outputs dynamic or static? Write a simple equation for each block if possible. Which signals are *disturbances*?

1.2 Sketch block-diagrams that can represent the following phenomena as *systems*:

(a) skin protection from sunscreen;
(b) money in a savings account;
(c) a chemical reaction.

Identify potential *input* and *output* signals that could be used to identify cause–effect relationships. Discuss the assumptions and limitations of your *model*. Is the relationship

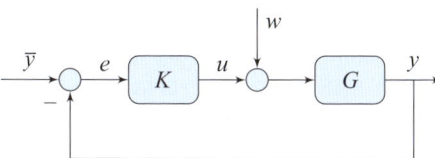

Figure 1.13 Closed-loop feedback configuration with input disturbance w.

14 Introduction

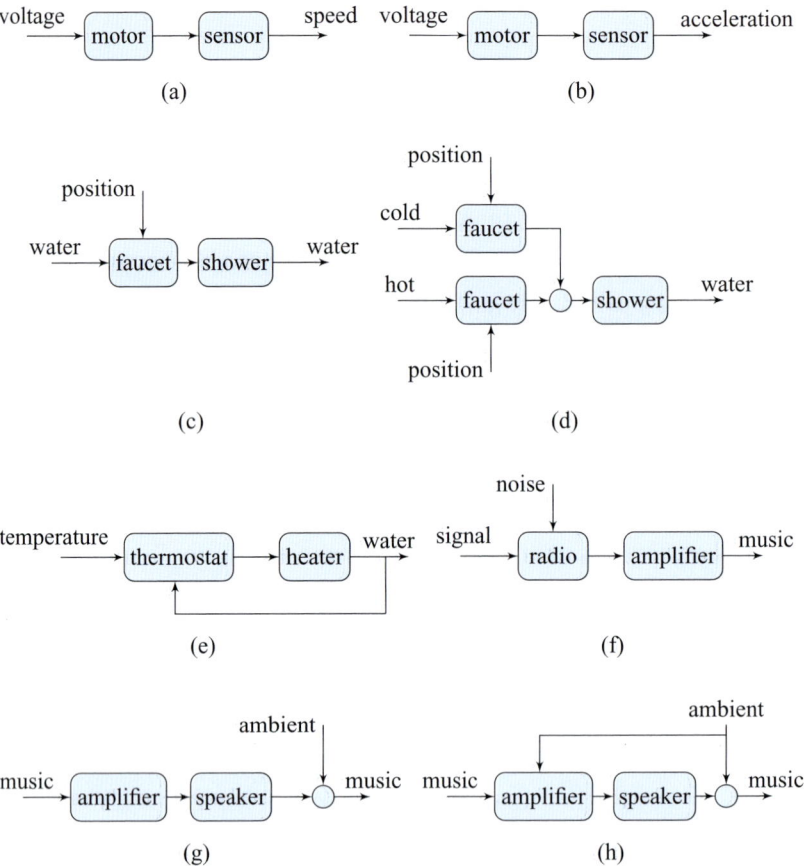

Figure 1.14 Block diagrams for P1.1.

between the inputs and outputs dynamic or static? Write simple equations for each block if possible.

1.3 Mammals are able to regulate their body temperature near 36.5 °C (~98 °F) despite fluctuations in the ambient temperature. Sketch a block-diagram that could represent a possible temperature control system in mammals. Identify disturbances, signals, and possible principles of sensing and actuation that could be used by mammals to lower or increase the body temperature. Compare possible *open-loop* and *closed-loop* solutions. Discuss the difficulties that need to be overcome in each case.

1.4 Most cars are equipped with an *anti-lock braking system* (ABS), which is designed to prevent the wheels from *locking up* when the driver actuates the brake pedal. It helps with emergencies and adverse road conditions by ensuring that traction is maintained on all wheels throughout breaking. An ABS system detects locking of a wheel by comparing the rotational speeds among wheels and modifies the pressure on the hydraulic brake actuator as needed. Sketch a block-diagram that could represent the signals and systems involved in ABS.

1.5 Humans learn to balance standing up early in life. Sketch a block-diagram that represents signals and systems required for standing up. Is there a sensor involved? Actuator? Feedback?

1.6 Sketch a block-diagram that represents the signals and systems required for a human to navigate across an unknown environment. Is there a sensor involved? Actuator? Feedback?

1.7 Repeat P1.5 and P1.6 from the perspective of a blind person.

1.8 Repeat P1.5 and P1.6 from the perspective of a robot or an autonomous vehicle.

1.9 For each block-diagram in Fig. 1.15 compute the transfer-function from the input u to the output y assuming that all blocks are linear.

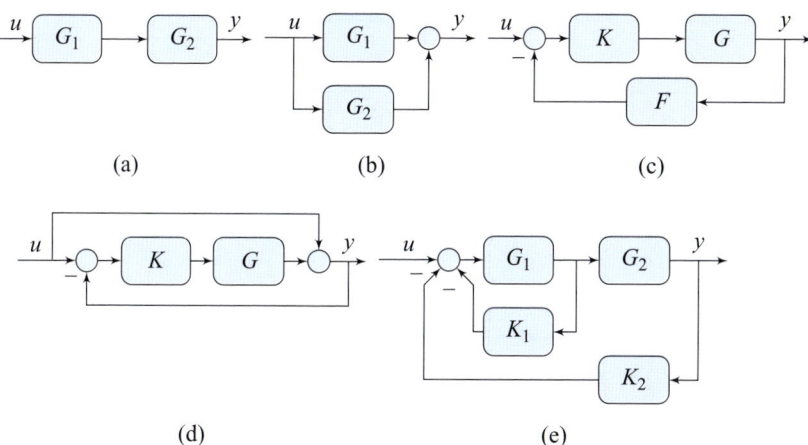

Figure 1.15 Block diagrams for P1.9.

1.10 Students participating in Rice University's Galileo Project [Jen14] set out to carefully reproduce some of Galileo's classic experiments. One was the study of projectile motion using an inclined plane, in which a ball accelerates down a plane inclined at a certain angle then rolls in the horizontal direction with uniform motion for a short while until falling off the edge of a table, as shown in Fig. 1.16. The distance the ball rolled along the inclined plane, ℓ in feet, and the distance from the end of the table to the landing site of the ball, d in inches, were recorded. Some of their data, five trials at two different angles, is reproduced in Table 1.2. Use MATLAB to plot and visualize the data. Fit simple equations, e.g. linear, quadratic, etc., to the data to relate the fall

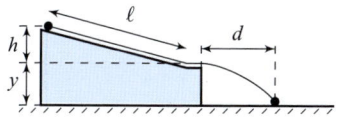

Figure 1.16 Galileo's inclined plane.

height, h, to the horizontal travel distance, d, given in Table 1.2. Justify your choice of equations and comment on the quality of the fit obtained in each case. Estimate using the given data the vertical distance y. Can you also estimate gravity?

Table 1.2 Data for P1.10

	\multicolumn{4}{c}{Ramp distance at 13.4°}
Try	1 ft	2 ft	4 ft	6 ft
1	$13\frac{15}{16}$	$19\frac{13}{16}$	$27\frac{11}{16}$	$33\frac{3}{8}$
2	$13\frac{7}{8}$	$19\frac{13}{16}$	$27\frac{3}{4}$	$33\frac{5}{16}$
3	$14\frac{1}{16}$	$19\frac{13}{16}$	$27\frac{3}{4}$	$33\frac{3}{16}$
4	14	$19\frac{3}{4}$	$27\frac{9}{16}$	$33\frac{7}{16}$
5	$13\frac{15}{16}$	$19\frac{3}{4}$	$27\frac{9}{16}$	$33\frac{5}{8}$

	\multicolumn{5}{c}{Ramp distance at 6.7°}
Try	1 ft	2 ft	4 ft	6 ft	8 ft
1	$10\frac{11}{6}$	$14\frac{1}{2}$	$20\frac{3}{4}$	$25\frac{7}{16}$	$29\frac{5}{8}$
2	$10\frac{11}{6}$	$14\frac{9}{16}$	$20\frac{3}{4}$	$25\frac{1}{2}$	$29\frac{1}{2}$
3	$10\frac{11}{6}$	$14\frac{1}{2}$	$20\frac{3}{4}$	$25\frac{3}{4}$	$29\frac{1}{2}$
4	$10\frac{11}{6}$	$14\frac{1}{2}$	$20\frac{3}{4}$	$25\frac{1}{2}$	$29\frac{5}{16}$
5	$10\frac{11}{6}$	$14\frac{9}{16}$	$20\frac{3}{16}$	$25\frac{5}{8}$	$29\frac{1}{2}$

1.11 Use MATLAB to determine the parameters α, β, and γ that produce the least-squares fit of the data in Fig. 1.4 to the curves $y(u) = \alpha \tan^{-1}(\beta u)$ and $y(u) = \gamma u$. Compare your answers with Figs. 1.5 and 1.6.

1.12 Compute the first-order Taylor series expansion of the function $y(u) = \alpha \tan^{-1}(\beta u)$ about $u = 0$ and use the solution to P1.11 to verify the value of the slope shown in Fig. 1.6.

1.13 Show that the sensitivity function in (1.3) is the one associated with the closed-loop transfer-function $H(G) = GK(1 + GK)^{-1}$.

2 Dynamic Systems

In Chapter 1 we contemplated solutions to our first control problem, a much simplified cruise controller for a car, without taking into account possible effects of *time*. System and controller models were *static* relations between the signals: the output signal, y, the input signal, u, the reference input, $\bar{y}$, and then the disturbance, w. Signals in block-diagrams flow *instantaneously*, and closed-loop solutions derived from such block-diagrams were deemed reasonable if they could be implemented *fast enough*. We drew encouraging conclusions from simple analysis but no rationale was given to support the conclusions if time were to be taken into consideration.

Of course, it is perfectly fine to construct a static mathematical model relating a car's pedal excursion with its terminal velocity, as long as we understand the model setup. Clearly a car does not reach its terminal velocity *instantaneously*! If we expect to implement the feedback cruise controller in a real car, we have to be prepared to say what happens between the time at which a terminal target velocity is set and the time at which the car reaches its terminal velocity. Controllers have to *understand* that it takes *time* for the car to reach its terminal velocity. That is, we will have to incorporate time not only into models and tools but also into controllers. For this reason we need to learn how to work with *dynamic* systems.

In the present book, mathematical models for dynamic systems take the form of ordinary[1] differential equations where signals evolve continuously in time. Bear in mind that this is not a course on differential equations, and previous exposure to the mathematical theory of differential equations helps. Familiarity with material covered in standard text books, e.g. [BD12], is enough. We make extensive use of the Laplace transform and provide a somewhat self-contained review of relevant facts in Chapter 3.

These days, when virtually all control systems are implemented in some form of digital computer, one is compelled to justify why not to discuss control systems directly from the point of view of discrete-time signals and systems. One reason is that continuous-time signals and systems have a long tradition in mathematics and physics that has established a *language* that most scientists and engineers are accustomed to. The converse is unfortunately not true, and it takes time to get comfortable with interpreting discrete-time models and making sense of some of the implied assumptions that come with them, mainly the effects of sampling and related practical issues, such as

---

[1] Ordinary, as opposed to *partial*, means that derivatives appear only with respect to one variable; in our case, time.

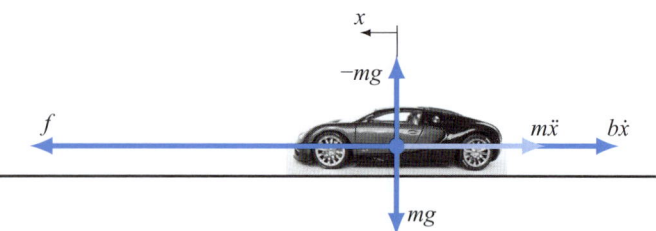

Figure 2.1 Free-body diagram showing forces acting in a car.

quantization and aliasing. In fact, for physical systems, it is impossible to appropriately choose an adequate sampling rate without having a good idea of the continuous-time model of the system being controlled. Finally, if a system is well modeled and a controller is properly designed in continuous-time, implementation in the form of a discrete-time controller is most of the time routine, especially when the available hardware sampling rates are fast enough.

2.1 Dynamic Models

Let us start with some notation: we denote time by the real variable t in the interval $[0, \infty)$, where 0 can be thought of as an arbitrary origin of time before which we are not interested in the behavior of the system or its signals. We employ functions of real variables to describe signals and use standard functional notation to indicate the dependence of signals on time. For example, the dynamic signals y and u are denoted as $y(t)$ and $u(t)$. At times, when no confusion is possible, we omit the dependence of signals on t.

We claimed in Chapter 1 that models should be rooted in well-defined experiments. Planning and performing experiments for dynamic systems is a much more complex task, which we do not have room to address in detail here. Instead, we will reach out to physics to help us introduce an abstract dynamic model, the parameters of which will later be determined through experiments.

In the tradition of simplified physical modeling, we use Newton's law to write equations for a car based on the free-body diagram shown in Fig. 2.1. The car is modeled as a particle with mass $m > 0$ and, after balancing all forces in the x-direction, we obtain the differential equation

$$m\ddot{x}(t) + b\dot{x}(t) = f(t),$$

where x is the linear coordinate representing the position of the car, $b \geq 0$ is the coefficient of friction, and f is a force, which we will use to put the car into motion. Much can be argued about the exact form of the friction force, which we have assumed to be *viscous*, that is of the form $-bv(t)$, linear, and opposed to the velocity $v(t) = \dot{x}(t)$. As we are interested in modeling the velocity of the car and not its position, it is convenient to rewrite the differential equation in terms of the velocity v, obtaining

$$m\dot{v}(t) + bv(t) = f(t). \qquad (2.1)$$

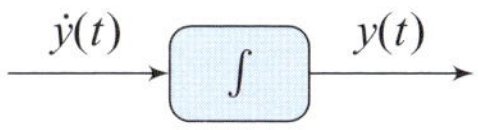

Figure 2.2 Block-diagram with integrator.

In order to complete our model we need to relate the car driving force, f, to the pedal excursion, u. Here we could resort to experimentation or appeal to basic principles. With simplicity in mind we choose a linear static model:

$$f = pu, \qquad (2.2)$$

where p represents a *pedal gain*, which can be determined experimentally by methods similar to the ones used in Chapter 1.

Of course, no one should believe that the simple force model (2.2) can accurately represent the response of the entire powertrain of the car in a variety of conditions. Among other things, the powertrain will have its own complex dynamic behaviors, which (2.2) gracefully ignores. Luckily, the validity[2] of such simplification depends not only on the behavior of the actual powertrain but also on the purpose of the model. In many cases, the time-constants[3] of the powertrain are much *faster* than the time-constant due to the inertial effects of the entire car. In this context, a simplified model can lead to satisfactory or at least insightful results when the purpose of the model is, say, predicting the velocity of the car. A human driver certainly does not need to have a deep knowledge of the mechanical behavior of an automobile for driving one!

Combining Equations (2.1) and (2.2), and labeling the velocity as the output of the system, i.e. $y(t) = v(t)$, we obtain the differential equation

$$\dot{y}(t) + \frac{b}{m} y(t) = \frac{p}{m} u(t), \qquad (2.3)$$

which is the mathematical dynamic model we will use to represent the car in the dynamic analysis of the cruise control problem.

2.2 Block-Diagrams for Differential Equations

In Chapter 1 we used block-diagrams to represent the interaction of signals, systems, and controllers. If we interpret differential equations as a relationship between signals and the signals' derivatives, it should be no surprise that ordinary differential equations can be represented in block-diagrams. The key is to use an *integrator* to relate the derivative to its primitive. An integrator block should be able to produce at its output the integral of its input. Alternatively, we can see the integrator input as the derivative of its output signal, as depicted in Fig. 2.2. We will study in Chapter 5 a number of physical devices that can be used to physically implement integrators.

Assuming that integrator blocks are available, all that is left to do is to rewrite the ordinary differential equation, isolating its highest derivative. For example, we

---

[2] A better word here may be *usefulness*. [3] More about that soon!

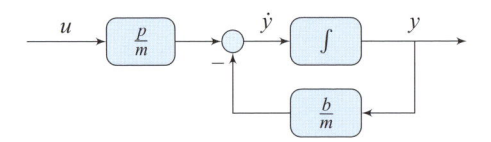

Figure 2.3 Dynamic model of the car: m is the mass, b is the viscous friction coefficient, p is the pedal gain, u is the pedal excursion, and y is the car's velocity.

rewrite (2.3) as

$$\dot{y}(t) = \frac{p}{m} u(t) - \frac{b}{m} y(t),$$

which can be represented by the block-diagram in Fig. 2.3.

Note the presence of a *feedback loop* in the diagram of Fig. 2.3! For this reason, tools for analyzing feedback loops often draw on the theory of differential equations and vice versa. We will explore the realization of differential equations using block-diagrams with integrators in detail in Chapter 5.

2.3 Dynamic Response

The differential equation (2.3) looks very different from the static linear models considered earlier in Chapter 1. In order to understand their differences and similarities we need to understand how the model (2.3) *responds* to inputs. Our experiment in Section 1.1 consisted of having a constant pedal excursion and letting the car reach a terminal velocity. We shall first attempt to emulate this setup using the differential equation (2.3) as a model.

A constant pedal excursion corresponds to the input function

$$u(t) = \tilde{u}, \quad t \geq 0,$$

where $\tilde{u}$ is constant. In response to this input we expect that the solution of the differential equation (2.3) approaches a constant terminal velocity, $\tilde{y}$, after some time has passed. The value of this terminal velocity can be calculated after noticing that

$$y(t) = \tilde{y}, \quad t \geq T \quad \Longrightarrow \quad \dot{y}(t) = 0, \quad t > T,$$

in which case (2.3) reduces to

$$\frac{b}{m}\tilde{y} = \frac{p}{m}\tilde{u} \quad \Longrightarrow \quad \tilde{y} = \frac{p}{b}\tilde{u}.$$

It is this relation that should be compared with the static model developed earlier. Experiments similar to the ones in Section 1.1 can be used to determine the value of the ratio p/b. In the language of differential equations the function

$$y_P(t) = \tilde{y} = \frac{p}{b}\tilde{u}$$

is a *particular solution* to the differential equation (2.3). See [BD12] for details.

2.3 Dynamic Response

The particular solution cannot, however, be a complete solution: if the initial velocity of the car at time $t = 0$ is not equal to $\tilde{y}$ then $y(0) = y_0 \neq \tilde{y} = y_P(0)$. The remaining component of the solution is found by solving the *homogeneous* version of Equation (2.3):

$$\dot{y}(t) + \frac{b}{m} y(t) = 0.$$

All solutions to this equation can be shown to be of the form

$$y_H(t) = e^{\lambda t}, \qquad (2.4)$$

where the constant λ is determined upon substitution of $y_H(t)$ into (2.3):

$$\dot{y}_H(t) + \frac{b}{m} y_H(t) = \left(\lambda + \frac{b}{m} \right) e^{\lambda t} = 0.$$

This is an algebraic equation that needs to hold for all $t \geq 0$, in particular $t = 0$, which will happen only if λ is a zero of the *characteristic equation*:

$$\lambda + \frac{b}{m} = 0 \quad \Longrightarrow \quad \lambda = -\frac{b}{m}. \qquad (2.5)$$

The complete solution to Equation (2.3) is a combination of the particular solution, $y_P(t)$, with all possible solutions to the homogeneous equation:[4]

$$y(t) = y_P(t) + \beta y_H(t) = \tilde{y} + \beta e^{\lambda t},$$

in which the constant β is calculated so that $y(t)$ matches the *initial condition*, y_0, at $t = 0$. That is,

$$y(0) = \tilde{y} + \beta = y_0 \quad \Longrightarrow \quad \beta = y_0 - \tilde{y}.$$

Putting it all together, the complete response is

$$y(t) = \tilde{y}\left(1 - e^{\lambda t}\right) + y_0 e^{\lambda t}, \qquad t \geq 0, \qquad (2.6)$$

where

$$\lambda = -\frac{b}{m}, \qquad \tilde{y} = \frac{p}{b} \tilde{u}. \qquad (2.7)$$

Plots of $y(t)$ for various values of λ and y_0 are shown in Fig. 2.4 for $\tilde{y} = 1$. Note how the responses *converge* to $\tilde{y}$ for all negative values of λ. The more negative the value of λ, the faster the convergence. When λ is positive the response does not converge to $\tilde{y}$, even when y_0 is very close to $\tilde{y}$.

It is customary to evaluate how fast the solution of the differential equation (2.3) converges to $\tilde{y}$ by analyzing its response to a zero initial condition $y_0 = 0$ and a nonzero $\tilde{y} \neq 0$. This is known as a *step response*. When $\lambda < 0$, the constant

$$\tau = -\frac{1}{\lambda} \qquad (2.8)$$

---

[4] In this simple example there is only one such solution, $y_H(t)$ given in (2.4).

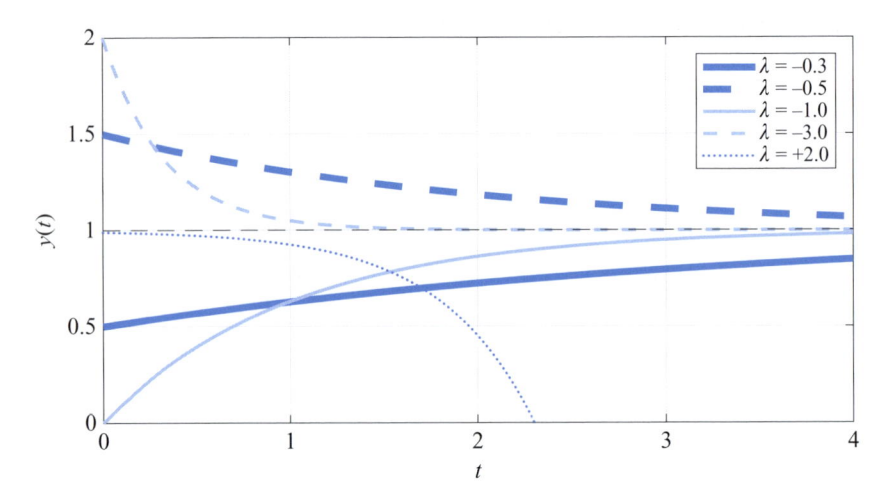

Figure 2.4 Plots of $y(t) = \tilde{y}\left(1 - e^{\lambda t}\right) + y_0 e^{\lambda t}$, $t \geq 0$, with $\tilde{y} = 1$; λ and y_0 are as shown.

is the *time-constant*. In P2.1 you will show that τ has units of time. At select times $t = \tau$ and $t = 3\tau$,

$$y(\tau) = \tilde{y}\left(1 - e^{-1}\right) \approx 0.63\tilde{y}, \qquad y(3\tau) = \tilde{y}\left(1 - e^{-3}\right) \approx 0.95\tilde{y}.$$

As τ depends only on λ and not on $\tilde{y}$, it is possible to compare the rate of convergence of different systems modeled by linear ordinary differential equations by comparing their time-constants. For differential equations more complex than (2.3), the time-constant is *defined* as the time it takes the step response to reach 63% of its terminal value. The smaller the time-constant, the faster the convergence.

Another measure of the rate of change of the response is the *rise-time*, t_r, which is the time it takes the step response to go from 10% to 90% of its final value.[5] Calculating

$$y(t_1) = \tilde{y}\left(1 - e^{\lambda t_1}\right) = 0.1\tilde{y}, \qquad y(t_2) = \tilde{y}\left(1 - e^{\lambda t_2}\right) = 0.9\tilde{y},$$

we obtain

$$t_r = t_2 - t_1 = \ln(1/9)\lambda^{-1} = \ln(9)\tau \approx 2.2\,\tau. \tag{2.9}$$

Again, the smaller the rise-time, the faster the convergence.

2.4 Experimental Dynamic Response

As shown in Section 2.3, the terminal velocity attained by the dynamic linear model, the differential equation (2.3), is related to the static linear model, the algebraic equation (1.1), through $\gamma = \tilde{y}/\tilde{u} = p/b$. This means that the ratio p/b can be determined in the same way as was done in Section 1.1. A new experiment is needed to determine the parameter $\lambda = -b/m$, which does not influence the terminal velocity but affects the rate at which the car approaches the terminal velocity.

---

[5] For systems in which the output develops only after a delay it is easier to measure the rise-time than the time-constant.

2.4 Experimental Dynamic Response

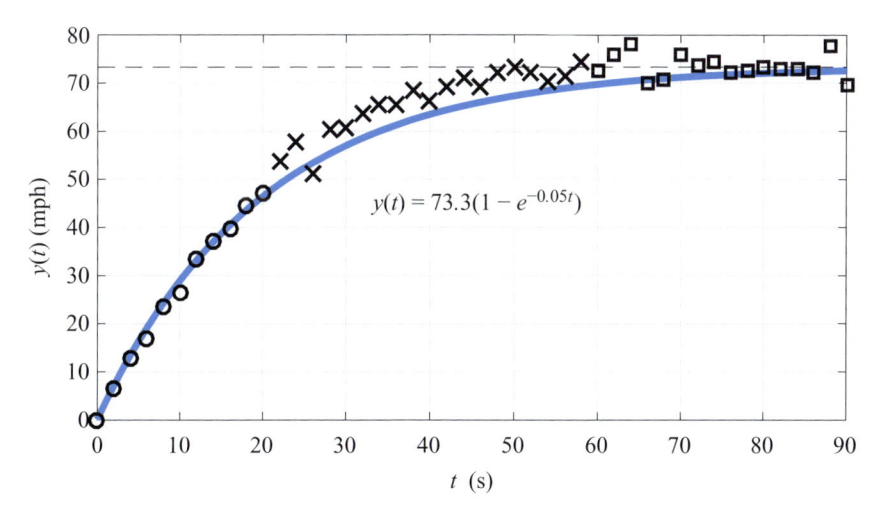

Figure 2.5 Experimental velocity response of a car to a constant pedal excursion, $u(t) = \tilde{u} = 1$ in, $t \geq 0$; samples are marked as circles, crosses, and squares.

First let us select a velocity around which we would like to *build* our model, preferably a velocity close to the expected operation of the cruise controller. Looking at Figs. 1.4 through 1.6, we observe that a pedal excursion of around 1 in will lead to a terminal velocity around 70 mph, which is close to highway speeds at which a cruise controller may be expected to operate. We perform the following dynamic experiment: starting at rest, apply constant pedal excursion, $u(t) = \tilde{u} = 1$ in, $t \geq 0$, and collect samples of the instantaneous velocity until the velocity becomes approximately constant. In other words, perform an *experimental step response*. The result of one such experiment may look like the plot in Fig. 2.5, in which samples (marked as circles, crosses, and squares) have been collected approximately every 2 s for 90 s.

We proceed by *fitting* the data in Fig. 2.5 to a function like (2.6) where the initial condition, y_0, is set to zero before estimating the parameters $\tilde{y}$ and λ. This fit can be performed in many ways. We do it as follows: we first average the samples over the last 30 s in Fig. 2.5 (squares) to compute an estimate of the terminal velocity. From the data shown in Fig. 2.5 we obtain the estimate $\tilde{y} \approx 73.3$ mph. If $y(t)$ is of the form (2.6) then

$$r(t) = 1 - y(t)/\tilde{y} = e^{\lambda t} \quad \Longrightarrow \quad \ln r(t) = \lambda t.$$

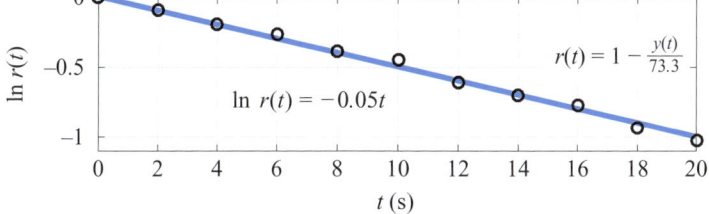

Figure 2.6 Plot of $\ln r(t)$ and fitted linear model.

That is, $\ln r(t)$ is a line with slope λ. With this in mind we plot $\ln r(t)$ in Fig. 2.6 using samples taken from the first 20 s from Fig. 2.5 (circles) and estimate the slope of the line $\ln r(t)$, that is $\lambda = -0.05$. The parameters b/m and p/m are then estimated based on the relationships

$$\frac{b}{m} = -\lambda \approx 0.05 \text{ s}^{-1},$$
$$\frac{p}{b} = \frac{\tilde{y}}{\tilde{u}} \approx 73.3 \text{ mph/in}, \qquad (2.10)$$
$$\frac{p}{m} = \frac{b}{m} \times \frac{p}{b} = -\lambda \times \frac{\tilde{y}}{\tilde{u}} = 3.7 \text{ mph/(in s)}.$$

Note that this model has a static gain of about 73.3 mph/in which lies somewhere between the two static linear models estimated earlier in Fig. 1.6. Indeed, this is the intermediate gain value that was used in Section 1.5 to calculate one of the static closed-loop transfer-functions in Table 1.1.

The estimation of the structure and the parameters of a dynamic system from experiments is known as *system identification*. The interested reader is referred to [Lju99] for an excellent introduction to a variety of useful methods.

2.5 Dynamic Feedback Control

We are now ready to revisit the feedback solution proposed in Section 1.3 for solving the cruise control problem. Let us keep the structure of the feedback loop the same, that is let the *proportional controller*

$$u(t) = Ke(t), \qquad e(t) = \bar{y} - y(t) \qquad (2.11)$$

be connected as in Fig. 1.8. Note a fundamental difference between this controller and the one analyzed before: in Section 1.5 the signals e and y were the terminal error and terminal velocity; controller (2.11) uses the *dynamic* error signal $e(t)$ and velocity $y(t)$. This dynamic feedback loop can be practically implemented if a sensor for the instantaneous velocity, $y(t)$, is used. Every vehicle comes equipped with one such sensor, the speedometer.[6]

In order to analyze the resulting dynamic feedback control loop we replace the system model, G, with the dynamic model, the differential equation (2.3), to account for the car's dynamic response to changes in the pedal excursion. In terms of block-diagrams, we replace G in Fig. 1.8 by the block-diagram representation of the differential equation (2.3) from Fig. 2.3. The result is the block-diagram shown in Fig. 2.7. Using Equations (2.3) and (2.11) we eliminate the input signal, $u(t)$, to obtain

$$\dot{y}(t) + \left(\frac{b}{m} + \frac{p}{m}K\right)y(t) = \frac{p}{m}K\bar{y}. \qquad (2.12)$$

---

[6] The speedometer measures the speed but it is easy to infer the direction, hence the velocity, in this simple one-dimensional setup.

2.5 Dynamic Feedback Control

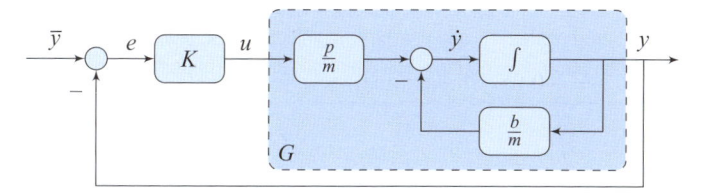

Figure 2.7 Dynamic closed-loop connection of the car model with proportional controller.

This linear ordinary differential equation governs the behavior of the closed-loop system. In the next chapters, you will learn to interpret this equation in terms of a closed-loop transfer-function using the Laplace transform. For now we proceed in the time domain and continue to work with differential equations.

Since Equation (2.12) has the same structure as Equation (2.3), its solution is also given by (2.6). That is

$$y(t) = \tilde{y}(1 - e^{\lambda t}) + y_0 e^{\lambda t}, \qquad t \geq 0,$$

with the constants

$$\lambda = -\frac{b}{m} - \frac{p}{m}K, \qquad \tilde{y} = \frac{(p/m)K}{(b/m) + (p/m)K}\bar{y} = \frac{(p/b)K}{1 + (p/b)K}\bar{y}.$$

When K is positive, the decay-rate λ is negative and hence $y(t)$ converges to $\tilde{y}$. In terms of the target velocity $\bar{y}$ we can write

$$\lim_{t \to \infty} y(t) = \tilde{y} = H(0)\bar{y}, \qquad H(0) = \frac{(p/b)K}{1 + (p/b)K}. \qquad (2.13)$$

We refer to components of the response of a dynamic system that persist as the time grows large as steady-state solutions. In this case, the closed-loop has a constant steady-state solution

$$y_{ss}(t) = \tilde{y} = H(0)\bar{y}.$$

Note that the value of $H(0)$ is equal to the static closed-loop transfer-function computed in Section 1.5 if G is replaced with p/b. As we will see in Chapter 4, this is not a mere coincidence. In terms of *tracking error*,

$$\lim_{t \to \infty} e(t) = \lim_{t \to \infty} \bar{y} - y(t) = S(0)\bar{y}, \qquad S(0) = \frac{1}{1 + (p/b)K}. \qquad (2.14)$$

The function $S(0) = 1 - H(0)$ is the static closed-loop *sensitivity function* computed before in Section 1.6. The reason for using the notation $H(0)$ and $S(0)$ will become clear in the next chapters.

For various values of $G = p/b$, including $p/b \approx 73.3$ which we estimated in Section 2.4, the steady-state closed-loop solution will track the reference $\bar{y}$ with accuracy $S(0) = 1 - H(0)$, which can be computed from the values listed in Table 1.1. In steady-state, the closed-loop behaves as predicted by the static analysis in Section 1.3. The dynamic analysis goes a step further: it predicts the rate at which convergence occurs,

Table 2.1 Open- and closed-loop steady-state transfer-function, $H(0)$, steady-state sensitivity, $S(0)$, steady-state limit, $\tilde{y}$, time-constant, τ, and rise-time, t_r, calculated for $b/m = 0.05$ and $p/b = 73.3$ and a constant target output of $\tilde{y} = 60$ mph. The open-loop solution is from Section 1.4.

K	$H(0)$	$S(0)$	$\tilde{y}$ (mph)	τ (s)	t_r (s)
Open-loop	1.00	0.00	60	20.0	43.9
0.02	0.60	0.40	36	8.1	17.8
0.05	0.79	0.21	47	4.3	9.4
0.50	0.97	0.03	58	0.5	1.2

which is related to the parameters λ and τ:

$$\lambda = -\frac{b}{m} - \frac{p}{m}K, \qquad \tau = -\lambda^{-1} = \frac{m}{b + pK}.$$

Notice that the time-constant, τ, becomes smaller as K grows. A numerical comparison for various values of gain, K, including the open-loop solution[7] (from Section 1.4), is given in Table 2.1. The corresponding dynamic responses calculated from zero initial conditions, $y(0) = 0$, are plotted in Fig. 2.8.

Some numbers in Table 2.1 and Fig. 2.8 look suspicious. Is it really possible to lower time-constants so much? Take, for example, the case of the largest gain $K = 0.5$: here we have almost perfect tracking (3% error) with a closed-loop rise-time that is more than 40 times faster than in open-loop. This kind of performance improvement is unlikely to be achieved by any controller that simply steps into the accelerator pedal. Surely there must be a catch! Indeed, so far we have been looking at the system output, the car's velocity, $y(t)$, and have paid little attention to the control input, the pedal excursion, $u(t)$. We shall now look at the control input in search of clues that could explain the impressive performance of the closed-loop controller.

The control inputs, $u(t)$, associated with the dynamic responses in Fig. 2.8 are plotted in Fig. 2.9. In Fig. 2.9 we see that the feedback controller is injecting into the system, the car, large inputs, pedal excursions, in order to achieve better tracking and faster response. The larger the control gain, K, the larger the required pedal excursion. Note that in this case the maximum required control signal happens at $t = 0$, when the tracking error is at a maximum, and

$$u(0) = Ke(0) = K(\tilde{y} - y(0)) = K\tilde{y}.$$

Clearly, the larger the gain, K, the larger the control input, u. For instance, with $K = 0.5$ the controller produces an input that exceeds the maximum possible pedal excursion of 3 in, which corresponds to *full throttle*. In other words, the control input is *saturated*. With $K = 0.05$ we have $u(0) = 3$, which is full throttle. Of course, any conclusions drawn for $K > 0.05$ will no longer be valid or, at least, not very accurate. It is not

---

[7] What we mean by open-loop solution is the simulation of the diagram in Fig. 1.7 where G is replaced by the differential equation model (2.3) but $K = (p/b)^{-1}$ is still the constant open-loop gain as calculated in Section 1.4. We are not yet equipped to speak of G^{-1} as the inverse of a dynamic model, which we will do in Sections 4.7, 5.1, and 8.6.

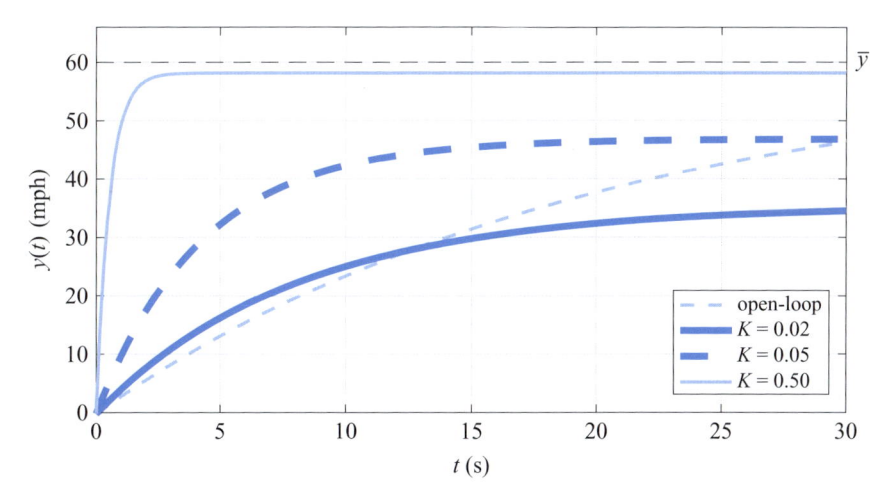

Figure 2.8 Open- and closed-loop dynamic response, $y(t)$, for the linear car velocity model (2.12) calculated for $b/m = 0.05$ and $p/b = 73.3$ and a constant target output of $\bar{y} = 60$ mph with proportional control (2.11) for various values of gain, K; the open-loop solution is from Section 1.4.

possible to achieve some of the predicted ultra-fast response times due to limitations in the system, in this case pedal and engine saturation, that were not represented in the linear models used to design and analyze the closed-loop. Ironically, the gain $K = 0.02$ is one for which the pedal excursion remains well below saturation, and is perhaps the one case in which the (poor) performance predicted by the linear model is likely to be accurate.

2.6 Nonlinear Models

In Section 2.5 we saw controllers that produced inputs that led to saturation of the system input, the car's pedal excursion. In some cases the required control input exceeded full

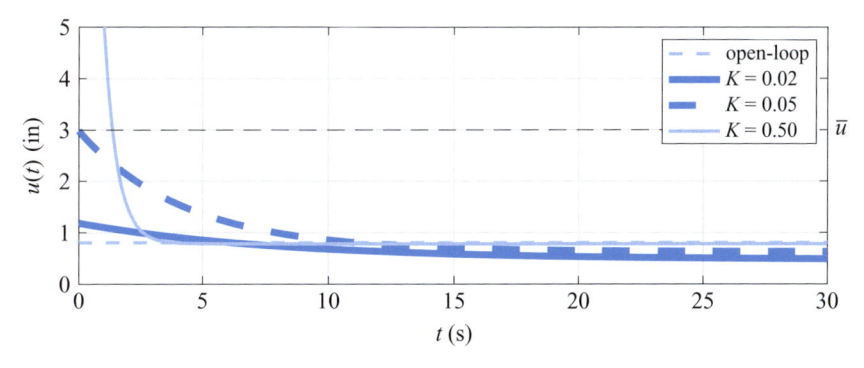

Figure 2.9 Open- and closed-loop control inputs (pedal excursion) corresponding to the dynamic responses in Fig. 2.8; the largest possible pedal excursion is 3 in.

throttle. In this section we digress a little to introduce a simple nonlinear model that can better predict the behavior of the system in closed-loop when saturation is present. This is not a course in nonlinear control, and the discussion will be kept at a very basic level. The goal is to be able to tell what happens in our simple example when the system reaches saturation.

In order to model the effect of saturation we will work with a nonlinear differential equation of the form

$$\dot{y}(t) + \frac{c}{m}\tan(\alpha^{-1}y(t)) = \frac{d}{m}u(t), \qquad u(t) \in [0, 3]. \tag{2.15}$$

When $u(t) = \tilde{u}$ is constant, one particular solution to (2.15) is $y(t) = \tilde{y}$, where

$$\frac{c}{m}\tan(\alpha^{-1}\tilde{y}) = \frac{d}{m}\tilde{u} \quad \Longrightarrow \quad \tilde{y} = \alpha \tan^{-1}(\beta\tilde{u}), \quad \beta = \frac{d}{c}.$$

This means that the steady-state response of the nonlinear differential equation (2.15) matches the empirical nonlinear fit performed earlier in Fig. 1.5. Moreover, at least for small values of $y(t)$, we should expect[8] that (see P2.2)

$$c\tan(\alpha^{-1}y(t)) \approx by(t), \qquad b = \alpha^{-1}c.$$

Intuitively, as long as $y(t)$ remains small and $u(t) \in [0, 3]$, the dynamic response of the nonlinear differential equation (2.15) should stay *close* to the dynamic response of the linear differential equation (2.3).

In order to estimate suitable parameters c/m, d/m, and α, we proceed as follows: first we borrow α and β from our previously computed nonlinear static fit ($\alpha = 82.8$ mph/in, $\beta = 1.2$ in^{-1}, see Fig. 1.5), then we estimate $b/m = -\lambda$ from the linear dynamic experiment described in Section 2.4, and calculate

$$\frac{c}{m} = \alpha \times \frac{b}{m} = 82.8 \times 0.05 = 4.1 \text{ mph/in},$$

$$\frac{d}{m} = \beta \times \frac{c}{m} = 1.2 \times 4.1 = 5.0 \text{ mph/in s}. \tag{2.16}$$

The resulting nonlinear model has a steady-state solution that matches the static fit from Fig. 1.5 and a time-constant close to that of the linear model from Section 2.1.

It is in closed-loop, however, that the nonlinear model will likely expose serious limitations of the controller based on the linear model (2.3). In order to capture the limits on pedal excursion we introduce the saturation nonlinearity:

$$\text{sat}_{(\underline{u},\bar{u})}(u(t)) = \begin{cases} \bar{u}, & u(t) > \bar{u}, \\ u(t), & \underline{u} \leq u(t) \leq \bar{u}, \\ \underline{u}, & u(t) < \underline{u}. \end{cases}$$

---

[8] This notion will be formalized in Section 5.4.

2.6 Nonlinear Models

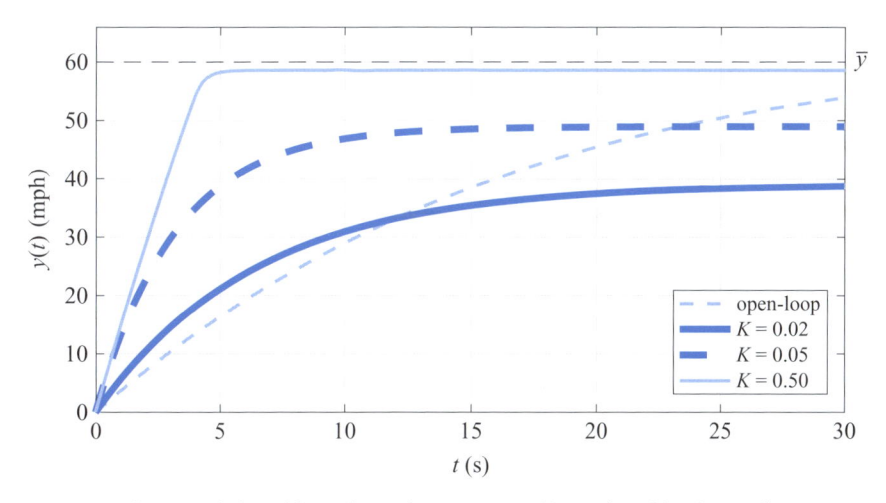

Figure 2.10 Open- and closed-loop dynamic response, $y(t)$, produced by the nonlinear car velocity model (2.17) calculated with $\alpha = 1.2$, $\beta = 82.8$, $c/m = 4.1$, and $d/m = 5.0$ and a constant target output of $\bar{y} = 60$ mph under proportional control (2.11) and various values of gain, K. Compare this with Fig. 2.8.

In the car model,[9] $\underline{u} = 0$ and $\bar{u} = 3$ in. The complete nonlinear model is

$$\dot{y}(t) + \frac{c}{m}\tan(\alpha^{-1}y(t)) = \frac{d}{m}\text{sat}_{(0,3)}(u(t)), \qquad u(t) = K(\bar{y} - y(t)).$$

Eliminating $u(t)$ we obtain

$$\dot{y}(t) + \frac{c}{m}\tan(\alpha^{-1}y(t)) = \frac{d}{m}\text{sat}_{(0,3)}(K(\bar{y} - y(t))). \qquad (2.17)$$

The above nonlinear ordinary differential equations cannot be solved analytically but can be simulated using standard numerical integration methods, e.g. one of the Runge–Kutta methods [BD12].

In order to see the effect of the nonlinearities on the closed-loop performance we repeat the simulations performed in Figs. 2.8 and 2.9, this time using the nonlinear feedback model (2.17). We show in Figs. 2.10 and 2.11 the nonlinear system closed-loop response, $y(t)$, and the control input, $u(t)$, for various values of the gain, K. These should be compared with Figs. 2.8 and 2.9. For values of $K = 0.05$ and $K = 0.02$, when the control input predicted using the linear model is within the linear region, i.e. $u(t) \in [0, 3]$, and the speeds are small, the nonlinearity has a minor impact. However, in the case of the larger gain $K = 0.5$, where the control input is heavily saturated (see Fig. 2.11), the response is significantly different. In particular, the extraordinarily fast response predicted by the linear model is not realized in the nonlinear model. In this simple example, the slower response seems to be the only apparent consequence. This will not always be the case, and severe nonlinearities will often negatively affect the performance of closed-loop systems. See Section 5.8.

---

[9] Note that by setting $\underline{u} = 0$ we prevent the model from applying any braking force.

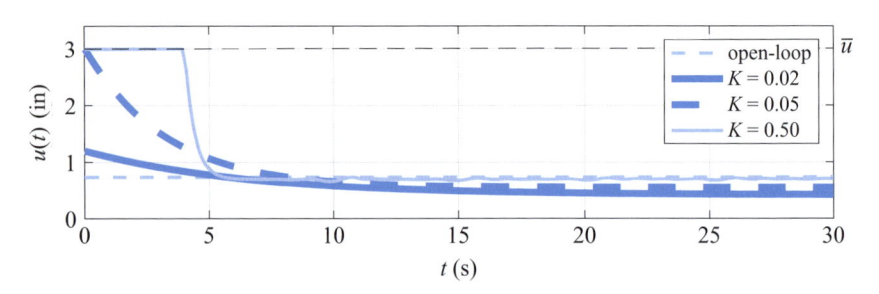

Figure 2.11 Open- and closed-loop control input, pedal excursion, $u(t)$, produced by the car velocity nonlinear model equation (2.17) under proportional control (2.11); the largest possible pedal excursion is 3 in; note the marked effect of pedal saturation in the case of the highest gain $K = 0.5$ and its impact in Fig. 2.10.

2.7 Disturbance Rejection

We now return to linear models to talk about a much desired feature of feedback control: *disturbance rejection*. Consider a modified version of the cruise control problem where the car is on a slope, as illustrated in Fig. 2.12. Newton's law applied to the car produces the differential equation[10]

$$m\dot{y}(t) + by(t) = f(t) - mg\sin(\theta(t)),$$

where $y = \dot{x}$ is the velocity of the car, θ is the angle the slope makes with the horizontal, and g is the gravitational acceleration.[11] When $\theta = 0$ the car is on the flat, and the model reduces to (2.1). As before, adoption of a linear model for the relationship between the force, f, and the pedal excursion, u, i.e. $f = pu$ from (2.2), produces the differential equation

$$\dot{y}(t) + \frac{b}{m}y(t) = \frac{p}{m}u(t) - g\sin(\theta(t)).$$

This is a linear differential equation except for the way in which $\theta(t)$ enters the equation. In most cases, the *signal* $\theta(t)$ is not known ahead of time, and can be seen as a *nuisance* or *disturbance*. Instead of working directly with θ, it is convenient to introduce the disturbance signal

$$w(t) = -\frac{mg}{p}\sin(\theta(t)) \qquad (2.18)$$

---

[10] Strictly speaking, this differential equation is true only if $\theta(t)$ is constant. When $\theta(t)$ is not constant, the car is subject to a (non-working) force that originates from changes in its frame of reference. This additional force can itself be treated as an additional disturbance if the changes in slope are moderate. No rollercoasters please!

[11] As we insist on using a non-standard unit for measuring velocity (mph), g will have to be expressed in mph/s, or $g = 9.8 \text{ m/s}^2 \approx 9.8 \times 3600/1609 \approx 21.9 \text{ mph/s}$. Ugh! That is ugly!

2.7 Disturbance Rejection

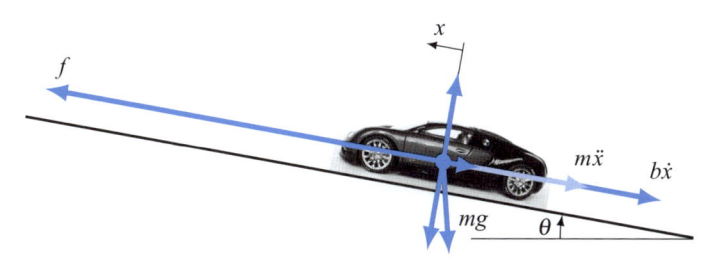

Figure 2.12 Free-body diagram showing forces acting on a car on a road slope.

as affecting the linear model

$$\dot{y}(t) + \frac{b}{m}y(t) = \frac{p}{m}u(t) + \frac{p}{m}w(t). \tag{2.19}$$

This differential equation is linear and can be analyzed with simpler tools.

The car model with the input disturbance is represented in closed-loop by the block-diagram in Fig. 2.13, which corresponds to the equations

$$\dot{y}(t) + \frac{b}{m}y(t) = \frac{p}{m}u(t) + \frac{p}{m}w(t), \qquad u(t) = K(\bar{y} - y(t)),$$

or, after eliminating u,

$$\dot{y}(t) + \left(\frac{b}{m} + \frac{p}{m}K\right)y(t) = \frac{p}{m}K\bar{y} + \frac{p}{m}w(t). \tag{2.20}$$

In order to understand how disturbances affect the closed-loop behavior we shall analyze the following scenario: suppose that the car is traveling on the flat, $w = \theta = 0$, with the cruise control in closed-loop at the steady-state velocity $y_{ss}(t) = H(0)\bar{y}$. At time $t = 0$ s, the car hits a 10% grade slope, $\bar{\theta} \approx 5.7°$. We use (2.18) to calculate the disturbance $w(t) = \bar{w} \approx -0.26, t \geq 0$.

The dynamic response of the car to the change in slope can be computed by formula (2.6) after setting

$$y_0 = H(0)\bar{y}, \qquad w(t) = \bar{w}, \quad t \geq 0$$

and calculating

$$\lambda = -\frac{b}{m} - \frac{p}{m}K, \qquad \tilde{y} = \frac{(p/m)K\bar{y} + (p/m)\bar{w}}{b/m + (p/m)K}.$$

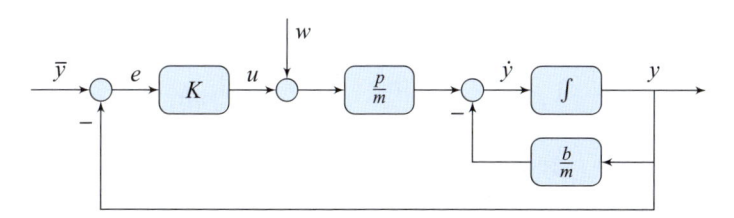

Figure 2.13 Closed-loop connection of the car showing the slope disturbance $w = -(mg/p)\sin(\theta)$.

Dynamic Systems

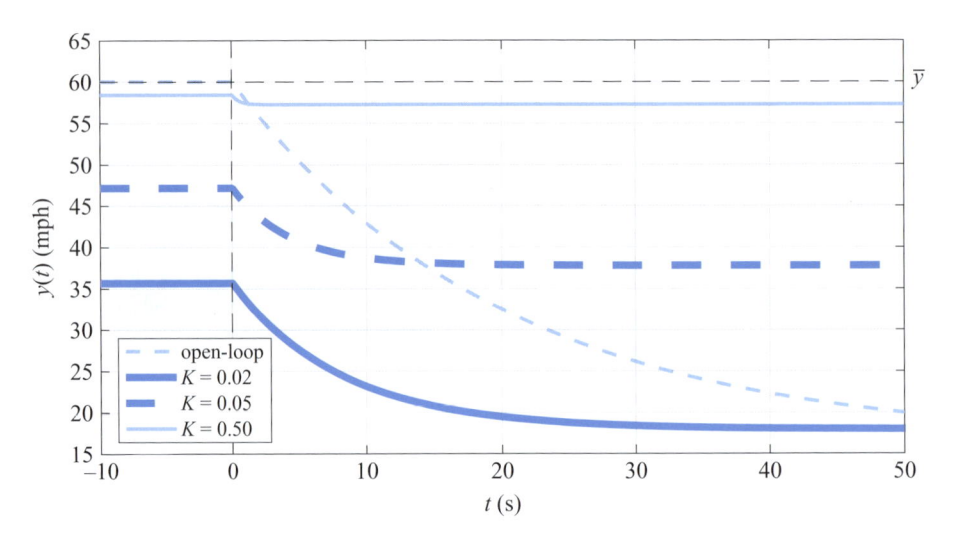

Figure 2.14 Closed-loop response of the velocity of the car with proportional cruise control (linear model (2.12), $b/m = 0.05$ and $p/b = 73.3$) to a change in road slope at $t = 0$, from flat to 10% grade for various values of the control gain.

It is useful to split $\tilde{y}$ into two components:

$$\tilde{y} = H(0)\bar{y} + D(0)\bar{w}, \qquad D(0) = \frac{p}{b}\frac{1}{1+(p/b)K}, \qquad (2.21)$$

where $H(0)$ is the same as in (2.13), and with which (2.6) becomes

$$y(t) = \tilde{y}\left(1 - e^{\lambda t}\right) + y_0 e^{\lambda t} = H(0)\bar{y} + \left(1 - e^{\lambda t}\right)D(0)\bar{w}.$$

The closed-loop response and the open-loop response are plotted in Fig. 2.14 for various values of the gain, K. The predicted change in velocity is equal to

$$\Delta y(t) = y(t) - y_0 = y(t) - H(0)\bar{y} = \left(1 - e^{\lambda t}\right)D(0)\bar{w}.$$

From (2.21), the larger K, the smaller $D(0)$, hence the smaller the change in velocity induced by the disturbance.

Compare the above analysis with the change in velocity produced by the open-loop solution (see P2.3):

$$\Delta y(t) = \left(1 - e^{-(b/m)t}\right)G(0)\bar{w}, \qquad G(0) = \frac{p}{b}. \qquad (2.22)$$

Because for any $K > 0$ we have

$$G(0) = \frac{p}{b} > \frac{p}{b}\frac{1}{1+(p/b)K} = D(0),$$

we conclude that the feedback solution always provides better *regulation* of the velocity in the presence of a road slope disturbance. Finally, large gains will bring down not only the tracking error but also the *regulation error* in response to a disturbance. Indeed, bigger Ks make both $S(0)$ and $D(0)$ small.

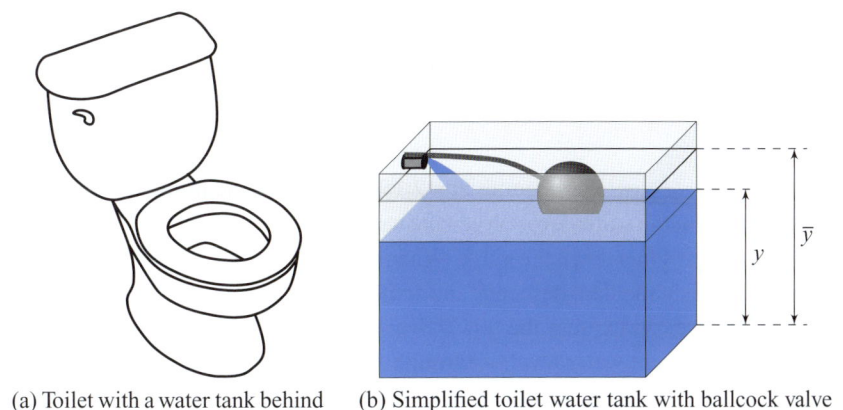

(a) Toilet with a water tank behind   (b) Simplified toilet water tank with ballcock valve

Figure 2.15 A toilet with a water tank and a simplified schematic diagram showing the ballcock valve. The tank is in the shape of a rectangular prism with cross-sectional area A. The water level is y and the fill line is $\bar{y}$.

2.8 Integral Action

We close this chapter with an analysis of a simple and familiar controlled system: a toilet water tank, Fig. 2.15. This system has a property of much interest in control, the so-called *integral action*. As seen in previous sections, large gains will generally lead to small tracking errors but with potentially damaging consequences, such as large control inputs that can lead to saturation and other nonlinear effects. In the examples presented so far, only an infinitely large gain could provide *zero* steady-state tracking error in closed-loop. As we will see in this section, integral action will allow closed-loop systems to track constant references with zero steady-state tracking error without resorting to infinite gains.

A schematic diagram of a toilet water tank is shown in Fig. 2.15(b). Assuming that the tank has a constant cross-sectional area, A, the amount of water in the tank, i.e. the volume, v, is related to the water level, y, by

$$v = Ay.$$

When the tank is closed, for instance, right after a complete flush, water flows in at a rate $u(t)$, which is controlled by the ballcock valve. Without leaks, the water volume in the tank is preserved, hence

$$\dot{v}(t) = u(t).$$

On combining these two equations in terms of the water level, y, we obtain the differential equation

$$\dot{y}(t) = \frac{1}{A}u(t), \qquad (2.23)$$

which reveals that the toilet water tank is essentially a *flow integrator*, as shown in the block-diagram representation in Fig. 2.16.

Dynamic Systems

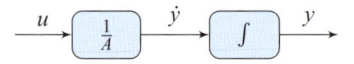

Figure 2.16 Block-diagram for water tank.

A *ballcock valve*, shown in Fig. 2.15(b), controls the inflow of water by using a float to measure the water level. When the water level reaches the *fill line*, $\bar{y}$, a lever connected to the float shuts down the valve. When the water level is below the fill line, such as right after a flush, the float descends and actuates the fill valve. This is a feedback mechanism. Indeed, we can express the flow valve as a function of the *error* between the fill line, $\bar{y}$, and the current water level, y, through

$$u(t) = K(\bar{y} - y),$$

where the profile of the function K is similar to the saturation curves encountered before in Figs. 1.4–1.6. The complete system is represented in the block-diagram Fig. 2.17, which shows that the valve is indeed a feedback element: the water level, y, tracks the reference level, fill line, $\bar{y}$.

With simplicity in mind, assume that the valve is linear. In this case, the behavior of the tank with the ballcock valve is given by the differential equation

$$\dot{y}(t) + \frac{K}{A}y(t) = \frac{K}{A}\bar{y}(t).$$

This equation is of the form (2.3) and has once again as solution (2.6), that is,

$$y(t) = \bar{y}(1 - e^{\lambda t}) + y_0 e^{\lambda t}, \quad \lambda = -\frac{K}{A}, \quad t \geq 0.$$

Note, however, the remarkable fact that

$$\lim_{t \to \infty} y(t) = \bar{y}.$$

In other words, the steady-state solution is always equal to the target fill line, $\bar{y}$, if $K/A > 0$, no matter what the actual values of K and A are! The toilet water tank level, y, tracks the fill line level, $\bar{y}$, exactly without a high-gain feedback controller. As will become clear in Chapter 4, the reason for this remarkable property is the presence of a pure integrator in the feedback loop. Of course, the values of K and A do not affect the steady-state solution but do influence the rate at which the system converges to it.

Integral action can be understood with the help of the closed-loop diagram in Fig. 2.17. First note that for the output of an integrator to converge to a constant

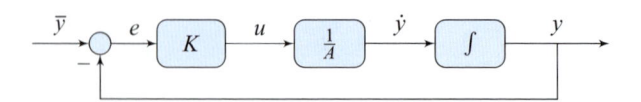

Figure 2.17 Block-diagram for water tank with ballcock valve.

value it is necessary that its input converge to zero. In Fig. 2.17, it is necessary[12] that $\lim_{t\to\infty} \dot{y}(t) = 0$. With that in mind, if $y(t)$ converges to anything other than $\bar{y}$, that is, $\lim_{t\to\infty} y(t) = \tilde{y} \neq \bar{y}$, then $\lim_{t\to\infty} e(t) = \tilde{y} - \bar{y} \neq 0$. But, if this is the case, $\lim_{t\to\infty} (A/K)\dot{y}(t) = \lim_{t\to\infty} e(t) = \tilde{y} - \bar{y} \neq 0$. Consequently $y(t)$ cannot converge to a constant other than $\bar{y}$. This is true even in the presence of some common nonlinearities in the loop. The ability to track constant references without high gains is the main reason behind the widespread use of integral control. We will analyze integral controllers in more detail in many parts of this book.

We conclude this section by revisiting the car example with linear model (2.3), by noting that when there is no damping, i.e. $b = 0$, then the car becomes a pure integrator. Indeed, in this case

$$\lim_{b\to 0} H(0) = \lim_{b\to 0} \frac{(p/b)K}{1+(p/b)K} = 1, \qquad \lim_{b\to 0} S(0) = \lim_{b\to 0} \frac{1}{1+(p/b)K} = 0,$$

which implies

$$\lim_{t\to\infty} y(t) = \tilde{y} = \lim_{b\to 0} H(0)\bar{y} = \bar{y},$$

independently of the value of K. We saw in Section 2.7 that large controller gains lead not only to small tracking errors but also to effective disturbance rejection. The same is true for an integrator *in the controller*, which leads to asymptotic tracking *and* asymptotic disturbance rejection. However, the position of the integrator in the loop matters: an integrator in the system but not in the controller will lead to zero tracking error but nonzero disturbance rejection error. For instance, in the example of the car we have seen that $b \to 0$ implies $S(0) \to 0$ but

$$\lim_{b\to 0} D(0) = \lim_{b\to 0} \frac{p}{b} \frac{1}{1+(p/b)K} = \frac{1}{K}, \qquad \lim_{t\to\infty} \Delta y(t) = D(0)\bar{w},$$

which is in general not zero. Nevertheless, it does gets smaller as the gain, K, gets large. By contrast, an integrator on the controller will generally lead to $S(0) = D(0) = 0$ independently of the loop gain, K. We will study this issue in more detail in Section 4.5.

Problems

2.1 Consider the solution (2.6) to the first-order ordinary differential equation (2.3) where the constant parameters m, b, and p are from the car velocity dynamic model developed in Section 2.1. Assign compatible units to the signals and constants in (2.3) and calculate the corresponding units of the parameter λ, from (2.7), and the time-constant τ, from (2.8).

---

[12] This is a necessary only condition, since $\dot{y}(t) = (1+t)^{-1}$ is such that $\lim_{t\to\infty} \dot{y}(t) = 0$ but $\lim_{t\to\infty} \int_0^t \dot{y}(\tau) d\tau = \ln(1+t) = \infty$.

Dynamic Systems

2.2 Use the Taylor series expansion of the function $f(x) = \tan^{-1}(x)$ to justify the approximation

$$c \tan(\alpha^{-1} y) \approx by$$

when $b = \alpha^{-1} c$.

2.3 Calculate the dynamic response, $y(t)$, of the open-loop car velocity model (2.19) when

$$y_0 = \bar{y}, \qquad u(t) = G(0)^{-1} \bar{y}, \qquad w(t) = \bar{w}, \qquad t \geq 0,$$

and $G(0) = p/b$. Calculate the change in speed $\Delta y(t) = y(t) - y_0$ and compare your answer with (2.22).

The next problems involve the motion of particle systems using Newton's law.

2.4 Show that the first-order ordinary differential equation

$$m\dot{v} + bv = mg$$

is a simplified description of the motion of an object of mass m dropping vertically under constant gravitational acceleration, g, and linear air resistance, $-bv$.

2.5 The first-order ordinary differential equation obtained in P2.4 can be seen as a dynamic system where the output is the vertical velocity, v, and the input is the gravitational force, mg. Calculate the solution to this equation. Consider $m = 1$ kg, $b = 10$ kg/s, $g = 10$ m/s². Sketch or use MATLAB to plot the response, $v(t)$, when $v(0) = 0$, $v(0) = 1$ m/s, or $v(0) = -1$ m/s.

2.6 Calculate the vertical position, $x(t)$, corresponding to the velocity, $v(t)$, computed in P2.5. How does the vertical position, $x(t)$, relate to the height measured from the ground, $h(t)$, of a free-falling object? Use the same data as in P2.5 and sketch or use MATLAB to plot the position $x(t)$ and the height $h(t)$, when $x(0) = 0$, $h(0) = 1$, and $v(0) = 0$, $v(0) = 1$ m/s, or $v(0) = -1$ m/s for 1 s.

2.7 The first-order nonlinear ordinary differential equation

$$m\dot{v} + bv^2 = mg, \qquad v > 0,$$

is a simplified description of the motion of an object of mass m dropping vertically under constant gravitational acceleration, g, and quadratic air resistance, bv^2. Verify that

$$v(t) = \frac{1 + \alpha e^{\lambda t}}{1 - \alpha e^{\lambda t}} \tilde{v}, \qquad \tilde{v} = \sqrt{\frac{mg}{b}}, \qquad \alpha = \frac{v(0) - \tilde{v}}{v(0) + \tilde{v}}, \qquad \lambda = -\frac{2b\tilde{v}}{m},$$

is a solution to this differential equation.

2.8 A sky diver weighing 70 kg reaches a constant vertical speed of 200 km/h during the free-fall phase of the dive and a vertical speed of 20 km/h after the parachute is opened. Approximate each phase of the fall by the ordinary differential equation obtained in P2.4 and estimate the resistance coefficients using the given information. Use $g = 10 \, \text{m/s}^2$. What are the time-constants in each phase? At what time and distance from the ground should the parachute be opened if the landing speed is to be less than or equal to 29 km/h? If a dive starts at a height of 4 km with zero vertical velocity at the moment of the jump and the parachute is opened 60 s into the dive, how long is the diver airborne?

2.9 Redo P2.8 using the nonlinear model from P2.7.

The next problems involve the planar rotation of a rigid body. Such systems can be approximately modeled by the first-order ordinary differential equation:

$$J \dot{\omega} = \tau,$$

where ω is the body's angular speed, J is the body's moment of inertia about its center of mass, and τ is the sum of all torques about the center of mass of the body. In constrained rotational systems, e.g. lever, gears, etc., the center of mass can be replaced by the center of rotation.

2.10 An (inextensible and massless) belt is used to drive a rotating machine without slip as shown in Fig. 2.18(a). The simplified motion of the inertia J_1 is described by

$$J_1 \dot{\omega}_1 = \tau + f_1 r_1 - f_2 r_1,$$

where τ is the torque applied by the driving motor and f_1 and f_2 are tensions on the belt. The machine is connected to the inertia J_2, which represents the sum of the inertias of all machine parts. The motion of the inertia J_2 is described by

$$J_2 \dot{\omega}_2 = f_2 r_2 - f_1 r_2.$$

(a) Rotating machine

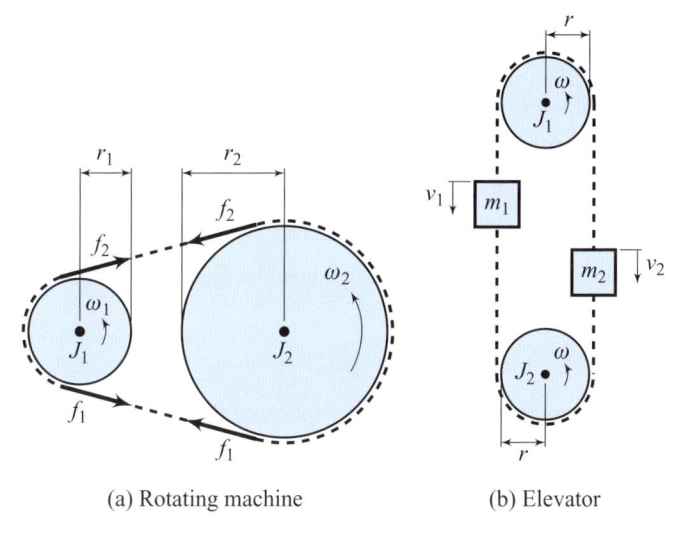

(b) Elevator

Figure 2.18 Diagrams for P2.10 and P2.18.

Show that the motion of the entire system can be described by the differential equation

$$(J_1 r_2^2 + J_2 r_1^2)\,\dot\omega_1 = r_2^2\,\tau, \qquad\qquad \omega_2 = (r_1/r_2)\,\omega_1.$$

2.11 Why is $f_1 \ne f_2$ in P2.10? Under what conditions are f_1 and f_2 equal?

2.12 Redo P2.10 in the presence of viscous friction torques, $-b_1\omega_1$ and $-b_2\omega_2$, on each inertia to obtain the differential equation

$$(J_1 r_2^2 + J_2 r_1^2)\dot\omega_1 + (b_1 r_2^2 + b_2 r_1^2)\omega_1 = r_2^2\,\tau.$$

2.13 Determine a first-order ordinary differential equation based on P2.10 and P2.12 to describe the rotating machine as a dynamic system where the output is the angular velocity of the inertia J_2, ω_2, and the input is the motor torque, τ. Calculate the solution to this equation. Consider $\tau = 1\,\text{N m}$, $r_1 = 25\,\text{mm}$, $r_2 = 500\,\text{mm}$, $b_1 = 0.01\,\text{kg m}^2/\text{s}$, $b_2 = 0.1\,\text{kg m}^2/\text{s}$, $J_1 = 0.0031\,\text{kg m}^2$, $J_2 = 25\,\text{kg m}^2$. Sketch or use MATLAB to plot the response, $\omega_2(t)$, when $\omega_2(0) = 0\,\text{rad/s}$, $\omega_2(0) = 3\,\text{rad/s}$, or $\omega_2(0) = 6\,\text{rad/s}$.

2.14 Calculate the (open-loop) motor torque, τ, for the rotating machine model in P2.10 and P2.12 so that the rotational speed of the mass J_2, ω_2, converges to $\bar\omega_2 = 4\,\text{rad/s}$ as t gets large. Use the same data as in P2.13 and sketch or use MATLAB to plot the response, $\omega_2(t)$, when $\omega_2(0) = 0\,\text{rad/s}$, $\omega_2(0) = 3\,\text{rad/s}$, or $\omega_2(0) = 6\,\text{rad/s}$.

2.15 What happens with the response in P2.14 if the actual damping coefficients b_1 and b_2 are 20% larger than the ones you used to calculate the open-loop torque?

2.16 The feedback controller

$$\tau(t) = K(\bar\omega_2 - \omega_2(t))$$

can be used to control the speed of the inertia J_2 in the rotating machine discussed in P2.10 and P2.12. Calculate and solve a differential equation that describes the closed-loop response of the rotating machine. Using data from P2.13, select a controller gain, K, with which the time-constant of the rotating machine is 3 s. Compare your answer with the open-loop time-constant. Calculate the closed-loop steady-state error between the desired rotational speed, $\bar\omega_2 = 4\,\text{rad/s}$, and $\omega_2(t)$. Sketch or use MATLAB to plot the response, $\omega_2(t)$, when $\omega_2(0) = 0\,\text{rad/s}$, $\omega_2(0) = 3\,\text{rad/s}$, or $\omega_2(0) = 6\,\text{rad/s}$.

2.17 What happens with the response in P2.16 if the actual damping coefficients b_1 and b_2 are 20% larger than the ones you used to calculate the closed-loop gain?

2.18 A schematic diagram of an elevator is shown in Fig. 2.18(b). Proceed as in P2.10 to show that

$$(J_1 + J_2 + r^2(m_1 + m_2))\,\dot\omega + (b_1 + b_2)\omega = \tau + gr(m_1 - m_2),$$

$v_1 = r\omega$, and $v_2 = -r\omega$, is a simplified description of the motion of the entire elevator system, where τ is the torque applied by the driving motor on the inertia J_1, and b_1 and b_2 are viscous friction torque coefficients at the inertias J_1 and J_2. If m_1 is the load to be lifted and m_2 is a counterweight, explain why is it advantageous to have the counterweight match the elevator load as closely as possible.

2.19 Determine a first-order ordinary differential equation based on P2.18 to describe the elevator as a dynamic system where the output is the vertical velocity of the mass m_1, v_1, and the inputs are the motor torque, τ, and the gravitational torque, $gr(m_1 - m_2)$. Calculate the solution to this equation. Consider $g = 10$ m/s^2, $\tau = 0$ N m, $r = 1$ m, $m_1 = m_2 = 1000$ kg, $b_1 = b_2 = 120$ kg m^2/s, $J_1 = J_2 = 20$ kg m^2. Sketch or use MATLAB to plot the response, $v_1(t)$, when $v_1(0) = 0$, $v_1(0) = 1$ m/s, or $v_1(0) = -1$ m/s.

2.20 Repeat P2.19 with $m_2 = 800$ kg.

2.21 Calculate the (open-loop) motor torque, τ, for the elevator model in P2.19 so that the vertical velocity of the mass m_1, v_1, converges to $\bar{v}_1 = 2$ m/s as t gets large. Use the same data as in P2.19 and sketch or use MATLAB to plot the response, $v_1(t)$, when $v_1(0) = 0$, $v_1(0) = 1$ m/s, or $v_1(0) = -1$ m/s.

2.22 Let $m_2 = 800$ kg and use the same motor torque, τ, you calculated in P2.21 and the rest of the data from P2.19 to sketch or use MATLAB to plot the response of the elevator mass m_1 velocity, $v_1(t)$, when $v_1(0) = 0$, $v_1(0) = 1$ m/s, or $v_1(0) = -1$ m/s. Did the velocity converge to $\bar{v}_1 = 2$ m/s? If not, recalculate a suitable torque. Plot the response with the modified torque, compare your answer with P2.21, and comment on the value of torque you obtained.

2.23 The feedback controller:

$$\tau(t) = K(\bar{v}_1 - v_1(t))$$

can be used to control the ascent and descent speed of the mass m_1 in the elevator discussed in P2.18 and P2.19. Calculate and solve a differential equation that describes the closed-loop response of the elevator. Using data from P2.19, select a controller gain, K, with which the time-constant of the elevator is approximately 5 s. Compare this value with the open-loop time-constant. Calculate the closed-loop steady-state error between the desired vertical velocity, $\bar{v}_1 = 2$ m/s, and $v_1(t)$. Sketch or use MATLAB to plot the response of the elevator mass m_1 velocity, $v_1(t)$, when $v_1(0) = 0$, $v_1(0) = 1$ m/s, or $v_1(0) = -1$ m/s. Compare the response with the open-loop control response from P2.22.

2.24 Repeat P2.23, this time setting the closed-loop time-constant to be about 0.5 s. What is the effect on the response? Do you see any problems with this solution?

2.25 Repeat P2.23 with $m_2 = 800$ kg. Treat the gravitational torque, $gr(m_1 - m_2)$, as a disturbance.

2.26 Repeat P2.25 this time setting the closed-loop time-constant to be about 0.5 s. What is the effect on the response? Do you see any problems with this solution?

Figures 2.19 through 2.21 show diagrams of mass–spring–damper systems. Assume that there is no friction between the wheels and the floor, and that all springs and dampers are linear: elongating a linear spring with rest length ℓ_0 by $\Delta\ell$ produces an opposing force $k\Delta\ell$ (Hooke's Law), where $k > 0$ is the spring stiffness; changing the length of a linear damper at a rate $\dot{\ell}$ produces an opposing force $b\dot{\ell}$, where b is the damper's damping coefficient.

2.27 Choose x_0 wisely to show that the ordinary differential equation

$$m\ddot{x} + b\dot{x} + kx = f$$

is a simplified description of the motion of the mass–spring–damper system in Fig. 2.19(a), where f is a force applied on mass m. Why does the equation not depend on the spring rest length ℓ_0?

2.28 Show that the ordinary differential equation

$$m\ddot{x} + b\dot{x} + kx = mg\sin\theta$$

is a simplified description of the motion of the mass–spring–damper system in Fig. 2.19(b), where g is the gravitational acceleration and x_0 is equal to the spring rest length ℓ_0.

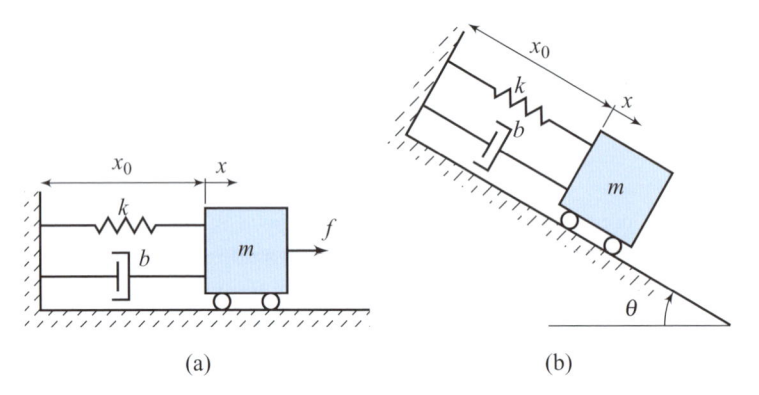

Figure 2.19 Diagrams for P2.27 and P2.28.

2.29 Rewrite the ordinary differential equation obtained in P2.28 as

$$m\ddot{y} + b\dot{y} + ky = 0, \qquad y = x - k^{-1}mg\sin\theta.$$

Relate this result to a different choice of x_0 and comment on your findings.

2.30 Show that the ordinary differential equation

$$m\ddot{x} + b\dot{x} + (k_1 + k_2)x = 0$$

is a simplified description of the motion of the mass–spring–damper system in Fig. 2.20(a). What does x represent? Why do the equations not depend either on the rest lengths $\ell_{0,1}$, $\ell_{0,2}$ or on the dimensions d and w? What is the difference between the cases $d \geq w + \ell_{0,1} + \ell_{0,2}$ and $d \leq w + \ell_{0,1} + \ell_{0,2}$?

2.31 Can you replace the two springs in P2.30 by a single spring and still obtain the same ordinary differential equation?

2.32 Show that the ordinary differential equations

$$m_1\ddot{x}_1 + (b_1 + b_2)\dot{x}_1 + (k_1 + k_2)x_1 - b_2\dot{x}_2 - k_2 x_2 = 0,$$
$$m_2\ddot{x}_2 + b_2(\dot{x}_2 - \dot{x}_1) + k_2(x_2 - x_1) = f_2$$

constitute a simplified description of the motion of the mass–spring–damper system in Fig. 2.20(b), where $x_1 = x_2 = 0$ when the length of both springs is equal to their rest lengths and f_2 is a force applied on mass m_2.

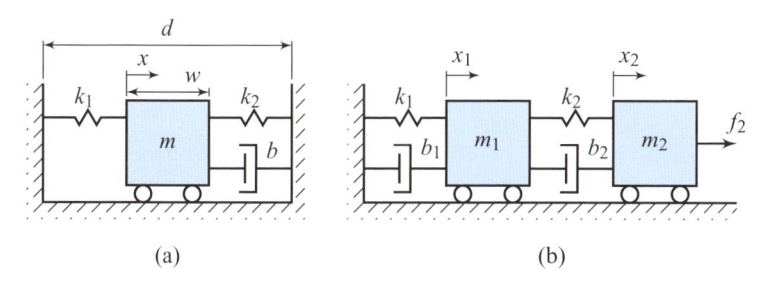

Figure 2.20 Diagrams for P2.30 and P2.32.

2.33 Show that the ordinary differential equations

$$m_1\ddot{x}_1 + b(\dot{x}_1 - \dot{x}_2) + k(x_1 - x_2) = 0,$$
$$m_2\ddot{x}_2 + b(\dot{x}_2 - \dot{x}_1) + k(x_2 - x_1) = 0,$$

constitute a simplified description of the motion of the mass–spring–damper system in Fig. 2.21 where $x_1 = x_2 = 0$ when the length of the spring is its rest length. Show that it is possible to *decouple* the equations if you write them in the coordinates

$$y_1 = \frac{m_1 x_1 + m_2 x_2}{m_1 + m_2}, \qquad y_2 = x_1 - x_2.$$

Use what you know from physics to explain why.

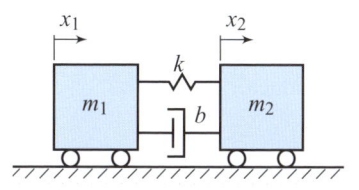

Figure 2.21 Diagram for P2.33.

The next problems have simple electric circuits. Electric circuits can be accurately modeled using ordinary differential equations.

2.34 An electric circuit in which a capacitor is in series with a resistor is shown in Fig. 2.22(a). In an electric circuit, the sum of the voltages around a loop must equal

Dynamic Systems

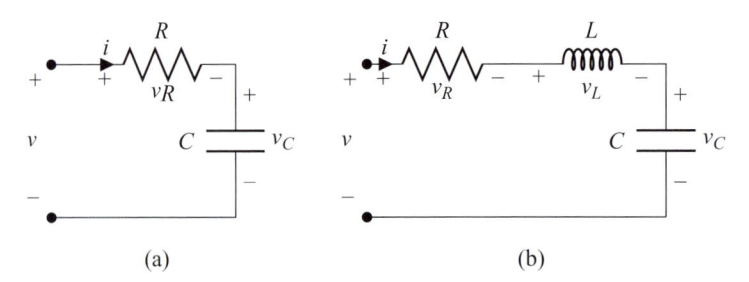

Figure 2.22 Diagrams for P2.34 and P2.36.

zero:
$$-v + v_R + v_C = 0.$$

This is Kirchhoff's voltage law. The voltage and the current on the capacitor and resistor satisfy

$$i_C = C\dot{v}_C, \qquad v_R = R i_R,$$

where C is the capacitor's *capacitance* and R is the resistor's *resistance*. In this circuit

$$i_R = i_C = i,$$

because all elements are *in series*. This is Kirchhoff's current law. Show that

$$RC\dot{v}_C + v_C = v$$

is the equation governing this *RC*-circuit.

2.35 Consider the *RC*-circuit from P2.34 where $R = 1\,\mathrm{M}\Omega$ and $C = 10\,\mathrm{\mu F}$. Assuming zero initial conditions, sketch the capacitor's voltage, $v_C(t)$, when a constant voltage $v(t) = 10\,\mathrm{V}, t \geq 0$, is applied to the circuit. Sketch also the circuit current $i(t)$.

2.36 An electric circuit in which an inductor, a capacitor, and a resistor are in series is shown in Fig. 2.22(b). As in P2.34, the sum of the voltages around a loop must equal zero:

$$-v + v_R + v_L + v_C = 0.$$

This is Kirchhoff's voltage law. The voltages and the currents on the capacitor and resistor are as in P2.34 and the voltage on the inductor is

$$v_L = L\dot{i}_L,$$

where L is the inductor's *inductance*. Because the elements are in series

$$i_R = i_C = i_L = i.$$

This is Kirchhoff's current law. Show that

$$LC\ddot{v}_c + RC\dot{v}_C + v_C = v$$

is the equation governing the *RLC*-circuit.

2.37 Consider the differential equation for the *RLC*-circuit from P2.36. Compare this equation with the equations of the mass–spring–damper system from P2.27 and explain how one could select values of the resistance, R, capacitance, C, inductance, L, and input voltage, v, to simulate the movement of the mass–spring–damper system in P2.27. The resulting device is an *analog computer*.

2.38 An approximate model for the electric circuit in Fig. 2.23, where the triangular element is an amplifier with a very large gain (operational amplifier, OpAmp), is obtained from

$$R_1 i_{R_1} = v - v_-, \qquad i_{C_1} = C_1(\dot{v} - \dot{v}_-), \qquad i_{C_2} = C_2(\dot{v}_- - \dot{v}_o),$$

and

$$v_- \approx v_+ = 0, \qquad i_{C_2} = i_{C_1} + i_{R_1}.$$

Show that

$$R_1 C_2 \dot{v}_o + R_1 C_1 \dot{v} + v = 0.$$

Solve the auxiliary differential equation:

$$\dot{z} + \frac{1}{R_1 C_2} v = 0$$

and show that

$$v_o(t) = R_1 C_1 \dot{z}(t) + z(t)$$

solves the original differential equation.

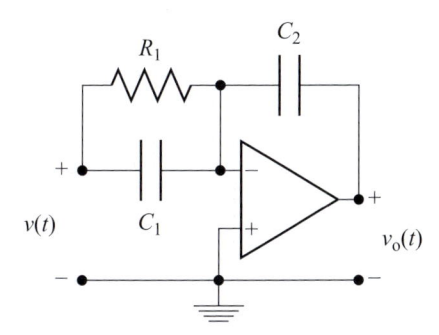

Figure 2.23 Diagram for P2.38.

2.39 Consider the OpAmp-circuit from P2.38 where $R_1 = 1\,\mathrm{M}\Omega$ and $C_1 = C_2 = 10\,\mu\mathrm{F}$. Assuming zero initial conditions, sketch the output voltage, $v_o(t)$, when a constant voltage $v(t) = 10\,\mathrm{V}, t \geq 0$, is applied to the circuit.

2.40 In P2.38, set $C_1 = 0$ and solve for $v_o(t)$ in terms of $v(t)$. Name one application for this circuit.

2.41 The mechanical motion of the rotor of a DC motor shown schematically in Fig. 2.24(a) can be described by the differential equation

$$J\dot{\omega} + b\omega = \tau,$$

where ω is the rotor angular speed, J is the rotor moment of inertia, b is the coefficient of viscous friction. The rotor torque, τ, is given by

$$\tau = K_t i_a$$

where i_a is the armature current and K_t is the motor torque constant. Neglecting the effects of the armature inductance ($L_a \approx 0$), the current is determined by the circuit in Fig. 2.24(b):

$$v_a = R_a i_a + K_e \omega,$$

where v_a is the armature voltage, R_a is the armature resistance, and K_e is the back-EMF constant. Combine these equations to show that

$$J\dot{\omega} + \left(b + \frac{K_e K_t}{R_a}\right)\omega = \frac{K_t}{R_a}v_a.$$

(a) Mechanical

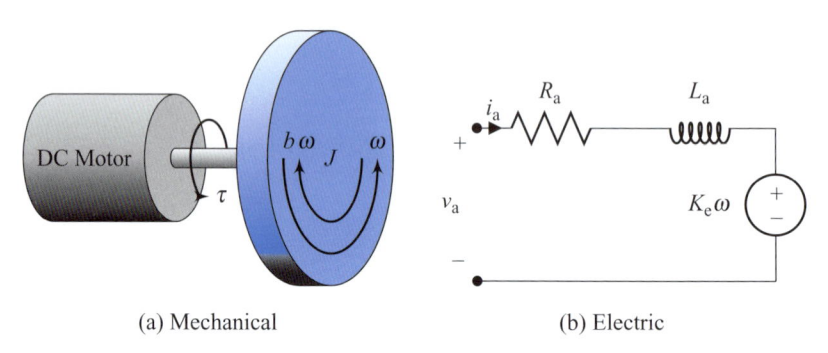

(b) Electric

Figure 2.24 Diagrams for P2.41.

2.42 Show that $K_t = K_e$. *Hint: Equate the mechanical power with the electric power.*

2.43 The first-order ordinary differential equation obtained in P2.41 can be seen as a dynamic system where the output is the angular velocity, ω, and the input is the armature voltage, v_a. Calculate the solution to this equation when v_a is constant. Estimate the parameters of the first-order differential equation describing a DC motor that achieves a steady-state angular velocity of 5000 RPM when $v_a = 12$ V and has a time-constant of 0.1 s. Can you also estimate the "physical" parameters J, b, K_t, K_e, and R_a with this information?

2.44 Can you estimate the parameters J, K_t, K_e, and b of the DC motor in P2.43 if you know $R_a = 0.2\ \Omega$ and the *stall torque* $\tau = 1.2$ N m at $v_a = 12$ V? *Hint: The stall torque is attained when the motor is held in place.*

2.45 DC motors with high-ratio gear boxes can be damaged if held in place. Can you estimate the parameters J, K_t, K_e, and b of the DC motor in P2.43 if you know that $R_a = 0.2\,\Omega$ and that after you attach an additional inertia $J' = 0.001$ kg m^2 the motor time-constant becomes 0.54 s?

2.46 Redo P2.41 using the equations in P2.36 to show that

$$JL_a\ddot{\omega} + (JR_a + bL_a)\dot{\omega} + (K_eK_t + bR_a)\omega = K_t v_a$$

when $L_a > 0$ is not negligible. Show that this equation reduces to the one in P2.41 if $L_a = 0$.

2.47 The feedback controller

$$v_a(t) = K(\bar{\omega} - \omega(t))$$

can be used to regulate the angular speed of the DC motor, $\omega(t)$, for which a model was developed in P2.41. Calculate and solve a differential equation that describes the closed-loop response of the DC motor. Using data from P2.43, select a controller gain, K, with which the closed-loop steady-state error between the desired angular speed, $\bar{\omega}$, and the actual angular speed, $\omega(t)$, is less than 10%. Calculate the resulting closed-loop time-constant and sketch or use MATLAB to plot the output $\omega(t)$ and the voltage $v_a(t)$ generated in response to a reference $\bar{\omega} = 4000$ RPM assuming zero initial conditions. What is the maximum value of $v_a(t)$?

2.48 Redo P2.47 but this time design K such that $v_a(0)$ is always smaller than 12 V when $\bar{\omega} = 4000$ RPM.

The next problems have simple examples of heat and fluid flow using ordinary differential equations. Detailed modeling of such phenomena often requires partial differential equations.

2.49 The temperature, T (in K or in °C), of a substance flowing in and out of a container kept at the ambient temperature, T_o, with an inflow temperature, T_i, and a heat source, q (in W), can be approximated by the differential equation

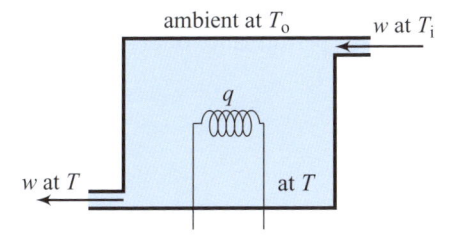

$$mc\dot{T} = q + wc(T_i - T) + \frac{1}{R}(T_o - T),$$

Figure 2.25 Diagram for P2.49.

where m and c are the substance's mass and specific heat, and R is the overall system's thermal resistance. The input and output flow mass rates are assumed to be equal to w. This differential equation model can be seen as a dynamic system where the output is the substance's temperature, T, and the inputs are the heat source, q, the flow rate, w, and the temperatures T_o and T_i. Calculate the solution to this equation when q, w, T_o, and T_i are constants.

2.50 Assume that water's density and specific heat are $997.1\,\text{kg/m}^3$ and $c = 4186\,\text{J/kg K}$. A 50 gal ($\approx 0.19\,\text{m}^3$) water heater is turned off full with water at $140\,°\text{F}$ ($\approx 60\,°\text{C}$). Use the differential equation in P2.49 to estimate the heater's thermal resistance, R, knowing that after 7 days left at a constant ambient temperature, $77\,°\text{F}$ ($\approx 25\,°\text{C}$), without turning it on, $q = 0$, or cycling any water, $w = 0$, the temperature of the water was about $80\,°\text{F}$ ($\approx 27\,°\text{C}$).

2.51 For the same conditions as in P2.50, calculate how much time it takes for the water temperature to reach $80\,°\text{F}$ ($\approx 27\,°\text{C}$) with a constant in/out flow of $20\,\text{gal/h}$ ($\approx 21 \times 10^{-6}\,\text{m}^3/\text{s}$) at ambient temperature. Compare your answer with the case when no water flows through the water heater.

2.52 Consider a water heater as in P2.50 rated at $40{,}000\,\text{BTU/h}$ ($\approx 12\,\text{kW}$). Calculate the time it takes to heat up a heater initially full with water at ambient temperature to $140\,°\text{F}$ ($\approx 60\,°\text{C}$) without any in/out flow of water, $w = 0$.

2.53 Repeat P2.52 for a constant in/out flow of $20\,\text{gal/h}$ at ambient temperature. Compare the solutions.

2.54 Most residential water heaters have a simple *on/off*-type controller: the water heater is turned on at full power when the water temperature, T, falls below a set value, $\underline{T}$, and is turned off when it reaches a second set point, $\overline{T}$. For a 50 gal ($\approx 0.19\,\text{m}^3$) heater as in P2.50 rated at $40{,}000\,\text{BTU/h}$ ($\approx 12\,\text{kW}$) and with thermal resistance $R = 0.27\,\text{K/W}$, sketch or use MATLAB to plot the temperature of the water during 24 hours for a heater with an on/off controller set with $\underline{T} = 122\,°\text{F}$ ($\approx 50\,°\text{C}$) and $\overline{T} = 140\,°\text{F}$ ($\approx 60\,°\text{C}$), without any in/out flow of water, $w = 0$. Assume that the heater is initially full with water a tad below $\underline{T}$. Compute the average water temperature and power consumption for a complete on/off cycle.

2.55 Repeat P2.54 for a constant in/out flow of $20\,\text{gal/h}$ at ambient temperature. Compare the solutions.

2.56 Repeat P2.54 with $\underline{T} = 129.2\,°\text{F}$ ($\approx 54\,°\text{C}$) and $\overline{T} = 132.8\,°\text{F}$ ($\approx 56\,°\text{C}$). What is the impact of the choice of $\underline{T}$ and $\overline{T}$ on the performance of the controller?

3 Transfer-Function Models

The dynamic models developed in Chapter 2 relate input and output signals evolving in the *time domain* through differential equations. In this chapter we will introduce transform methods that can relate input and output signals in the *frequency domain*, establishing a correspondence between a *time-invariant* linear system model and its frequency-domain *transfer-function*. A linear system model may not be time-invariant but, for linear time-invariant systems, the frequency domain provides an alternative vantage point from which to perform calculations and interpret the behavior of the system and the associated signals. This perspective will be essential to many of the control design methods to be introduced later in this book, especially those of Chapter 7.

Frequency-domain models can be obtained experimentally or derived formally from differential equations using the Laplace or Fourier transforms. In controls we work mostly with the Laplace transform. This chapter starts with a brief but not completely elementary review of the Laplace transform before returning to signals, systems, and controls. From this chapter on, transfer-functions become functions of complex variables and we use the symbol j to represents the imaginary unit, that is $j = \sqrt{-1}$. We use $\dot{f}(t)$ and $f'(s)$ to distinguish between differentiation with respect to the real variable $t \in \mathbb{R}$ or differentiation with respect to the complex variable $s \in \mathbb{C}$.

3.1 The Laplace Transform

The Laplace transform is essentially a sophisticated *change-of-variables* that associates a function[1] $f(t)$ of the *real-valued* time variable $t \in \mathbb{R}$, $t \geq 0$, with a function $F(s)$ of a *complex-valued* frequency variable $s \in \mathbb{C}$. Formally, the Laplace transform of the function $f(t)$ is the result of the integral:[2]

$$F(s) = \mathscr{L}\{f(t)\} = \int_{0^-}^{\infty} f(t) e^{-st} dt. \qquad (3.1)$$

This integral may not converge for every function $f(t)$ or every s, and when it converges it may not have a *closed-form* solution. However, for a large number of common

---

[1] A formal setup that is comfortable is that of piecewise continuous or piecewise smooth (continuous and infinitely differentiable) functions with only a discrete set of discontinuities, such as the one adopted in [LeP10].

[2] The notation 0^- means the one-side limit $\lim_{\epsilon \uparrow 0} \epsilon$, which is used to accommodate possible discontinuities of $f(t)$ at the origin.

Table 3.1 Laplace transform pairs

Function	$f(t)$, $t \geq 0$	$F(s)$
Impulse	$\delta(t)$	1
Step	1, $1(t)$	$\dfrac{1}{s}$
Ramp	t	$\dfrac{1}{s^2}$
Monomial	t^m	$\dfrac{m!}{s^{m+1}}$, $m \geq 1$
Exponential	e^{-at}	$\dfrac{1}{s+a}$
	$t^m e^{-at}$	$\dfrac{m!}{(s+a)^{m+1}}$, $m \geq 1$
Sine	$\sin(at)$	$\dfrac{a}{s^2+a^2}$
Cosine	$\cos(at)$	$\dfrac{s}{s^2+a^2}$
	$\cos(at + \phi)$	$\dfrac{s\cos\phi - a\sin\phi}{s^2+a^2}$
	$2|k|e^{-bt}\cos(at + \angle k)$	$\dfrac{k}{s+b-ja} + \dfrac{k^*}{s+b+ja}$

functions, some of which are given in Table 3.1, the integral converges and has a closed-form solution. This short collection will be enough to get us through all applications considered in this book. A sufficient condition for convergence is *exponential growth*. A function f has exponential growth if there exist $M \geq 0$ and $\alpha \in \mathbb{R}$ such that

$$|f(t)| \leq Me^{\alpha t}, \qquad t \geq 0.$$

Indeed, for $s = a + jb$ and $t \geq 0$,

$$|f(t)e^{-st}| = |f(t)e^{-jbt}||e^{-at}| = |f(t)|e^{-at} \leq Me^{(\alpha-a)t}. \tag{3.2}$$

Therefore, for all $s = a + jb$ such that $\text{Re}(s) = a > \alpha$,

$$|F(s)| = \int_{0^-}^{\infty} |f(t)e^{-st}|dt \leq M \int_{0^-}^{\infty} e^{(\alpha-a)t}dt = \left.\frac{Me^{(\alpha-a)t}}{\alpha - a}\right|_{0^-}^{\infty} = \frac{M}{\alpha - a}.$$

In other words, the Laplace transform integral (3.1) converges and is bounded for all s such that $\text{Re}(s) > \alpha$. The smallest possible value of α, labeled α_c, for which the Laplace transform converges for all s such that $\text{Re}(s) > \alpha_c$, is called the *abscissa of convergence*. Because the complex-valued function $F(s)$ is bounded it does not have any singularities in its region of convergence, that is for all s such that $\text{Re}(s) > \alpha_c$. In fact $F(s)$ is *analytic*[3] in its region of convergence. See [LeP10] or [Chu72] for a more thorough discussion of these issues.

---

[3] If a function is analytic at a point s_0 then the function and all its derivatives exist in a neighborhood of s_0. See Section 3.4 and [BC14, Section 25] for details.

For a concrete example, consider the Laplace transform of the constant function $f : \mathbb{R} \to \mathbb{R}$, $f(t) = 1$. This function has exponential growth ($\alpha = 0$, $M = 1$) and its Laplace transform is

$$F(s) = \int_{0^-}^{\infty} e^{-st} dt = \left. \frac{-e^{-st}}{s} \right|_{0^-}^{\infty} = 0 - \frac{-1}{s} = \frac{1}{s}, \qquad \text{Re}(s) > 0. \qquad (3.3)$$

As expected, $F(s) = s^{-1}$ does not have any singularities in the region of convergence $\text{Re}(s) > \alpha_c = 0$.

From a practical standpoint, one might find it useful to think of the Laplace transform as a *calculator*:[4] it takes up functions in the time domain and transforms them into the frequency domain where operations are quicker and easier. If needed, the results can be transformed back into the time domain. This can be done with the help of the *inverse Laplace transform*:

$$\mathcal{L}^{-1}\{F(s)\} = \frac{1}{2\pi j} \lim_{\omega \to \infty} \int_{\alpha - j\omega}^{\alpha + j\omega} F(s) e^{st} ds, \qquad \alpha > \alpha_c, \qquad (3.4)$$

which takes the form of a line integral in the complex plane. Note that the line $\alpha + j\omega$ must lie in the region of convergence of $F(s)$. Fortunately it will not be necessary to explicitly compute the integral (3.4) to find inverse Laplace transforms. The key is to use powerful results from complex analysis to convert the integral into a sum of simpler integrals, the *residues*, which need be evaluated only at points where $F(s)$ is singular. We will study residues later in Section 3.4, but before we delve further into the details, let us bring up a cautionary note about the use of the particular form[5] of Laplace transform in (3.1).

First, the transform uses information from $f(t)$ only on $t \geq 0$, and we expect the inverse Laplace transform (3.4) to be zero when $t < 0$, a fact we will prove later in Section 3.4. Because of the integration, discrete discontinuities do not contribute to the value of the integral. Consequently, the inverse Laplace transform is generally not unique, as piecewise continuous functions that differ only on a discrete set of discontinuities will have the same Laplace transform. Indeed, uniqueness of the Laplace transform has to be interpreted *up to a discrete set of discontinuities*. With this caveat in mind, each row in Table 3.1 is a Laplace transform *pair*.

As an example, consider the *step function*:[6]

$$1(t) = \begin{cases} 0, & t < 0; \\ 1, & t \geq 0. \end{cases} \qquad (3.5)$$

---

[4] A much older engineering student would be familiar with a *slide rule*, which was a popular instrument used to quickly perform multiplications and divisions by converting numbers to logarithms. Slide rules became extinct after the advent of cheap handheld electronic calculators.

[5] Versions of the Laplace transform that operate on functions on the entire real axis also exist but will be of no use to us [LeP10].

[6] The step function is sometimes denoted by the symbol $u(t)$. We use $1(t)$ in order to avoid confusion with the input signal $u(t)$.

Table 3.2 Laplace transform properties

Operation	t-domain (time)	s-domain (frequency)
Linearity	$\alpha f(t) + \beta g(t)$	$\alpha F(s) + \beta G(s)$
Integration	$\int_{0^-}^{t} f(\tau)d\tau$	$\dfrac{F(s)}{s}$
Differentiation in time	$\dot{f}(t) = \dfrac{df(t)}{dt}$	$sF(s) - f(0^-)$
	$f^{(n)}(t) = \dfrac{d^n f(t)}{dt^n}$	$s^n F(s) - s^{n-1} f(0^-) \cdots - f^{(n-1)}(0^-)$
Differentiation in frequency	$(-t)^n f(t)$	$\dfrac{d^n F(s)}{ds^n}$
Convolution	$\int_{0^-}^{t} f(\tau)g(t-\tau)d\tau$	$F(s)G(s)$
Shift in time	$f(t-\tau)1(t-\tau)$	$e^{-\tau s} F(s)$
Shift in frequency	$e^{-at} f(t)$	$F(s+a)$
Initial-value	$\lim_{t \to 0^+} f(t)$	$\lim_{|s| \to \infty} sF(s)$
Final-value[a]	$\lim_{t \to \infty} f(t)$	$\lim_{s \to 0} sF(s)$

[a] The final-value property may not hold if $F(s)$ has a singularity on the imaginary axis or the right-hand side of the complex plane. For example $f(t) = e^{at}$, $a > 0$, is such that $\lim_{t \to \infty} f(t) = \infty$. However, $F(s) = (s-a)^{-1}$ and $\lim_{s \to 0} sF(s) = 0$.

Because one cannot distinguish the step function from the constant function $f(t) = 1$ in $t \geq 0$, they have the same Laplace transform. However, the integration in (3.1) is performed in the interval $(0^-, \infty)$, which makes one wonder whether the difference $1(0^-) = 0 \neq 1$ matters. The issue is delicate and has deceived many [LMT07]. Its importance is due to the fact that the choice of the interval $(0^-, \infty)$ impacts many useful properties of the Laplace transform which depend on $f(0^-)$. One reason for the trouble with $f(0^-)$ is a remarkable property of the Laplace transform: it is capable of handling the differentiation of functions at points which are not continuous by cleverly *encoding* the size of the discontinuities with the help of *impulses*, that is the function $\delta(t)$ appearing on the first row of Table 3.1, for which $\mathscr{L}\{\delta(t)\} = 1$. In order to understand this behavior we need to learn how to compute integrals and derivatives using the Laplace transform, which we do by invoking the *differentiation* and *integration* properties listed in Table 3.2. All properties in Table 3.2 will be proved by you in P3.1–P3.14 at the end of this chapter. Key properties are *linearity* and *convolution*, which we will discuss in more detail in Section 3.2. For now we will continue on differentiation and integration.

Consider the constant function $f(t) = 1$ and the unit step $f(t) = 1(t)$. Their values agree for all $t \geq 0$ and they have the same Laplace transform, $F(s) = s^{-1}$. The constant function $f(t) = 1$ is differentiable everywhere and has $\dot{f}(t) = 0$. Application of the formal differentiation property from Table 3.2 to the Laplace transform of the constant function produces

$$sF(s) - f(0^-) = 1 - 1 = 0 \quad \Longrightarrow \quad \dot{f}(t) = 0, \quad t \geq 0,$$

which is what one would have expected. The step function $f(t) = 1(t)$ is differentiable everywhere except at the origin. Wherever it is differentiable $\dot{f}(t) = 0$. The same formal differentiation property applied to the step function yields

$$sF(s) - f(0^-) = 1 - 0 = 1 \quad \Longrightarrow \quad \dot{f}(t) = \delta(t), \quad t \geq 0,$$

which is an *impulse* at the origin, obtained from Table 3.1. How should this be interpreted? The Laplace transform can formally compute (generalized) derivatives of piecewise smooth functions even at points of discontinuity and it indicates that by formally adding impulses to the derivative function. If $\tau \geq 0$ is a point where the function $f(t)$ is not continuous, the term

$$[f(\tau^+) - f(\tau^-)]\delta(t - \tau),$$

is added to the generalized derivative computed by the Laplace transform at $t = \tau$. See [LMT07] for more examples and details.

Things get complicated if one tries too hard to *understand* impulses. The first complication is that there exists no piecewise smooth function $f(t)$ for which $L\{f(t)\} = 1$. The impulse is not a function in the strict sense.[7] Nevertheless, some see the *need* to define the impulse by means of the limit of a regular piecewise smooth function.[8] Choices are varied, and a common one is

$$\delta(t) = \lim_{\epsilon \to 0} p_\epsilon(t),$$

where $p_\epsilon(t)$ is the finite *pulse*:

$$p_\epsilon(t) = \epsilon^{-1}[1(t) - 1(t - \epsilon)]. \tag{3.6}$$

These *definitions* of the impulse are unnecessary if one understands the role played by the impulse in encoding the size and order of the discontinuities of piecewise smooth functions. A definition of the impulse based on the formal inverse Laplace transform[9] formula (3.4),

$$\delta(t) = \frac{1}{2\pi j} \lim_{\rho \to \infty} \int_{-\rho}^{\rho} e^{j\omega t} j \, d\omega = \lim_{\rho \to \infty} \frac{\sin(\rho t)}{\pi t} = \lim_{\epsilon \to 0} \frac{\sin(t/\epsilon)}{\pi t}, \tag{3.7}$$

is popular when the impulse is introduced using Fourier rather than Laplace transforms. All these attempts to define the impulse leave a bitter taste. For example, both definitions require careful analysis of the limit as $\epsilon \to 0$ and the functions used in (3.7) are not zero for $t \leq 0$, a fact that causes trouble for the unilateral definition of the Laplace transform

---

[7] Formally it is a *generalized function* or *distribution*. See [KF75, Chapter 21] and [LMT07].
[8] See P3.15–P3.26.
[9] The Laplace transform of the impulse $F(s) = 1$ converges everywhere and has no singularities, and hence α can be chosen to be 0 in (3.4).

in (3.1). See also Section 5.1 for a discussion of a physical system, an electric circuit with a capacitor, whose response approximates an impulse as the losses in the capacitor become zero.

The properties of the impulse are also consistent with integration. If $f(t)$ is a piecewise smooth function then

$$g(t) = \int_{0^-}^{t} f(\tau)d\tau \quad \text{and} \quad g(0^-) = g(0^+) = 0,$$

even if $f(t)$ has a discontinuity at $t = 0$. Try with the step function $f(t) = 1(t)$! However, one might suspect that $g(t)$ can be the result of integration and yet $g(0^+) \neq 0$. Take the impulse, $f(t) = \delta(t)$, for which $F(s) = 1$, and the step function, $g(t) = 1(t)$, for which $G(s) = s^{-1}$. In this case,

$$\mathscr{L}\{1(t)\} = \frac{1}{s}\mathscr{L}\{\delta(t)\}, \quad 1(t) = \int_{0^-}^{t} \delta(\tau)d\tau, \quad 0 = 1(0^-) \neq 1(0^+) = 1.$$

The reason for this apparent inconsistency is the fact that the impulse $\delta(t)$ is *not* a piecewise smooth function! The role of the impulse here, as seen before, is to encode a discontinuity at the origin. This behavior is consistent with the formal properties of Table 3.2 outside the origin as well. For example, a step at time $\tau > 0$, that is $g(t) = 1(t - \tau)$, for which $G(s) = s^{-1}e^{-s\tau}$, is related to an impulse at τ, that is $f(t) = \delta(t - \tau)$, for which $F(s) = e^{-s\tau}$, by integration: $G(s) = s^{-1}F(s)$ and $0 = g(\tau^-) \neq g(\tau^+) = 1$.

Impulses can also encode information about discontinuities on higher-order derivatives. For an example, consider the function $f(t) = t$, $t \geq 0$, which is known as a *ramp* and has $F(s) = s^{-2}$. A ramp is continuous at $t = 0$, i.e. $f(0^-) = f(0^+) = 0$, but not differentiable. Application of the differentiation rule yields

$$sF(s) - f(0^-) = s\frac{1}{s^2} - 0 = \frac{1}{s} \quad \Longrightarrow \quad \dot{f}(t) = 1(t).$$

Recall that $0 = \dot{f}(0^-) \neq \dot{f}(0^+) = 1$ and differentiate to obtain

$$s^2 F(s) - sf(0^-) - \dot{f}(0^-) = s^2\frac{1}{s^2} - 0 - 0 = 1 \quad \Longrightarrow \quad \ddot{f}(t) = \delta(t),$$

which correctly indicates the discontinuity of the first derivative. Formal differentiation of the impulse can be interpreted as *higher-order* impulses [LeP10], but these will have little use in the present book.

3.2 Linearity, Causality, and Time-Invariance

We are now ready to formalize a notion of linearity that can be used with dynamic systems. Roughly speaking,[10] a dynamic system with input $u(t)$ and output $y(t)$ described by the mathematical model

$$y(t) = G(u(t), t)$$

---
[10] See [Kai80, Chapter 1] for some not so obvious caveats.

3.2 Linearity, Causality, and Time-Invariance

is *linear* whenever the following property holds: any linear combination of input signals $u_1(t), u_2(t)$ in the form

$$u(t) = \alpha_1 u_1(t) + \alpha_2 u_2(t), \qquad \alpha_1 \in \mathbb{R}, \quad \alpha_2 \in \mathbb{R}, \tag{3.8}$$

generates the output signal

$$y(t) = G(u(t), t) = \alpha_1 y_1(t) + \alpha_2 y_2(t), \tag{3.9}$$

where

$$y_1(t) = G(u_1(t), t), \quad y_2(t) = G(u_2(t), t),$$

must hold for all α_1, α_2 and also all inputs $u_1(t), u_2(t)$.

Another relevant property is *time-invariance*. A dynamic system is time-invariant if it passes the following test. Apply the input signal $u(t)$ and record

$$y(t) = G(u(t), t).$$

Then apply the same input signal delayed by $\tau > 0$, $u(t - \tau)$, and record

$$y(t, \tau) = G(u(t - \tau), t).$$

The system is time-invariant if

$$y(t, \tau) = y(t - \tau) \tag{3.10}$$

for any $u(t)$ and $\tau > 0$. Time-invariant systems respond the same way to the same input regardless of when the input is applied.

It is important to note that there are systems which are linear but not time-invariant. For example, an amplitude modulator is a linear system for which

$$y(t) = 2\cos(\omega_f t) u(t), \qquad \omega_f > 0. \tag{3.11}$$

Modulators are the basic building block in virtually all communication systems. You will show in P3.50 that a modulator is linear but not time-invariant. Of course, there are also time-invariant systems which are not linear. See P3.50 through P3.52 for more examples of time-varying linear systems.

In this book we are mostly interested in linear time-invariant systems. While some of the results to be presented will hold true for linear time-varying systems, many do not. See [Bro15] for more details on time-varying linear systems. Most notably many conclusions to be obtained in the frequency domain will not hold for time-varying systems.

For time-invariant or time-varying linear systems, it is possible to take advantage of the properties of the impulse, $\delta(t)$, to obtain the following compact and powerful representation in terms of an integral known as the *convolution integral*.

Start by decomposing a given input signal $u(t), t \geq 0$, as a finite sum of given signals, $b_k(t), t \geq 0, k = 1, \ldots, n$, as in (3.8):

$$u(t) = \sum_{k=1}^{n} \alpha_k b_k(t).$$

Transfer-Function Models

The set of signals $b_k(t)$, $t \geq 0$, $k = 1, \ldots, n$, can be thought of as a *basis* in which the signal $u(t)$ can be expressed. A reader familiar with Fourier series or Fourier transforms will recall that continuous-time signals can rarely be expressed as a *finite* sum of other signals. With that in mind, replace the finite summation above by the infinite summation, that is, the integration

$$u(t) = \int_{-\infty}^{\infty} \alpha(\tau) b(t, \tau) d\tau, \qquad (3.12)$$

in which the "coefficient" $\alpha(\tau)$ becomes a continuous function of the "index" τ. The same metamorphosis occurs with the "basis signals" $b(t, \tau)$. Given a signal $u(t)$, $t \geq 0$, and a suitable set of basis signals, $b(t, \tau)$, our next task is to calculate the coefficients, $\alpha(\tau)$, such that $u(t)$ is indeed represented by (3.12). This calculation, which can be a difficult task, is tremendously simplified if one chooses as the basis signals the set of delayed impulses

$$b(t, \tau) = \delta(t - \tau).$$

In this very special case, in view of the *sifting property* of the impulse that you will prove in P3.12, one can simply set

$$\alpha(\tau) = u(\tau)$$

to obtain

$$u(t) = \int_{-\infty}^{\infty} \alpha(\tau) b(t, \tau) d\tau = \int_{-\infty}^{\infty} u(\tau) \delta(t - \tau) d\tau = \int_{0^-}^{t} u(\tau) \delta(t - \tau) d\tau.$$

The limits of integration can be made finite because $u(t) = 0$ for $t < 0$, and $\delta(t - \tau) = 0$ for any $t \neq \tau$, including $\tau > t$.

Equipped with the above representation of the input signal $u(t)$ we calculate[11] the response of the linear system to each component of the basis of delayed impulses,

$$g(t, \tau) = y(t, \tau) = G(\delta(t - \tau), t).$$

This is the system's *impulse response*. If the system model is linear, that is, if (3.8) and (3.9) hold, then

$$y(t) = \int_{-\infty}^{\infty} \alpha(\tau) y(t, \tau) d\tau = \int_{0^-}^{\infty} u(\tau) g(t, \tau) d\tau,$$

where the lower limit of integration can be changed because $u(t) = 0$, $t < 0$. The upper limit cannot be changed without the following additional assumption: $g(t, \tau) = 0$, $t < \tau$. For reasons that will be clear soon, a system for which this property holds is known as *causal* or *non-anticipatory*. With linearity and causality,

$$y(t) = \int_{0^-}^{t} g(t, \tau) u(\tau) d\tau.$$

---

[11] No one is suggesting physically recording the response of the system to the unrealizable impulse function. This is simply a useful mathematical construction.

This is the general form of the response of a linear time-varying system, in which the impulse response, $g(t, \tau)$, may change depending on the time the impulse is applied, τ. Note that a consequence of causality is that $y(t) = 0$, $t < 0$. More importantly, the output, $y(t)$, depends only on past values of the input, that is $u(\tau)$, $0 \leq \tau \leq t$. In other words, a causal or non-anticipatory system cannot anticipate or predict its input.

For a causal linear time-invariant system the impulse response can be described in terms of a single signal:

$$g(t) = g_0(t) = G(\delta(t), t),$$

because (3.10) implies

$$g(t, \tau) = g(t - \tau).$$

This leads to the well-known representation of a linear time-invariant system through the *convolution integral*:

$$y(t) = \int_{0^-}^{t} g(t - \tau) u(\tau) d\tau. \tag{3.13}$$

Application of the convolution property of the Laplace transform produces

$$Y(s) = G(s) U(s), \tag{3.14}$$

which shows that the Laplace transform of the output, $Y(s) = \mathscr{L}\{y(t)\}$, is simply the product of the Laplace transform of the input, $U(s) = \mathscr{L}\{u(t)\}$ and the Laplace transform of the impulse response, $G(s) = \mathscr{L}\{g(t)\}$.

The function $G(s)$ is known as the *transfer-function* of a causal, linear, and time-invariant system. This representation is key to most of the methods studied in the next chapters. In this book, with rare exceptions, we are mainly interested in the case when $U(s)$, $G(s)$, and $Y(s)$ are *rational*. In particular, rational transfer-functions can be obtained directly from linear ordinary differential equations, as shown in the next section.

3.3 Differential Equations and Transfer-Functions

In this section we illustrate how the Laplace transform can be used to compute the transfer-function and, more generally, the response of linear models to arbitrary inputs directly from linear ordinary differential equation models. Recall a familiar example: the car model in the form of the differential equation (2.3). Application of the differentiation and linear properties of the Laplace transform from Table 3.2 yields

$$\mathscr{L}\left\{\dot{y}(t) + \frac{b}{m} y(t) - \frac{p}{m} u(t)\right\} = sY(s) - y(0^-) + \frac{b}{m} Y(s) - \frac{p}{m} U(s) = 0,$$

where $Y(s)$ is the Laplace transform of the output $y(t)$, and $U(s)$ is the Laplace transform of the input $u(t)$. When the Laplace transform is applied to linear ordinary differential equations, impulses encode the *initial conditions*. In the above example $\mathscr{L}\{\dot{y}(t)\} = sY(s) - y(0^-)$ and $\mathscr{L}^{-1}\{y(0^-)\} = y(0^-)\delta(t)$. After solving for $Y(s)$, we

Transfer-Function Models

obtain

$$Y(s) = G(s)U(s) + F(s)y(0^-). \quad (3.15)$$

The function $Y(s)$ has two terms involving the rational functions:

$$G(s) = \frac{p/m}{s + b/m}, \qquad F(s) = \frac{1}{s + b/m}. \quad (3.16)$$

The first term is related to the Laplace transform of the input, $U(s)$, through $G(s)$; the second term is related to the initial condition, $y(0^-)$, through $F(s)$. The rational functions $G(s)$ and $F(s)$ are similar but not equal. Because they share[12] the same denominator, they will display similar properties, as will soon become clear. Indeed, most of the time we can safely assume zero initial conditions and focus on $G(s)$. Some exceptions to this rule are discussed in Sections 4.7 and 5.3.

From the discussion in Section 3.2, the rational function $G(s)$ is precisely the system *transfer-function*. When initial conditions are zero, the system response can be computed from the transfer-function through the formula (3.14) or, equivalently, by the impulse response and the time-domain convolution formula (3.13). The impulse response associated with the transfer-function (3.16) is

$$g(t) = \mathscr{L}^{-1}\left\{\frac{p/m}{s + b/m}\right\} = \frac{p}{m}e^{-(b/m)t}, \quad t \geq 0,$$

which was obtained directly from Table 3.1. Because $g(t) = 0$ for $t < 0$, the ordinary differential equation model (2.3) is causal, linear, and time-invariant.

Analysis of the impulse response can provide valuable information on the dynamic behavior of a system and clues on the response to actual input functions. Some might be tempted to *record* the impulse response, $g(t)$, experimentally after applying an input that resembles the infinitesimal pulse (3.6) to obtain an approximate transfer-function, a practice that we discourage due to the many difficulties in realizing the impulse and collecting the resulting data. Better results can be obtained by system identification methods, such as the ones described in [Lju99].

The above procedure for calculating transfer-functions directly from linear ordinary differential equations is not limited to first-order equations. Assume for the moment zero initial conditions, e.g. $y(0^-) = 0$ in (3.15), and consider a linear system modeled by the generic linear ordinary differential equation

$$y^{(n)}(t) + a_1 y^{(n-1)}(t) + \cdots + a_n y(t) = b_0 u^{(m)}(t) + b_1 u^{(m-1)}(t) + \cdots + b_m u(t). \quad (3.17)$$

Application of the Laplace transform under zero initial conditions[13] produces

$$(s^n + a_1 s^{n-1} + \cdots + a_n)Y(s) = (b_0 s^m + b_1 s^{m-1} + \cdots + b_m)U(s),$$

---

[12] Unless pole–zero cancellations occur. See Sections 4.7 and 5.3 for more details.
[13] The initial conditions here are $y(0^-), \dot{y}(0^-), \ldots, y^{(n-1)}(0^-)$.

leading to the rational transfer-function

$$G(s) = \frac{Y(s)}{U(s)} = \frac{b_0 s^m + b_1 s^{m-1} + \cdots + b_m}{s^n + a_1 s^{n-1} + \cdots + a_n}. \quad (3.18)$$

The integer n, the degree of the polynomial in the denominator, is the *order* of the transfer-function, which we might by extension refer to as the order of the system. When $m \leq n$ we say that the transfer-function is *proper* and if $m < n$ the transfer-function is *strictly proper*. The transfer-function (3.16) is rational, of order $n = 1$, and strictly proper, $m = 0 < 1$.

3.4 Integration and Residues

Given the Laplace transform of any input signal, $U(s)$, and the system transfer-function $G(s)$, formula (3.14) computes the Laplace transform of the output response signal, $Y(s)$. From $Y(s)$ one can compute the output response, $y(t)$, using the inverse Laplace transform (3.4). So far we have been able to compute the inverse Laplace transform by simply looking up Table 3.1. This will not always be possible. For example, Table 3.1 does not tell one how to invert the generic rational transfer-function (3.18) if $n > 2$. If we want to take full advantage of the frequency-domain formula (3.14) we need to learn more sophisticated methods.

Our main tool is a result that facilitates the computation of integrals of functions of a single complex variable known as Cauchy's residue theorem. The theory leading to the theorem is rich and beyond the scope of this book. However, the result itself is pleasantly simple and its application to the computation of the inverse Laplace transform is highly practical. One does not need to talk about integration and residues at all to be able to *calculate* inverse Laplace transforms. However, as we shall also use residues in connection with Chapter 7, we choose to briefly review some of the key results. Readers who might feel intimidated by the language can rest assured that practical application of the theory will be possible even without complete mastery of the subject. Of course, a deeper knowledge is always to be preferred, and good books on complex variables, e.g. [BC14; LeP10], are good complementary sources of information.

We will use the concepts of an *analytic function* and *singular points*. A function f of a single complex variable, s, is said to be *analytic* at a given point s_0 if f and f', the complex derivative of f, exist at s_0 *and* at every point in a neighborhood[14] of s_0. Asking for a function to be analytic at a point is no small requirement and one requires more than the mere existence of the derivative.[15] Indeed, it is possible to show that if f is analytic at s_0 then not only do f and f' exist but also derivatives of all orders exist and are themselves analytic at s_0 [BC14, Section 57]. Furthermore, the function's Taylor series expansion exists and is guaranteed to converge in a neighborhood of s_0. By extension, we say that f is analytic in an open set S if f is analytic at all points of S.

---

[14] A neighborhood is a sufficiently small open disk $S = \{s : |s - s_0| < R\}$, with $R > 0$.
[15] If $f(x + jy) = u(x, y) + jv(x, y)$, where $x, y \in \mathbb{R}$ and $u, v : \mathbb{R}^2 \to \mathbb{R}$ then f' exists only if $\partial u/\partial x = \partial v/\partial y$ and $\partial u/\partial y = \partial v/\partial x$. These are the Cauchy–Riemann equations [BC14, Section 21].

For elementary functions, such as polynomial, exponential, trigonometric, and hyperbolic functions, the calculation of f' can be formally done as if they were real functions. For example, the derivative of the function $f_1(s) = e^{-s}$ exists and is equal to $f_1'(s) = -e^{-s}$ at any point $s \in \mathbb{C}$. Likewise the derivative of the monomial $f_2(s) = s^m$ exists and is equal to $f_2'(s) = ms^{m-1}$ at any point $s \in \mathbb{C}$. Both f_1 and f_2 are *analytic in* $\mathbb{C}$. Functions that are analytic everywhere in $\mathbb{C}$ are called *entire*.

If the function f fails to be analytic at s_0 but is analytic at some point in every neighborhood of s_0 then s_0 is a *singular point*. The rational function $f_3(s) = (s-1)^{-1}$ is analytic everywhere except at the singular point $s_0 = 1$. Its derivative exists and is equal to $f_3'(s) = -(s-1)^{-2}$ at any $s \neq 1$. A rational function is analytic everywhere except at the roots of its denominator. If f is singular at s_0 but analytic at all points in a neighborhood of s_0 then s_0 is an *isolated singularity*.[16] Functions that have only a finite number of singularities, such as rational functions, have only isolated singularities. In the case of rational functions, the singular points are the finite number of roots of a polynomial, the denominator, and all such singularities are *poles*.[17] When we say singularity in this book we mean isolated singularity, and the vast majority of such singularities will be poles.

A complex number is a two-dimensional entity, therefore integration of a complex function must be done over a prescribed curve or area in the complex plane. Some of the most powerful results in complex analysis arise from integration along a *closed path* or *contour*. We say that a closed contour is *simple* when it does not intersect itself. It is *positively oriented* when the traversal of the contour is done in the counter-clockwise direction and *negatively oriented* if the traversal is clockwise. Figure 3.1 illustrates some simple closed contours and their orientation is indicated by arrows along the path. Integration of the function f along the simple closed contour C is denoted by

$$\int_C f(s)ds.$$

The actual integration requires a suitable path. The contour C_3 in Fig. 3.1 is a positively oriented circle that can be parametrized as

$$s = s_0 + \rho e^{j\theta}, \quad 0 \le \theta \le 2\pi, \tag{3.19}$$

where s_0 is the center of the circle, $\rho > 0$ its radius, and θ a parameter describing the traversal of the circle in the counter-clockwise direction. The parameters used to plot C_3 in Fig. 3.1 were $s_0 = 0.5 + j$, $\rho = 0.75$. The same parametrization can describe the negatively oriented circle C_2 in Fig. 3.1 on replacing "θ" with "$-\theta$" to reverse the direction of travel.

---

[16] The function $f(s) = \sin(\pi/s)^{-1}$ is singular at $s = 0$ and every $s = k^{-1}$ where $k \neq 0$ is integer. The singular points $s = k^{-1}$ are all isolated. However, $s = 0$ is not isolated since for any $\epsilon > 0$ such that $S_\epsilon = \{s : |s| < \epsilon\}$ there exists a large enough k such that $s = k^{-1}$ is in S_ϵ.

[17] The isolated singularity s_0 of the function f is a *pole* if the number of terms with negative exponents in its Laurent series expansion is finite [BC14, Section 66].

3.4 Integration and Residues

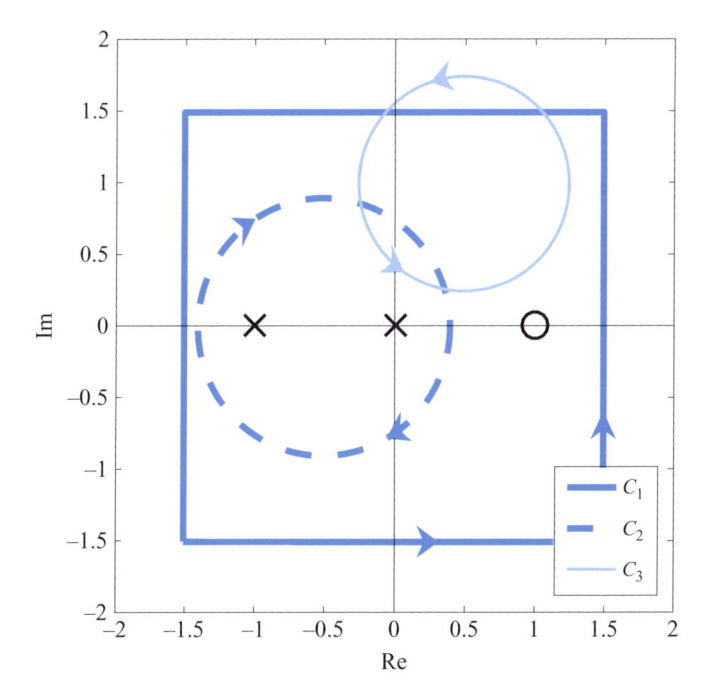

Figure 3.1 Simple closed contours in the complex plane. The rectangle (thick) C_1 and the circle (thin) C_3 contours are *positively oriented* and the circle (dashed) C_2 is *negatively oriented*.

Integration of the function $f(s) = 1$ along any positively oriented circular contour C parametrized by (3.19) can be performed after evaluating

$$ds = j\rho e^{j\theta} d\theta$$

and substituting

$$\int_C f(s)ds = j\rho \int_0^{2\pi} e^{j\theta} d\theta = \rho e^{j\theta}\Big|_0^{2\pi} = 0. \quad (3.20)$$

Note that one obtains zero after integration along any circle since the result depends neither on s_0 nor on ρ. It gets better than that: the integral is zero even if C is replaced by any simple closed contour and f is replaced by any function that is analytic on and inside the contour. This incredibly strong statement is a consequence of the following key result [BC14, p. 76]:

THEOREM 3.1 (Cauchy's residue theorem) *If a function f is analytic inside and on the positively oriented simple closed contour C except at the singular points p_k, $k = 1, \ldots, n$, inside C, then*

$$\int_C f(s)ds = 2\pi j \sum_{k=1}^n \operatorname*{Res}_{s=p_k} f(s).$$

The notation $\operatorname*{Res}_{s=p_k} f(s)$ indicates the *residue* of f at $s = p_k$. We will discuss residues in more detail soon, but notice that when f is analytic on and inside C then it has no

Transfer-Function Models

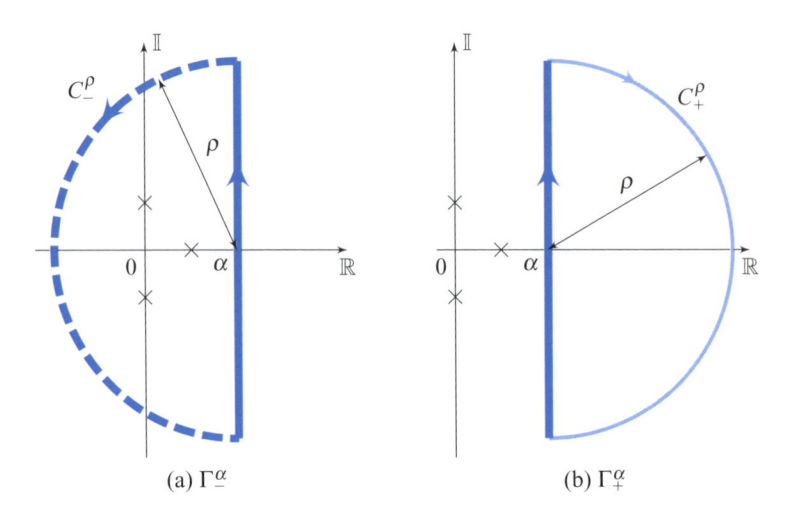

Figure 3.2 Simple closed contours used to compute the inverse Laplace transform; crosses indicate singular points of $F(s)$; C_-^ρ and C_+^ρ denote the semi-circular part of the contours; Γ_-^α encloses all singularities, Γ_+^α is free of singularities, and $F(s)$ vanishes on the semi-circular parts of the contours as $\rho \to \infty$.

singular points inside C and the integral reduces to zero. This is what is at work in (3.20), since $f(s) = 1$ is analytic everywhere.

Application of Cauchy's residue theorem to calculate the integral (3.20) may not seem very impressive. A more sophisticated and useful application is the computation of the inverse Laplace transform. The main idea is to perform the line integral (3.4) as part of a suitable contour integral, such as the ones shown in Fig. 3.2, where the linear path is made closed by traversing a semi-circular path of radius ρ centered at α, labeled C_-^ρ and C_+^ρ in the figure.

For example, integration along the simple positively oriented contour Γ_-^α can be split into

$$\int_{\Gamma_-^\alpha} F(s)e^{st}\,ds = \int_{\alpha-j\rho}^{\alpha+j\rho} F(s)e^{st}\,ds + \int_{C_-^\rho} F(s)e^{st}\,ds.$$

If F is such that

$$\lim_{\rho \to \infty} \int_{C_-^\rho} F(s)e^{st}\,ds = 0, \quad t > 0, \tag{3.21}$$

for instance, if F vanishes for large s, then it is possible to evaluate the inverse Laplace transform by calculating residues, that is,

$$\begin{aligned}
\mathcal{L}^{-1}\{F(s)\} &= \frac{1}{2\pi j} \lim_{\rho \to \infty} \int_{\alpha-j\rho}^{\alpha+j\rho} F(s)e^{st}\,ds \\
&= \frac{1}{2\pi j} \lim_{\rho \to \infty} \int_{\Gamma_-^\alpha} F(s)e^{st}\,ds \\
&= \sum_{k=1}^{n} \operatorname*{Res}_{s=p_k}\left(F(s)e^{st}\right), \quad t > 0,
\end{aligned} \tag{3.22}$$

3.4 Integration and Residues

where the p_ks are the singular points of $F(s)e^{st}$ located inside the contour, which in this case coincides with the half-plane $\operatorname{Re}(s) < \alpha$ as $\rho \to \infty$. Moreover, if α is chosen such that $\alpha > \alpha_c$ then *all* singular points of F are enclosed by Γ_-^ρ as $\rho \to \infty$. Because e^{st} is analytic in $\mathbb{C}$, the singular points of $F(s)e^{st}$ are simply the singular points of $F(s)$. The above integral formula is known as the *Bromwich integral* [BC14, Section 95]. The main advantage of this reformulation is that evaluation of the Bromwich integral reduces to a calculus of residues.

Sufficient conditions for (3.21) to hold exist but are often restrictive. Roughly speaking, it is necessary that F becomes small as $|s|$ gets large, i.e.

$$\lim_{|s| \to \infty} F(s) = 0. \tag{3.23}$$

More precisely, if there exists $M > 0$, $R > 0$, and an integer $k \geq 1$ such that

$$|F(s)| \leq \frac{M}{|s|^k}, \quad \text{for all } s \in C_-^\rho, \text{ and } \rho > R, \tag{3.24}$$

then (3.21) will hold for all $t > 0$. See P3.27–P3.30. Additionally, if $k \geq 2$ then it also holds for $t = 0$. This result is [Chu72, Theorem 5, Section 67]. Note that if F is rational and strictly proper, i.e. the degree of the numerator is smaller than the degree of the denominator, then k is at least one.

That k needs to be such that $k \geq 2$ in (3.24) for the semi-circular path integral in (3.21) to converge to zero also at $t = 0$ can be attributed to a loss of continuity of the inverse Laplace transform. Indeed, the inverse Laplace transform of a rational function with a simple pole at p_k, i.e. $F(s) = (s - p_k)^{-1}$, is of the form $f(t) = e^{p_k t}$, $t > 0$. Because the inverse Laplace transform must be zero for all $t < 0$, this means that there is a discontinuity at the origin. For continuity at the origin, the initial-value property in Table 3.2 suggests that $\lim_{|s| \to \infty} s F(s) = 0$, which requires $k \geq 2$ in (3.24). This situation can be partially remedied by using a slightly more sophisticated setup using the inverse Fourier transform [Chu72, Theorem 6, Section 68] that allows $k = 1$ at the expense of averaging at discontinuities. A key result is unicity of the inverse Laplace transform [Chu72, Theorem 7, Section 69], which allows one to be content with inverse Laplace transforms obtained by any reasonable method, be it table lookup, calculus of residues, or direct integration.

When F is only bounded, that is,

$$\lim_{|s| \to \infty} |F(s)| \leq M < \infty, \tag{3.25}$$

but M cannot be made zero, it is usually possible to split the calculation as a sum of a bounded function, often a constant, and a convergent function satisfying (3.23). We will illustrate this procedure with rational functions in Section 3.5. If impulses at the origin are present one needs to explicitly compute the integral (3.4) at $t = 0$. The simplest possible example is $F(s) = 1$. Since $F(s) = 1$ is analytic everywhere, $f(t) = \delta(t) = 0$ for all $t \neq 0$. In some special cases when (3.25) does not hold, it may still be possible to compute the inverse Laplace transform using residues. For example, $F(s) = e^{s\tau}/s$ does not satisfy (3.25) but $f(t) = 1(t - \tau)$, $\tau \geq 0$, can be computed after shifting the origin of time (see the time-shift property in Table 3.2). Of course $f(t) \neq 0$ for $-\tau \leq t < 0$ if $\tau < 0$ so one should be careful when applying this result. Care is also required when the function $F(s)$ is multivalued. See [LeP10, Section 10-21] for a concrete example.

Another satisfying application of Cauchy's residue theorem is in establishing that the inverse Laplace transform is causal,[18] in other words, that it is always zero when $t < 0$. We shall use the contour Γ_+^ρ for that purpose. If (3.23) holds, or there exists $M > 0$, $R > 0$, and an integer $k \geq 1$ such that

$$|F(s)| \leq \frac{M}{|s|^k}, \quad \text{for all } s \in C_+^\rho, \text{ and } \rho > R, \tag{3.26}$$

then

$$\lim_{\rho \to \infty} \int_{C_+^\rho} F(s)e^{st}\, ds = 0, \quad t < 0. \tag{3.27}$$

This time, however,

$$\mathcal{L}^{-1}\{F(s)\} = \frac{1}{2\pi j} \lim_{\rho \to \infty} \int_{\alpha - j\rho}^{\alpha + j\rho} F(s)e^{st}\, ds$$

$$= \frac{1}{2\pi j} \lim_{\rho \to \infty} \int_{\Gamma_+^\alpha} F(s)e^{st}\, ds = 0, \quad t < 0, \tag{3.28}$$

because no singular points of F lie in the half-plane $\text{Re}(s) > \alpha$ as $\rho \to \infty$ if α is chosen so that $\alpha > \alpha_c$.

In order to take advantage of the Bromwich integral formula (3.22) we need to learn how to calculate residues. If s_0 is an isolated singular point the residue at s_0 is the result of the integration:

$$\operatorname*{Res}_{s=s_0} f(s) = \int_{C \in D} f(s)\, ds, \quad D = \{s : 0 < |s - s_0| < R\},$$

where C is any positively oriented simple closed contour in D. The advantage is of course that it is often easier to calculate the residue than it is to calculate an integral on an arbitrary contour. In general, if s_0 is a pole of order m then

$$f(s) = \frac{g(s)}{(s - s_0)^m}, \quad g(s_0) \neq 0,$$

where $g(s)$ is some appropriate function which is analytic at s_0. The analytic function g is often calculated from a Laurent series expansion of f around s_0. A Laurent series is a special type of power series involving positive and negative exponents [BC14, Section 66]. After computing g, the residue at s_0 is [BC14, Section 80]:

$$\operatorname*{Res}_{s=s_0} f(s) = \begin{cases} g(s_0), & m = 1, \\ \dfrac{g^{(m-1)}(s_0)}{(m-1)!}, & m \geq 2. \end{cases} \tag{3.29}$$

Application of this formula will be illustrated in the next sections.

---

[18] This is one case in which the conditions (3.23) and (3.25) may be too strong. In fact, if $F(s)$ is such that $F(j\omega)$ is the Fourier transform of the corresponding $f(t)$, for example whenever $|F(j\omega)|$ is square integrable, then a classical necessary and sufficient condition for causality is that the Paley–Wiener integral $\int_{-\infty}^{\infty} |\ln|F(j\omega)||/(1 + \omega^2)\, d\omega$ be finite [PW34].

3.5 Rational Functions

We saw in Section 3.3 that transfer-functions for a large class of linear system models are rational. The Laplace transforms of many input functions of interest are also rational or can be approximated with rational functions. For this reason, we often have to compute the inverse Laplace transform of rational functions. For example, in Section 3.3 we computed the inverse Laplace transform of a rational transfer-function, $G(s)$, to obtain the impulse response, $g(t) = \mathscr{L}^{-1}\{G(s)\}$. Many calculations discussed in Section 3.4 also become much simpler when the functions involved are rational.

A rational function of the single complex variable s is one which can be brought through algebraic manipulation to a ratio of polynomials in s. When $F(s)$ is a rational function its *poles* are the roots of the denominator, that is solutions to the polynomial equation

$$s^n + a_1 s^{n-1} + \cdots + a_n = 0. \tag{3.30}$$

This polynomial equation is known as the *characteristic equation*. Characteristic equations appeared earlier when we solved differential equations in Chapter 2. For instance, the transfer-function (3.16) has as characteristic equation

$$s + \frac{b}{m} = 0.$$

Compare this equation with (2.5). The *zeros* of a rational transfer-function are the roots of the numerator. We will have more to say about zeros later.

If a rational function $F(s)$ is obtained as a result of the Laplace transform formula (3.1), then it has a non-empty region of convergence, $\operatorname{Re}(s) > \alpha_c = \max_k \operatorname{Re}(p_k)$, where the p_ks are the poles of F. Furthermore, a rational function $F(s)$ is always analytic in its region of convergence. If $F(s)$ is strictly proper then (3.23) holds and it is possible to compute the inverse Laplace transform of $F(s)$ using the residue formula (3.22). For that we need first to compute the poles of $F(s)$.

By the fundamental theorem of algebra [Olm61, § 1515], we know that the polynomial (3.30) has exactly n possibly complex roots; that is, a rational function $F(s)$ of order n will have n poles. In this book we are interested in models that correspond to physical systems, where the coefficients of the rational function $F(s)$ are real numbers. Indeed, when we do not say anything about a complex-valued function it is to be assumed that the coefficients are real. A consequence of the fundamental theorem of algebra is that the denominator of $F(s)$ can be *factored* as the product of its n roots, i.e. the poles

$$s^n + a_1 s^{n-1} + \cdots + a_n = (s - p_1)(s - p_2) \cdots (s - p_n).$$

Roots may appear more than once, according to their algebraic multiplicity. Having obtained[19] the poles of a rational function we are ready to compute the residues in (3.22).

---

[19] One of the most remarkable results in abstract algebra is the Abel–Ruffini theorem [Wae91, Section 8.7] which states that no general "simple formula" exists that can express the roots of polynomials of order 5 or higher. In this sense, it is not possible to factor such polynomials exactly and we shall rely on numerical methods and approximate roots.

Consider first the case of a simple pole s_0, i.e. a pole with multiplicity one. Because $e^{st} \neq 0$ and e^{st} is analytic for all $s \in \mathbb{C}$ we can write

$$F(s)e^{st} = \frac{G(s)}{s - s_0}, \qquad G(s) = (s - s_0)F(s)e^{st}, \qquad G(s_0) \neq 0,$$

and straightforward application of the residue formula (3.29) yields

$$\operatorname*{Res}_{s=s_0}\left(F(s)e^{st}\right) = G(s_0) = k_0 e^{s_0 t}, \qquad k_0 = \lim_{s \to s_0}(s - s_0)F(s). \tag{3.31}$$

For example,

$$F(s) = \frac{b_1}{s + a_1}$$

is rational and strictly proper so it satisfies (3.23). Hence

$$G(s) = b_1 e^{st}, \qquad \operatorname*{Res}_{s=-a_1}\left(F(s)e^{st}\right) = G(-a_1) = b_1 e^{-a_1 t}$$

and

$$f(t) = \mathscr{L}^{-1}\left\{\frac{b_1}{s + a_1}\right\} = \operatorname*{Res}_{s=-a_1}\left(F(s)e^{st}\right) = b_1 e^{-a_1 t}, \qquad t > 0.$$

Of course one could have computed this simple inverse Laplace transform just by looking up Table 3.1. The next example, which we borrow from Section 2.3, is slightly more involved.

We repeat the calculation of the step response performed in Section 2.3, this time using Laplace transforms. Recall the unit step function, $1(t)$, and its Laplace transform, $\mathscr{L}(1(t)) = s^{-1}$. Now let

$$u(t) = \tilde{u}, \quad t \geq 0,$$

be a step of amplitude $\tilde{u}$. Substituting $U(s) = \tilde{u} s^{-1}$ into (3.15) we obtain

$$Y(s) = \frac{\tilde{u}}{s} \times \frac{p/m}{s + b/m} + \frac{1}{s + b/m} y_0.$$

The function $Y(s)$ is rational and its poles are

$$p_1 = 0, \qquad p_2 = -\frac{b}{m}.$$

Both have multiplicity one. Applying (3.22) and (3.31) the response is

$$y(t) = \sum_{k=1}^{2} \operatorname*{Res}_{s=p_k}\left(Y(s)e^{st}\right) = k_1 e^{p_1 t} + k_2 e^{p_2 t}, \qquad t > 0.$$

Evaluating

$$k_1 = \lim_{s \to 0} s Y(s) = \frac{p}{b}\tilde{u}, \qquad k_2 = \lim_{s \to -b/m}\left(s + \frac{b}{m}\right)Y(s) = y_0 - \frac{p}{b}\tilde{u},$$

3.5 Rational Functions

and substituting we obtain the complete response:

$$y(t) = \frac{p}{b}\tilde{u} + \left(y_0 - \frac{p}{b}\tilde{u}\right)e^{-\frac{b}{m}t}, \qquad t > 0,$$

which coincides with (2.6) and (2.7) as expected.

Converse application of the formula (3.22) in the case of a rational function provides an alternative perspective on the above calculation. Combining (3.22) and (3.31) we obtain

$$F(s) = \mathscr{L}\left\{\sum_{k=1}^{n} \operatorname*{Res}_{s=p_k}\left(F(s)e^{st}\right)\right\} = \mathscr{L}\left\{\sum_{k=1}^{n} k_k e^{p_k t}\right\} = \sum_{k=1}^{n} \frac{k_k}{s - p_k},$$

which is known as an expansion in *partial fractions*. From the partial-fraction expansion one can go back to the time domain by applying Table 3.1. For the example above, the function $Y(s)$ expanded in partial fractions is

$$Y(s) = \frac{k_1}{s} + \frac{k_2}{s + b/m},$$

which leads to the same inverse Laplace transform as the one obtained using residues. It is through expansion in partial fractions that most introductory systems and control books motivate the calculation of the inverse Laplace transform for rational functions, e.g. [FPE14; DB10]. We took a detour to highlight its connection with the theory of integration and the calculus of residues, which we will benefit from later.

Formula (3.22) requires the convergence condition (3.23), which means that $F(s)$ needs to be strictly proper when $F(s)$ is rational. If $F(s)$ is proper but not strictly proper, then the boundedness condition (3.25) holds and it is possible to split the calculation as discussed in Section 3.4. For example

$$F(s) = \frac{s}{s+1} = F_1(s) + F_2(s), \qquad F_1(s) = 1, \qquad F_2(s) = -\frac{1}{s+1},$$

where we used polynomial division to split F into a sum of a polynomial, F_1, and a strictly proper function, F_2. The inverse Laplace transform of a polynomial will always lead to a sum of impulses and higher-order impulses which contribute only to the inverse Laplace transform at $t = 0$. The strictly proper part, F_2, satisfies the convergence condition (3.23) and its contribution to the inverse Laplace transform can be computed using residues. The complete inverse Laplace transform is therefore

$$f(t) = \mathscr{L}^{-1}\{F_1(s)\} + \mathscr{L}^{-1}\{F_2(s)\} = \delta(t) - e^{-t}, \qquad t \geq 0.$$

When F is proper but not strictly proper the polynomial term F_1 is simply a constant which can be calculated directly by evaluating the limit

$$F_1 = \lim_{|s| \to \infty} F(s). \qquad (3.32)$$

The case of poles with multiplicity greater than one is handled in much the same way. Let s_0 be a pole with multiplicity $m \geq 2$. In this case

$$F(s)e^{st} = \frac{G(s)}{(s-s_0)^m}, \qquad G(s) = (s-s_0)^m F(s)e^{st}, \qquad G(s_0) \neq 0.$$

Because $m \geq 2$ we need to use the second formula in (3.29), which requires differentiating $G(s)$. In the case of rational functions it is possible to calculate the following general formula:

$$\frac{G^{(m-1)}(s_0)}{(m-1)!} = \lim_{s \to s_0} \frac{d^{m-1}}{ds^{m-1}}\left((s-s_0)^m F(s)e^{st}\right) = \left(k_{0,1} + k_{0,2}t + \cdots + k_{0,m}t^{m-1}\right)e^{s_0 t},$$

(3.33)

where

$$k_{0,j} = \frac{1}{(j-1)!}\lim_{s \to s_0}\frac{d^{j-1}}{ds^{j-1}}(s-s_0)^m F(s), \qquad j = 1,\ldots, m. \qquad (3.34)$$

Note the appearance of terms in $t^{j-1}e^{s_0 t}$, $j = 1,\ldots, m$ in the response. An example illustrates the necessary steps.

Let us compute the closed-loop response of the car linear model, Equation (2.12) in Section 2.5, to a reference input of the form

$$\bar{y}(t) = \mu t, \quad t \geq 0.$$

This is a *ramp* of slope μ. For the car, the coefficient μ represents a desired acceleration. For example, this type of reference input could be used instead of a constant velocity reference to control the closed-loop vehicle acceleration when a user presses a "resume" button on the cruise control panel and the difference between the current and the set speed is too large. In Chapter 4 we will show how to compute the closed-loop transfer-function directly by combining the open-loop transfer-function with the transfer-function of the controller. For now we apply the Laplace transform to the previously computed closed-loop differential equation (2.12) to calculate the closed-loop transfer-function

$$Y(s) = H(s)\bar{Y}(s), \qquad H(s) = \frac{b_1}{s+a_1}, \qquad (3.35)$$

where

$$a_1 = \frac{b}{m} + \frac{p}{m}K, \qquad b_1 = \frac{p}{m}K, \qquad (3.36)$$

The rational function $H(s)$ is the closed-loop transfer-function. From Table 3.1 the Laplace transform of the ramp input $\bar{y}(t)$ is

$$\bar{Y}(s) = \frac{\mu}{s^2}.$$

Multiplying $\bar{Y}(s)$ by the closed-loop transfer-function $H(s)$ we obtain

$$Y(s) = H(s)\bar{Y}(s) = \frac{\mu}{s^2} \times \frac{b_1}{s+a_1},$$

3.5 Rational Functions

which is also rational with poles at

$$p_1 = -a_1, \qquad p_2 = p_3 = 0.$$

The resulting $Y(s)$ is rational and strictly proper, satisfying (3.23). The pole p_1 is simple and the pole $p_2 = p_3 = 0$ has multiplicity 2, and the complete response is computed using (3.22), (3.31), (3.33), and (3.34) to obtain

$$y(t) = \sum_{k=1}^{2} \operatorname*{Res}_{s=p_k} \left(Y(s)e^{st}\right) = k_1 e^{p_1 t} + (k_{2,1} + k_{2,2}t)e^{p_2 t}, \qquad t \geq 0,$$

where

$$k_1 = \lim_{s \to -a_1}(s + a_1)Y(s) = \lim_{s \to -a_1}\frac{\mu b_1}{s^2} = \frac{\mu b_1}{a_1^2},$$

$$k_{2,2} = \lim_{s \to 0} s^2 Y(s) = \lim_{s \to 0}\frac{\mu b_1}{s + a_1} = \frac{\mu b_1}{a_1},$$

$$k_{2,1} = \lim_{s \to 0}\frac{d}{ds}s^2 Y(s) = \lim_{s \to 0}\frac{-\mu b_1}{(s + a_1)^2} = -\frac{\mu b_1}{a_1^2}.$$

The complete response is

$$y(t) = \frac{\mu b_1}{a_1^2}\left[e^{-a_1 t} - 1 + a_1 t\right], \qquad t \geq 0.$$

Combined application of (3.22), (3.31), and (3.33) leads to a generalized form for the expansion of rational functions in partial fractions in which poles with higher multiplicities appear in powers up to their multiplicities. In our example

$$Y(s) = \frac{\mu b_1/a_1^2}{s + a_1} - \frac{\mu b_1/a_1^2}{s} + \frac{\mu b_1/a_1}{s^2},$$

is the partial-fraction expansion of $Y(s)$.

In anticipation of things to come, we substitute the car parameters (3.36) into the response and rewrite it as

$$y(t) = \frac{\mu}{Kp/m}H(0)^2\left[e^{\lambda t} - 1\right] + \mu H(0)t, \qquad t \geq 0,$$

where $\lambda = -(b/m + Kp/m)$ and $H(0)$ is the same as in (2.13). After a small transient, the car response to a ramp of slope μ will be another ramp of slope $\mu H(0)$. Indeed, the steady-state solution is

$$y_{\text{ss}}(t) = \mu H(0)t - \frac{\mu}{Kp/m}H(0)^2, \qquad (3.37)$$

which can be interpreted as a "delayed" ramp. As studied in Section 2.5, the higher the control gain K the closer $H(0)$ gets to one, in other words, the closer the response slope

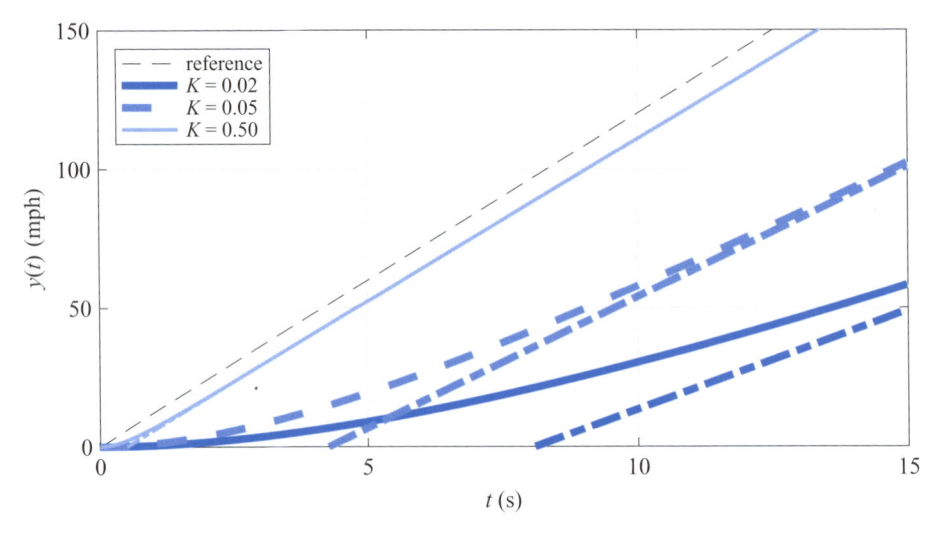

Figure 3.3 Closed-loop response of the velocity of the car cruise control according to the linear model (2.12) to a ramp with slope $\mu = 12$ mph/s (reference, dashed line), calculated for $b/m = 0.05$ and $p/b = 73.3$ and various values of the control gain K. The figure shows the reference ramp (dashed line), and the complete response and the steady-state response (dash–dotted lines) for various values of K.

is to μ. Note that the steady-state tracking error is

$$e_{ss}(t) = \bar{y}(t) - y_{ss}(t) = \mu(1 - H(0))t - \frac{\mu}{Kp/m} H(0)^2$$

$$= \mu S(0)t - \frac{\mu}{Kp/m} H(0)^2 \qquad (3.38)$$

and this error grows unbounded unless $S(0) = 0$. We will have more to say about this in Section 4.1. The closed-loop ramp response for the car model is plotted in Fig. 3.3 for various values of the gain K using p/m and b/m estimated previously.

The last case we shall cover is that of complex poles. Even though no new theory is necessary to handle complex poles, an example is in order to illustrate simplifications that will lead to a real signal, $f(t)$, even when the Laplace transform, $F(s)$, has complex poles. Consider the strictly proper transfer-function

$$F(s) = \frac{2}{s^2 + 2s + 2}.$$

The poles are

$$p_1 = -1 + j, \qquad p_2 = -1 - j = p_1^*,$$

where the symbol x^* denotes the complex conjugate of x. Because both poles are simple, the inverse Laplace transform is computed by straightforward application of formulas (3.22) and (3.31):

$$f(t) = \operatorname*{Res}_{s=p_1}(F(s)e^{st}) + \operatorname*{Res}_{s=p_2}(F(s)e^{st}) = k_1 e^{p_1 t} + k_2^* e^{p_2 t},$$

where

$$k_1 = \lim_{s \to p_1} (s - p_1)F(s), \qquad k_2 = \lim_{s \to p_2} (s - p_2)F(s).$$

A simplification is possible whenever $F(s)$ has real coefficients. In this case $F(s^*) = F^*(s)$ and

$$k_1^* = \lim_{s \to p_1} (s^* - p_1^*)F^*(s) = \lim_{s^* \to p_2} (s^* - p_2)F(s^*) = k_2$$

are complex-conjugates. The actual calculation leads to

$$k_1 = \lim_{s \to -1+j} \frac{2}{s+1+j} = \frac{2}{2j} = -j, \qquad k_2 = k_1^* = j,$$

and

$$f(t) = -j\,e^{(-1+j)t} + j\,e^{(-1-j)t} = j(e^{-jt} - e^{jt})e^{-t} = 2\sin(t)e^{-t}, \quad t \geq 0,$$

which is a real signal, as expected.

3.6 Stability

We now return to systems. Stability of a system is essentially a certificate of the system's *good behavior*. A number of different *definitions* of stability exist, each offering advantages in particular contexts. Fortunately, for linear systems, most notions of stability can be related to *asymptotic stability*. For a linear time-invariant system, asymptotic stability translates into location of the poles of the system's transfer-function, which can easily be checked. Even though we will not be fully equipped to comprehend asymptotic stability until Chapter 5, we can begin discussing stability in the time domain using the notion of *bounded-input–bounded-output stability*.

Consider the convolution formula (3.13) which we reproduce here:

$$y(t) = \int_{0^-}^{t} g(t-\tau)u(\tau)d\tau = \int_{0^-}^{t} g(\tau)u(t-\tau)d\tau.$$

A measure of good behavior of a system with impulse response $g(t)$ is that the output, $y(t)$, be bounded whenever the input, $u(t)$, is bounded. That is, we would like to verify the implication

$$|u(t)| \leq M_u < \infty \quad \Longrightarrow \quad |y(t)| \leq M_y < \infty \qquad (3.39)$$

for all possible input functions $u(t)$ at all $t \geq 0$. If (3.39) holds we say that the system is *bounded-input–bounded-output stable* (BIBO stable). Because

$$|y(t)| = \int_{0^-}^{t} |g(\tau)||u(t-\tau)|d\tau \leq M_u \int_{0^-}^{t} |g(\tau)|d\tau$$

one suspects that the implication (3.39) is true if

$$\|g\|_1 = \int_{0^-}^{\infty} |g(\tau)|d\tau < M_g < \infty. \qquad (3.40)$$

The above condition depends only on the *system*, in this case represented by the impulse response g(t), and not on a particular input, u(t), which is reassuring. The reason for using the symbol $\|g\|_1$ will be explained in Section 3.9. Boundedness of $\|g\|_1$ is not only sufficient but also necessary for BIBO stability: for any given $T > 0$ the input signal[20]

$$u_T(t) = \begin{cases} \text{sign}(g(T-t)), & 0 \le t \le T \\ 0, & t < 0 \text{ or } t > T \end{cases}$$

is such that $|u(t)| \le 1$ and

$$y_T(t) = \int_{0^-}^t g(t-\tau) u_T(\tau) d\tau = \int_{0^-}^t |g(\tau)| d\tau$$

which becomes arbitrarily close to $\|g\|_1$ when t and T are made large. The fact that the stability condition (3.40) does not depend on the size of $u(t)$ or $y(t)$ is certainly due to the fact that the system is linear.

One should be careful when generalized functions take part in the impulse response. Without getting into details on how to evaluate the resulting integrals, note that derivatives of the impulse cannot be present since[21]

$$\int_{0^-}^t \dot\delta(\sigma-\tau) u(t-\sigma) d\sigma = \int_{0^-}^t \delta(\sigma-\tau) \dot u(t-\sigma) d\sigma = \dot u(t-\tau),$$

which suggests an infinite response $y(t)$ when $u(t)$ is bounded but has a discontinuity. The same applies to higher-order derivatives of generalized functions. Note that impulses and shifted impulses are fine. Readers looking for a rigorous discussion are referred to [Vid93, Section 6.4].

For linear time-invariant systems, it is possible to verify the stability condition (3.40) in the frequency domain. A linear time-invariant system with impulse response $g(t)$ is asymptotically stable if and only if its transfer-function, $G(s) = \mathscr{L}\{g(t)\}$, converges and is bounded for all $\text{Re}(s) \ge 0$; in other words, if $G(s)$ does not have poles on the imaginary axis or on the right-hand side of the complex plane. The meaning of the terminology *asymptotic stability* will be discussed later in Chapter 5. To show one side of this statement, recall that $|e^{-st}| \le 1$ for all s such that $\text{Re}(s) \ge 0$. From (3.1)

$$|G(s)| = |\mathscr{L}\{g(t)\}| = \left|\int_{0^-}^\infty g(t) e^{-st} dt\right| \le \int_{0^-}^\infty |g(t)||e^{-st}| dt \le \int_{0^-}^\infty |g(t)| dt = \|g\|_1,$$

which shows that boundedness of $\|g\|_1$ implies boundedness and convergence of $G(s)$ for all s such that $\text{Re}(s) \ge 0$. Proving the converse, that is, that convergence of $G(s)$ on $\text{Re}(s) \ge 0$ implies boundedness of $\|g\|_1$, is more complicated, and we will show it for the special case of rational functions in the next section. Note that derivatives of

---

[20] $\text{sign}(x) = \{1, x > 0;\ 0, x = 0;\ -1, x < 0\}$.
[21] In this form, this formula follows formally from integration by parts. See P3.13. Alternatively, one can use the derivative property of the Laplace transform in Table 3.2.

generalized functions are naturally excluded by the requirement that $G(s)$ be bounded on the imaginary axis.

3.7 Transient and Steady-State Response

In many parts of this text we identify a *persistent* component of the response of a dynamic system, which we label the *steady-state response*, as opposed to the *transient response*, which vanishes with time. In this section we formalize the notions of transient and steady-state response.

The ability to rewrite the inverse Laplace transform of the output response function $Y(s)$ using residues (3.22) has another remarkable implication: when $Y(s)$ satisfies (3.25) it is possible to split $Y(s)$ into a sum according to the location of the singular points of $Y(s)$. A useful splitting is

$$Y(s) = Y_-(s) + Y_0(s) + Y_+(s), \tag{3.41}$$

where all singularities of $Y_-(s)$ are located on the left-hand side of the complex plane, i.e. they are such that $\mathrm{Re}(s) < 0$; all singularities of $Y_+(s)$ are located on the right-hand side of the complex plane, i.e. $\mathrm{Re}(s) > 0$; and all singularities of $Y_0(s)$ are on the imaginary axis.[22] When $Y(s)$ is rational, the components of (3.41) can be obtained from a partial-fraction expansion. The corresponding *components* of the output,

$$y(t) = y_-(t) + y_0(t) + y_+(t), \tag{3.42}$$

can then be classified according to their behavior in time.

The component

$$y_{\mathrm{tr}}(t) = y_-(t) = \mathscr{L}\{Y_-(s)\}$$

is called the *transient*. In order to justify this name we use the notion of asymptotic stability. From Section 3.6, the function $Y_-(s)$ is asymptotically stable and therefore $\|y_{\mathrm{tr}}\|_1 = \|y_-\|_1$ is bounded. It would be great if we could at this point conclude from asymptotic stability that

$$\lim_{t \to \infty} y_{\mathrm{tr}}(t) = 0. \tag{3.43}$$

Unfortunately, it is possible to come up with counterexamples of signals $y(t)$ which have bounded $\|y\|_1$ but for which $\lim_{t \to \infty} y(t) \neq 0$.[23] However, when $Y_-(s)$ is rational, proper, and with simple[24] poles we have $Y_-(s) = F_\mathrm{p} + F_\mathrm{sp}(s)$, where F_p is given by the

---

[22] If $Y(s)$ has a component without singularities (entire) it can be grouped with $Y_-(s)$. As discussed at the end of Section 3.4, these terms affect the response only at $t = 0$.
[23] See [GO03, p. 45] for an example where $y(t)$ is continuous and bounded and [Che98, p. 122] for an example where $y(t)$ is continuous and unbounded.
[24] If there are repeated poles one must use the more complicated expressions in Section 3.5.

formula (3.32) and

$$y_{\text{tr}}(t) = \mathscr{L}^{-1}\{Y_-(s)\} = F_p\delta(t) + \sum_{k=1}^{n} \operatorname*{Res}_{s=p_k}(F_{\text{sp}}(s)e^{st})$$

$$= F_p\delta(t) + \sum_{k=1}^{n} G_k(p_k)e^{p_k t}, \qquad t \geq 0.$$

Since the poles, p_k, and the residues $G_k(p_k)$ are bounded and $\operatorname{Re}(p_k) < 0$, the function $y_{\text{tr}}(t)$ is comprised of an impulse at the origin and exponentials that vanish as time grows, from which (3.43) follows.

When $Y(s)$ is proper and rational it is possible to go even further and prove the converse implication of asymptotic stability mentioned at the end of Section 3.6. Taking absolute values:

$$|y_{\text{tr}}(t)| \leq \sum_{k=1}^{n} |G_k(p_k)||e^{p_k t}| = \sum_{k=1}^{n} |G_k(p_k)|e^{\operatorname{Re}(p_k)t}, \qquad t > 0.$$

After integration we obtain

$$\int_{0^-}^{\infty} |y_{\text{tr}}(t)|dt \leq F_p + \sum_{k=1}^{n} |G_k(p_k)| \int_{0^-}^{\infty} e^{\operatorname{Re}(p_k)t}\, dt = F_p + \sum_{k=1}^{n} \frac{|G_k(p_k)|}{-\operatorname{Re}(p_k)} < \infty,$$

which implies $\|y_{\text{tr}}\|_1$ is bounded. Having poles with multiplicity greater than one does not affect the above conclusion since any additional terms involving

$$\int_{0^-}^{\infty} t^m e^{\operatorname{Re}(p_k)t}\, dt = \frac{(m+1)!}{(-\operatorname{Re}(p_k))^{m+1}} < \infty, \qquad 1 \leq m < \infty,$$

are also bounded and converge to zero when $\operatorname{Re}(p_k) < 0$.

Conversely, $Y_+(s)$ has all singularities on the right-hand side of the complex plane and, because of an argument similar to the one used with $Y_-(s)$, the response $y_+(t)$ will display unbounded growth. That is,

$$\lim_{t \to \infty} y_+(t) \to \infty.$$

This growth is at least exponential. In the case that $Y_+(s)$ is rational the output is comprised of terms of the form $t^{j-1}e^{\operatorname{Re}(p_k)t}$, $j = 1, \ldots, m_k$, stemming from each pole p_k, $\operatorname{Re}(p_k) > 0$, with multiplicity m_k.

Finally, the component

$$y_{\text{ss}}(t) = y_0(t) = \mathscr{L}^{-1}\{Y_0(s)\}$$

is what we refer to as the *steady-state response*. The signal $y_{\text{ss}}(t)$ may not necessarily be bounded. For example, if $s = j\omega_0$ is a pole with multiplicity higher than one then $y_{\text{ss}}(t)$ will grow unbounded. However, this growth is polynomial. For example, if $Y_0(s)$ is rational and $s = j\omega_0$ is a pole with multiplicity m then $y_0(t)$ will display growth of the order of t^{m-1}. Note that one might argue that it makes no sense to talk about steady-state or even transient response if the response grows unbounded. One can, however, always speak of transient and steady-state *components* of the response even when the overall response diverges, as done in this section.[25]

We shall revisit some earlier examples. In the calculation of the step response of the car in Section 3.3,

$$Y_-(s) = \frac{y_0 - (p/b)\tilde{u}}{s + b/m}, \qquad Y_0(s) = \frac{(p/b)\tilde{u}}{s}, \qquad Y_+(s) = 0,$$

from which

$$y_{\text{tr}}(t) = \left(y_0 - \frac{p}{b}\tilde{u}\right)e^{-\frac{b}{m}t}, \qquad y_{\text{ss}}(t) = \frac{p}{b}\tilde{u}, \qquad t > 0.$$

In the calculation of the response to a ramp of slope μ considered in Section 3.5,

$$Y_-(s) = \frac{\mu b_1/a_1^2}{s + a_1}, \qquad Y_0(s) = \frac{\mu b_1/a_1}{s^2} - \frac{\mu b_1/a_1^2}{s}, \qquad Y_+(s) = 0,$$

from which

$$y_{\text{tr}}(t) = \frac{\mu b_1}{a_1^2}e^{-a_1 t}, \qquad y_{\text{ss}}(t) = \frac{\mu b_1}{a_1^2}(a_1 t - 1), \qquad t > 0.$$

Note the linear growth of the steady-state response because of the double pole at the origin.

3.8 Frequency Response

The key idea behind the notion of frequency response is the calculation of the *steady-state* response of a linear time-invariant system to the family of inputs:

$$u(t) = A\cos(\omega t + \phi), \qquad t \geq 0. \tag{3.44}$$

As we will see shortly, for a linear time-invariant system without poles on the imaginary axis,

$$y_{\text{ss}}(t) = A|G(j\omega)|\cos(\omega t + \phi + \angle G(j\omega)), \tag{3.45}$$

where $G(s)$ is the system's transfer-function. That is, the steady-state response is another cosine function with the same frequency, ω, but different amplitude and phase. The amplitude and phase can be computed by evaluating the transfer-function, $G(s)$, at $s = j\omega$. We stress that the assumption that the linear system is time-invariant should not be

---

[25] Our definition of steady-state is indeed a bit unorthodox. One might prefer to further split the steady-state response in order to separate imaginary poles with multiplicity greater than one and enforce boundedness. This kind of exercise is not free from flaws. For example, what are the steady-state components of the signal with Laplace transform

$$\frac{s-1}{s^2(s+1)} = \frac{2}{s} - \frac{1}{s^2} - \frac{2}{s+1}?$$

Should one distinguish $\mathcal{L}^{-1}\{2s^{-1} - 2(s+1)^{-1}\} = 2 - 2e^{-t}$ from $\mathcal{L}^{-1}\{-s^{-2}\} = -t$?

taken for granted. For example, the modulator (3.11) is linear but not time-invariant. In particular, its steady-state response to the input (3.44) is

$$y_{ss}(t) = A\cos[(\omega + \omega_f)t] + A\cos[(\omega - \omega_f)t]$$

which is not of the form (3.45)!

The function $G(j\omega)$ is a complex-valued function of the transfer-function, $G(s)$, and of the input frequency, ω, and is known as the *frequency response*. The frequency response can also be obtained experimentally. The notion of steady-state solution introduced in Section 3.7 does not require asymptotic stability. However, a steady-state response like (3.45) can be *measured* from an actual system only if it is asymptotically stable. In Chapter 7 we will study the frequency response of stable and unstable systems in detail, learning how to sketch and interpret the magnitude and phase components of the frequency response and how to use it for closed-loop analysis and controller design.

We start by Laplace transforming the input (3.44). From Table 3.1,

$$U(s) = A\frac{s\cos\phi - \omega\sin\phi}{s^2 + \omega^2}.$$

Assuming zero initial conditions, this input is applied to a linear time-invariant system with transfer-function $G(s)$ to produce an output with transform

$$Y(s) = AG(s)\frac{s\cos\phi - \omega\sin\phi}{s^2 + \omega^2}.$$

As in Section 3.7, the steady-state response is

$$y_{ss}(t) = \mathcal{L}^{-1}\{Y_0(s)\}.$$

Assume[26] that $G(s)$ does not have poles on the imaginary axis, i.e. $G_0(s) = 0$. In this case the only poles in $Y_0(s)$ will be the pair originating from the input:

$$p_1 = j\omega, \qquad p_2 = -j\omega.$$

Using the residue calculations from Section 3.4 we obtain[27]

$$y_{ss}(t) = \mathcal{L}^{-1}\{Y_0(s)\} = ke^{j\omega t} + k^*e^{-j\omega t}, \qquad k = \lim_{s \to j\omega}(s - j\omega)Y_0(s),$$

where the coefficient k is from (3.31). Evaluating,

$$k = \lim_{s \to j\omega} AG(s)\frac{s\cos\phi - \omega\sin\phi}{s + j\omega} = \frac{A}{2}G(j\omega)(\cos\phi + j\sin\phi) = \frac{A}{2}G(j\omega)e^{j\phi}.$$

Formula (3.45) comes after using trigonometry[28] to rewrite

$$y_{ss}(t) = 2|k|\cos(\omega t + \angle k) = A|G(j\omega)|\cos(\omega t + \phi + \angle G(j\omega)).$$

The frequency response is key to understanding linear time-invariant systems.

---

[26] When $G(s)$ does have poles on the imaginary axis the steady-state response will have contributions from these imaginary poles as well. However, as we will see in Chapter 7, the frequency response is still useful in the presence of imaginary poles.
[27] The coefficient associated with the residue at the conjugate pole $-j\omega$ is simply k^*.
[28] See also the last entry in Table 3.2.

More abstractly, the frequency response of a linear system can be understood as steady-state *eigenvalues* of the linear system. The response of a linear system with impulse response $g(t)$ to a possibly complex input *eigenfunction*

$$u(t) = e^{s_0 t}, \quad t \geq 0,$$

is given by

$$y(t) = \int_{0-}^{t} g(t-\tau)e^{s_0 \tau}\, d\tau,$$

$$= e^{s_0 t} \int_{0-}^{t} g(t-\tau)e^{-s_0(t-\tau)}\, d\tau,$$

$$= e^{s_0 t} \int_{0-}^{t} g(\sigma)e^{-s_0 \sigma}\, d\sigma.$$

Therefore, for any s_0 in the region of convergence of the transfer-function $G(s)$, the output converges to the steady-state solution

$$y_{ss}(t) = e^{s_0 t} \lim_{t\to\infty} \int_{0-}^{t} g(\sigma)e^{-s_0 \sigma}\, d\sigma = G(s_0)e^{s_0 t}.$$

In particular when $G(s)$ is asymptotically stable, s_0 can be taken on the imaginary axis, $s_0 = j\omega$, in which case $G(s_0) = G(j\omega)$ is the frequency response and the expression above is the complex version of the steady-state solution derived earlier.

3.9 Norms of Signals and Systems

The notion of stability discussed in Section 3.6 highlights the importance of asserting the boundedness of signals that interact with systems. Signals can be measured by computing *norms*. A variety of signal norms exist, spanning a large number of applications. We do not have room to go into the details here, so the interested reader is referred to [DFT09] for a more advanced yet accessible discussion.

Some commonly used signal norms are the select p-norms:[29,30,31]

$$\|y\|_1 = \int_{0-}^{\infty} |y(t)|\, dt, \quad \|y\|_2 = \left(\int_{0-}^{\infty} y(t)^2\, dt\right)^{1/2}, \quad \|y\|_\infty = \sup_{t\geq 0} |y(t)|.$$

We have seen these norms at work in Section 3.6. For example, the stability condition (3.40) is equivalent to $\|g\|_1$ bounded and BIBO stability implies

$$\|y\|_\infty \leq \|g\|_1 \|u\|_\infty. \tag{3.46}$$

---

[29] $\|y\|_p = \left(\int_{0-}^{\infty} |y(t)|^p\, dt\right)^{1/p}, p = 1, 2, \ldots, \infty.$

[30] We have not paid much attention to what form of integral to use so far. For most of our developments the Lebesgue integral [KF75] is the one assumed implicitly.

[31] sup is a fancy replacement for max. When the maximum of a certain function is not attained at any point in its domain we use sup to indicate its least upper bound. For example, the function $f(x) = x/(1+x)$ is such that $f(x) < 1$ for all $x \geq 0$, hence $\max_{x\geq 0} f(x) < 1$. On the other hand, $\sup_{x\geq 0} f(x) = \min_y \{y : y \geq f(x); x \geq 0\}$ is equal to 1.

Any norm has the following useful properties: $\|y\| \geq 0$ is never negative; moreover, $\|y\| = 0$ if and only if $y(t) = 0$ for all $t \geq 0$; and finally, the *triangle inequality*

$$\|y_1 + y_2\| \leq \|y_1\| + \|y_2\| \tag{3.47}$$

must hold for any two signals y_1 and y_2.

Some signals can be bounded when measured by a norm while being unbounded according to a different norm. For instance, the signal

$$y(t) = (t+1)^{-1}, \qquad t \geq 0,$$

is unbounded in the 1-norm but $\|y\|_2 = \|y\|_\infty = 1$. Even simpler, the step function $y(t) = 1(t)$ is bounded only in the ∞-norm. However, for linear systems, input–output stability measured with respect to any one p-norm implies input–output stability in all other p-norms. For example, it is possible to establish that [Kha96, p. 266]

$$\|y\|_p \leq \|g\|_1 \|u\|_p, \qquad p = 1, 2, \ldots, \infty.$$

This bound is tight in the case $p = \infty$, i.e. (3.46), as we have shown in Section 3.6. For other values of p there might be tighter bounds. One example is $p = 2$, which we address next.

Some time-domain norms have frequency-domain counterparts. Assume that $y(t)$, $t \geq 0$, is such that its Laplace transform, $Y(s)$, is asymptotically stable. Then

$$\|y\|_2^2 = \int_{0^-}^{\infty} y(t)^2 \, dt = \int_{-\infty}^{\infty} y^*(t) y(t) \, dt.$$

Note that the limit of the integral has been changed because we assume that $y(t) = 0$ for $t < 0$. Since $Y(s)$ is asymptotically stable, we can represent $y(t)$ in terms of the inverse Laplace transform (3.4) where α is set to zero, i.e.

$$y(t) = \mathcal{L}^{-1}\{Y(s)\} = \frac{1}{2\pi j} \lim_{\omega \to \infty} \int_{-j\omega}^{j\omega} Y(s) e^{st} \, ds = \frac{1}{2\pi} \int_{-\infty}^{\infty} Y(j\omega) e^{j\omega t} \, d\omega.$$

Substituting this signal in the previous calculation gives

$$\|y\|_2^2 = \int_{-\infty}^{\infty} y^*(t) y(t) \, dt,$$

$$= \frac{1}{4\pi^2} \int_{-\infty}^{\infty} \int_{-\infty}^{\infty} Y(-j\sigma) e^{-j\sigma t} \, d\sigma \int_{-\infty}^{\infty} Y(j\omega) e^{j\omega t} \, d\omega \, dt,$$

$$= \frac{1}{2\pi} \int_{-\infty}^{\infty} Y(j\omega) \int_{-\infty}^{\infty} Y(-j\sigma) \frac{1}{2\pi} \int_{-\infty}^{\infty} e^{j(\omega - \sigma)t} \, dt \, d\sigma \, d\omega,$$

after rearranging the order of the integrals. The key step now is to use the formula (3.7) to write

$$\frac{1}{2\pi} \int_{-\infty}^{\infty} e^{j(\omega - \sigma)t} \, dt = \delta(\omega - \sigma)$$

as an impulse. Using the sifting property of the impulse (P3.12) we calculate

$$\|y\|_2^2 = \frac{1}{2\pi} \int_{-\infty}^{\infty} Y(j\omega) \int_{-\infty}^{\infty} Y(-j\sigma)\delta(\omega - \sigma) d\sigma \, d\omega,$$

$$= \frac{1}{2\pi} \int_{-\infty}^{\infty} Y(j\omega) Y(-j\omega) d\omega = \frac{1}{2\pi} \int_{-\infty}^{\infty} |Y(j\omega)|^2 \, d\omega = \|Y\|_2^2,$$

where

$$\|Y\|_2 = \left(\frac{1}{2\pi} \int_{-\infty}^{\infty} |Y(j\omega)|^2 \, d\omega \right)^{1/2}$$

is the 2-norm of $Y(s)$ evaluated at $s = j\omega$. The equivalence

$$\|y\|_2 = \|Y\|_2$$

is the subject of the remarkable theorems of *Plancherel* and *Parseval* [PW34, p. 2] (see also [LeP10, p. 279]), which allow the computation of the time-domain norm to be carried out in the frequency domain.

We can use Plancherel's theorem to bound the output of a linear time-invariant system. If $U(s)$ and $G(s)$ are asymptotically stable, then

$$\|y\|_2^2 = \frac{1}{2\pi} \int_{-\infty}^{\infty} |Y(j\omega)|^2 \, d\omega = \frac{1}{2\pi} \int_{-\infty}^{\infty} |G(j\omega)|^2 |U(j\omega)|^2 \, d\omega$$

$$\leq \left(\sup_{\omega \in \mathbb{R}} |G(j\omega)| \right)^2 \frac{1}{2\pi} \int_{-\infty}^{\infty} |U(j\omega)|^2 \, d\omega = \|G\|_\infty^2 \|u\|_2^2.$$

The quantity

$$\|G\|_\infty = \sup_{\omega \in \mathbb{R}} |G(j\omega)|$$

is the H_∞ norm of the transfer-function $G(s)$, in analogy with the ∞-norm of a time signal discussed earlier. This inequality is the same as

$$\|y\|_2 \leq \|G\|_\infty \|u\|_2, \tag{3.48}$$

which is of the form (3.46) with the 2-norm replacing the ∞-norm for the input and output signals. It is possible to show that this bound is tight, in the sense that there exists an input signal $u(t)$ such that $\|u\|_2$ is bounded and equality holds in (3.48). The proof of this is rather technical though [Vid81, Chapter 3].

Another application of Plancherel and Parseval's theorem is in establishing the correspondence

$$\|g\|_2 = \|G\|_2 = \left(\frac{1}{2\pi} \int_{-\infty}^{\infty} |G(j\omega)|^2 \, d\omega \right)^{1/2}.$$

The quantity $\|G\|_2$ is known as the H_2 norm of the transfer function $G(s)$. When $G(s)$ is rational, boundedness of $\|G\|_2$ implies that $G(s)$ must be strictly proper. This is because the 1-norm of the impulse is bounded,[32] that is, an impulse is asymptotically stable, but

---

[32] Impulse derivatives of any order have unbounded p-norms, and hence are not asymptotically stable. See P3.19 and P3.20.

the 2-norm is unbounded. A rational $G(s)$ needs to be proper but not strictly proper if $\|G\|_\infty$ is to be bounded. The H_2 norm also provides the useful bound

$$\|y\|_\infty \leq \|G\|_2 \|u\|_2. \tag{3.49}$$

You will show that this bound is tight in P3.31 and will learn how to compute the H_2 norm using residues in P3.32.

Problems

3.1 Show that the Laplace transform, (3.1), and the inverse Laplace transform, (3.4), are linear operators. *Hint: The operator L is linear when $L(\alpha f_1 + \beta f_2) = \alpha L(f_1) + \beta L(f_2)$ for all functions f_1, f_2 and scalars α and β.*

In P3.2–P3.14, assume that all functions of time and their derivatives exist and are continuous. Assume also that they are zero when $t < 0$.

3.2 Assume that the function f has exponential growth. Prove the integration property in Table 3.2. *Hint: Evaluate the integral in $\mathcal{L}\{\int_{0^-}^{t} f(\tau)d\tau\}$ by parts.*

3.3 Assume that the function f has exponential growth and that the Laplace transform of $\dot{f}$ exists. Prove the first differentiation property in Table 3.2 for the first derivative. *Hint: Evaluate the integral in $\mathcal{L}\{\dot{f}(t)\}$ by parts.*

3.4 Use the differentiation property for the first-order derivative to formally prove the time differentiation property in Table 3.2 for higher-order derivatives.

3.5 Prove the frequency differentiation property in Table 3.2. *Hint: Calculate $F'(s)$ by differentiating inside the Laplace integral.*

3.6 Change variables under the Laplace integral to prove the time-shift property in Table 3.2.

3.7 Change variables under the Laplace integral to prove the frequency-shift property in Table 3.2.

3.8 Evaluate the product $F(s)G(s)$ first as $\mathcal{L}\{f(t)\}G(s)$ then use the shift property to establish the convolution property in Table 3.2. *Hint: Assume that $f(t) = g(t) = 0$, for all $t < 0$ and that you can switch the order of integration.*

3.9 Show that

$$\mathcal{L}\left\{\frac{\sin(at)}{t}\right\} = \tan^{-1}\left(\frac{a}{s}\right).$$

Hint: Use $\int_0^a \cos(at)da = \sin(at)/t$.

3.10 Use P3.9, the time integration property, and the final value property to show that

$$\lim_{t\to\infty}\int_{0-}^{t}\frac{\sin(at)}{t}\,dt = \text{sign}(a)\frac{\pi}{2}.$$

3.11 Use P3.10 and (3.7) to show that

$$\int_{-\infty}^{\infty}\delta(t)\,dt = 1.$$

3.12 Use the convolution property to prove the *sifting property* of the impulse:

$$\int_{0-}^{t}f(\tau)\delta(t-\tau)\,d\tau = f(t).$$

3.13 Formally integrate by parts and use P3.12 to show that

$$\int_{0-}^{t}f(\tau)\dot{\delta}(\tau)\,d\tau = -\int_{0-}^{t}\dot{f}(\tau)\delta(\tau)\,d\tau = \dot{f}(0).$$

3.14 Use P3.12 and P3.13 to show that

$$\int_{0-}^{t}f(t-\tau)\dot{\delta}(\tau)\,d\tau = \int_{0-}^{t}\dot{f}(t-\tau)\delta(\tau)\,d\tau = \dot{f}(t).$$

Interpret the result in terms of the differentiation property of the Laplace transform.

In the following questions assume that $p_\epsilon(t)$ is the continuous pulse in Fig. 3.4(a).

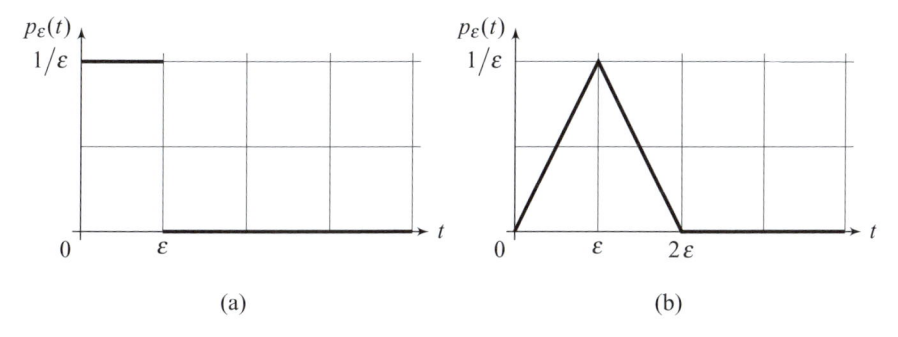

Figure 3.4 Waveforms for P3.15–P3.25.

3.15 Show that

$$\int_{0-}^{\infty}p_\epsilon(\tau)\,d\tau = \lim_{\epsilon\to 0}\int_{0-}^{\epsilon}p_\epsilon(\tau)\,d\tau = 1.$$

3.16 Show that

$$\lim_{\epsilon\to 0}\int_{0-}^{t}f(t-\tau)p_\epsilon(\tau)\,d\tau = f(t),$$

when f is continuous and differentiable. *Hint:* Use the mean-value theorem to write $f(t) - f(t-\tau) = \tau \dot{f}(\xi)$, where $\xi \in [t-\tau, t]$.

3.17 Show that

$$\lim_{\epsilon \to 0} \int_{0^-}^{t} f(\tau)p_\epsilon(\tau)d\tau = f(0),$$

when f is continuous and differentiable.   *Hint: Proceed as in P3.16.*

3.18 Show that

$$\lim_{\epsilon \to 0} \mathscr{L}\{p_\epsilon(t)\} = 1.$$

Hint: Use P3.17.

3.19 Calculate

$$\lim_{\epsilon \to 0} \int_{0^-}^{\infty} |p_\epsilon(\tau)|d\tau.$$

3.20 Calculate

$$\lim_{\epsilon \to 0} \int_{0^-}^{\infty} p_\epsilon(\tau)^2 \, d\tau.$$

3.21 Can one switch the order of the limit and integration in P3.15–P3.20?

In the following questions assume that $p_\epsilon(t)$ is the continuous pulse in Fig. 3.4(b).

3.22 Redo problems P3.15–P3.21.

3.23 Show that

$$\lim_{\epsilon \to 0} \int_{0^-}^{t} f(t-\tau)\dot{p}_\epsilon(\tau)d\tau = \dot{f}(t),$$

when f is continuous and differentiable.   *Hint: Apply the mean-value theorem on each segment where $\dot{p}_\epsilon(t)$ is not zero.*

3.24 Calculate

$$\lim_{\epsilon \to 0} \int_{0^-}^{\infty} |\dot{p}_\epsilon(\tau)|d\tau.$$

3.25 Calculate

$$\lim_{\epsilon \to 0} \int_{0^-}^{\infty} \dot{p}_\epsilon(\tau)^2 \, d\tau.$$

3.26 Replace $p_\epsilon(t)$ by a continuous and differentiable pulse and redo P3.15–P3.25. *Hint: Use a higher-order polynomial to replace each leg of the triangle in Fig. 3.4(b).*

3.27 Show that

$$\int_{-\pi/2}^{\pi/2} e^{-R\cos\theta} \, d\theta < \frac{\pi}{R}, \quad R > 0.$$

Hint: Use the fact that $\cos\theta \geq 1 - 2|\theta|/\pi$.

3.28 Assume that $F(s)$ satisfies (3.26). Let the parametrization of C_+^ρ be $s : \alpha + \rho e^{j\theta}$, $-\pi/2 \leq \theta \leq \pi/2$, and show that

$$|F(s)e^{st}| \leq \frac{Me^{\alpha t}}{(\rho - |\alpha|)^k} e^{\rho t \cos \theta}.$$

Hint: Recall that $|x + y| \geq |x| - |y|$.

3.29 Assume that $F(s)$ satisfies (3.26). Show that if $t < 0$ then

$$\left| \int_{C_+^\rho} F(s)e^{st}\, ds \right| \leq \frac{\pi M e^{\alpha t}}{|t|(\rho - |\alpha|)^k}.$$

Use this inequality to prove (3.27). *Hint: Use P3.27 and P3.28.*

3.30 Modify the arguments in P3.27–P3.29 to prove (3.21) assuming (3.24) is true and $t > 0$.

3.31 Assume that the causal impulse response $g(t)$ is such that the transfer-function $G(s)$ is asymptotically stable and $\|G\|_2$ is bounded. If

$$u(t) = u_\sigma(t) = \frac{g(\sigma - t)}{\|G\|_2}, \quad \tau \in [0, \sigma],$$

show that

$$\lim_{\sigma \to \infty} \|u_\sigma\|_2^2 = \lim_{\sigma \to \infty} \int_{0^-}^\infty u_\sigma(\tau)^2\, d\tau = 1$$

and

$$\|y\|_\infty \geq \lim_{t \to \infty} |y(t)| = \|G\|_2 \|u\|_2,$$

to conclude that the inequality (3.49) is tight.

3.32 Let $G(s)$ be asymptotically stable and satisfying (3.23). Show that

$$\|G\|_2^2 = \frac{1}{2\pi} \int_{-\infty}^\infty |G(j\omega)|\, d\omega = \frac{1}{2\pi j} \lim_{\rho \to \infty} \int_{\Gamma_-^0} G(-s)G(s)\, ds$$

where Γ_-^0 is the contour Γ_-^α from Fig. 3.2 with $\alpha = 0$. Explain how to compute this integral using Cauchy's residue theorem (Theorem 3.1). Use this method to verify that

$$\|G\|_2^2 = \frac{1}{2a}$$

when $G(s) = (s + a)^{-1}$.

The next problems are inspired by a list of useful facts in [Bro+07].

3.33 Show that if $f : [0, \infty) \to \mathbb{R}$ is non-decreasing and $f(t) \leq M < \infty$ for all t then $\lim_{t \to \infty} f(t)$ exists, that is, f converges. *Hint: Show that if f does not converge it cannot be bounded.*

Transfer-Function Models

3.34 Show that if $\int_0^t |f(\tau)| d\tau$ converges then $\int_0^t f(\tau) d\tau$ also converges. *Hint: Partition the integration interval into two disjoint sets of intervals in which either $f(t) \geq 0$ or $f(t) < 0$ to show that each part converges.*

3.35 Show that if $\int_0^\infty |\dot{f}(\tau)| d\tau < \infty$ then f converges. *Hint: Use P3.33 to show that $\int_0^t |\dot{f}(\tau)| d\tau$ converges and then P3.34 to show that $\int_0^t \dot{f}(\tau) d\tau$ converges.*

3.36 Show that
$$2 \int_0^t |f(\tau)\dot{f}(\tau)| d\tau \leq \int_0^t f(\tau)^2 d\tau + \int_0^t \dot{f}(\tau)^2 d\tau.$$
Hint: Use the fact that $[f(t) - \dot{f}(t)]^2 \geq 0$.

3.37 Show that if $\int_0^\infty f(\tau)^2 d\tau < \infty$ and $\dot{f}^2$ converges then f converges to zero. *Hint: Look at what happens to the integral as t grows if $\dot{f}^2$ converges.*

3.38 Show that if $\int_0^\infty f(\tau)^2 d\tau < \infty$ and $\int_0^\infty \dot{f}(\tau)^2 d\tau < \infty$ then $\dot{f}$ is bounded and $\lim_{t \to \infty} f(t) = 0$. *Hint: Let $g(t) = (d/dt) f(t)^2$ and use P3.36 to show that $\int_0^\infty |g(\tau)| d\tau < \infty$. Then use P3.35 to show that f^2 converges and P3.37 to show that f must converge to zero.*

3.39 Compute the Laplace transform of the following signals $f(t)$ defined for $t \geq 0$:

(a) $1 - e^{-t}$;
(b) $e^{-t} - e^{-2t}$;
(c) $\sinh(t) - \cosh(2t)$;
(d) $\sin(t) - \cos(t)$;
(e) $t + te^{-t}$;
(f) $\sin(t) - e^{-t}\cos(2t)$;
(g) $t \sin(t) + \cos(2t)$;
(h) $e^{-t}\cos(t + \pi/4)$;
(i) $1(t) + 1(t-1)$;
(j) $e^{-t} + 1(t-1)e^{t-1}$;
(k) $\delta(t) + te^{-t}\sin(t)$;
(l) $1 + t - t^2$.

3.40 Compute the expansion in partial fractions of the following rational functions:

(a) $\dfrac{1}{(s+1)(s+2)}$;
(b) $\dfrac{s-1}{(s+1)(s+2)}$;
(c) $\dfrac{s(s-1)}{(s+1)(s+2)}$;
(d) $\dfrac{s^2-1}{(s+1)(s+2)}$;
(e) $\dfrac{s^2(s-1)}{(s+1)(s+2)}$;
(f) $\dfrac{1}{s(s+1)(s+2)}$;
(g) $\dfrac{1}{s^2(s+1)}$;
(h) $\dfrac{s-1}{s^2(s+1)}$;
(i) $\dfrac{1}{s^2+2s+1}$;
(j) $\dfrac{s}{s^2+2s+1}$;
(k) $\dfrac{1}{s^2+2s+2}$;
(l) $\dfrac{s+1}{(s^2+2s+2)^2}$.

3.41 Compute the inverse Laplace transform of the following complex-valued functions:

(a) $\dfrac{1}{s(s+1)}$;
(b) $\dfrac{s-1}{s+1}$;
(c) $\dfrac{s-1}{s(s+1)}$;
(d) $\dfrac{s}{s^2+2s+1}$;
(e) $\dfrac{1}{s^2-1}$;
(f) $\dfrac{1}{s^2(s+2)^2}$;
(g) $\dfrac{1}{(s+1)^2+1}$;
(h) $\dfrac{1-e^{-s}}{s}$;
(i) $\dfrac{s+1-e^{-s}}{s(s+1)}$;
(j) $\dfrac{1+s+s^2}{(1+s)^3}$;
(k) $\dfrac{1}{s+1} - \dfrac{s}{(s+1)^2} + \dfrac{s^2}{(s+1)^3}$.

3.42 Use the Laplace transform to solve the following linear ordinary differential equations:

(a) $\dot{y}+y=0,$ $\qquad$ $y(0)=1;$
(b) $\dot{y}+y=1,$ $\qquad$ $y(0)=0;$
(c) $\dot{y}-y=0,$ $\qquad$ $y(0)=1;$
(d) $\dot{y}-y=1,$ $\qquad$ $y(0)=0;$
(e) $\ddot{y}+\dot{y}=0,$ $\qquad$ $y(0)=1, \dot{y}(0)=-1;$
(f) $\ddot{y}+\dot{y}=e^{-t},$ $\qquad$ $y(0)=\dot{y}(0)=0;$
(g) $\ddot{y}+y=0,$ $\qquad$ $y(0)=1, \dot{y}(0)=-1;$
(h) $\ddot{y}+y=\cos(t),$ $\qquad$ $y(0)=\dot{y}(0)=0;$
(i) $\ddot{y}+y=\cos(2t),$ $\qquad$ $y(0)=\dot{y}(0)=0;$
(j) $\ddot{y}+y=\cos(t),$ $\qquad$ $y(0)=1, \dot{y}(0)=0;$
(k) $\ddot{y}+y=\cos(2t),$ $\qquad$ $y(0)=1, \dot{y}(0)=0;$
(l) $\ddot{y}+2\dot{y}+y=0,$ $\qquad$ $y(0)=1, \dot{y}(0)=0;$
(m) $\ddot{y}+2\dot{y}+y=\sin(t),$ $\qquad$ $y(0)=\dot{y}(0)=0;$
(n) $\ddot{y}+2\dot{y}+2y=0,$ $\qquad$ $y(0)=1, \dot{y}(0)=0;$
(o) $\ddot{y}+2\dot{y}+2y=\cos(t),$ $\qquad$ $y(0)=\dot{y}(0)=0.$

3.43 Prove that, when $f(t), t \geq 0$, is such that $f(t) = 0$, for all $t \geq T$, then its Laplace transform is analytic in $\mathbb{C}$ (entire) or its singularities are *removable*. Verify this by showing that the following functions have only removable singularities:

(a) $1-e^{-s}+e^{-2s};$
(b) $\dfrac{1-e^{-s}}{s};$
(c) $\dfrac{1-2e^{-s}+e^{-2s}}{s};$
(d) $\dfrac{1-2e^{-s}+e^{-2s}}{s^2};$
(e) $\dfrac{1-2e^{-s}+2e^{-3s}-e^{-4s}}{s^2};$
(f) $\dfrac{1-2e^{-s}+2e^{-3s}-e^{-4s}}{s^3};$
(g) $\dfrac{e^{-s-1}(e^s-e)}{s-1}.$

Compute the inverse Laplace transform, $f(t) = \mathcal{L}^{-1}\{F(s)\}$, and sketch the plot of $f(t)$, $t \geq 0$. *Hint:* Roughly speaking, $F(s)$ is singular at a removable singular point s_0 if $\lim_{s \to s_0} F(s_0)$ exists and is bounded and $F'(s_0)$ and all higher-order derivatives exist, e.g. $s^{-1}\sin(s)$ at 0, i.e. it admits a Taylor series expansion at 0.

3.44 The functions

(a) $\dfrac{1}{s};$
(b) $\dfrac{1}{s+1};$
(c) $\dfrac{1}{s^2};$
(d) $\dfrac{1}{(s+1)^2};$
(e) $\dfrac{s}{(s+1)^2};$
(f) $\dfrac{e^{-s}}{s};$
(g) $\dfrac{e^{-s}}{s+1};$
(h) $\dfrac{1-e^{-2\pi s}}{s^2+1};$
(i) $\dfrac{(1-e^{-2\pi s})s}{s^2+1}$

are Laplace transforms obtained from functions $f(t), t \geq 0$. Assume that $f(0^-) = 0$ and use the formal derivative property and the inverse Laplace transform to calculate the time-derivative function, $\dot{f}(t), t \geq 0$, without evaluating $f(t)$. Locate any discontinuities of the original function $f(t)$.

3.45 Let C be the unit circle traversed in the counter-clockwise direction and calculate the contour integral

$$\int_C f(s)ds$$

using Cauchy's residue theorem (Theorem 3.1) for the following functions:

(a) $1 - e^{-s}$;

(b) $\dfrac{1 + e^{-s}}{s}$;

(c) $\dfrac{1}{s^2}$;

(d) $\dfrac{1}{s(s + 1/2)}$;

(e) $\dfrac{s}{(s + 1/2)(s + 2)}$;

(f) $\dfrac{1}{s(s + 1/2)}$;

(g) $\dfrac{1}{s^2 + 1/4}$.

3.46 A linear time-invariant system has as impulse response

$$g(t) = e^{-t}, \qquad t \geq 0.$$

Compute the system's transfer-function. What is the order of the system? Is the system asymptotically stable? Assuming zero initial conditions, calculate and sketch the response to a constant input $u(t) = 1, t \geq 0$. Identify the transient and steady-state components of the response.

3.47 Repeat P3.46 for the following impulse responses:

(a) $e^{-2t} - e^{-t}$;
(b) $e^{t} - e^{-t}$;
(c) $-te^{-2t}$;
(d) $e^{-t}\cos(t)$;
(e) $e^{-t}(\cos(t) - \sin(t))$;

(f) $\cos(t + \pi/6)$;
(g) $t\cos(t)$;
(h) $\delta(t) + te^{-t}\sin(t)$;
(i) $1(t) - 1(t - 1)$;
(j) $t - 2(t - 1)1(t - 1) + (t - 2)1(t - 2)$.

3.48 Let $u(t), t \geq 0$, be such that $u(t) = 0$ for all $t \geq T$ and let $U(s)$ be its Laplace transform. Show that

$$y(t) = \mathcal{L}^{-1}\{Y(s)\}, \qquad Y(s) = \dfrac{U(s)}{1 - e^{-sT}},$$

is a periodic function with period T. Use this fact to calculate and sketch the plot of the inverse Laplace transform of

$$Y(t) = \dfrac{1 - e^{-sT/2}}{s(1 - e^{-sT})}.$$

Explain what happens when $u(t) \neq 0$, $t \geq T$. Repeat for

$$Y(t) = \frac{1 - e^{-s3T/2}}{s(1 - e^{-sT})}.$$

Hint: Recall that $(1-x)^{-1} = 1 + x + x^2 + \cdots$, for all $|x| < 1$ and use P3.43.

3.49 Calculate $Y(s)$ and the associated transfer-function in the block-diagram in Fig. 3.5. Compare with P3.48. Is this system asymptotically stable?

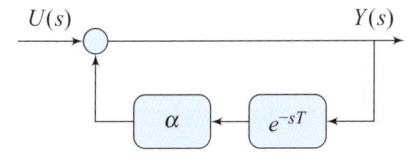

Figure 3.5 Block diagram for P3.49.

3.50 A modulator is the basic building block of any radio. A modulator converts signals from one frequency range to another frequency range. Given an input $u(t)$, $t \geq 0$, an *amplitude modulator* (AM modulator) produces the output

$$y(t) = 2\cos(\omega_f t)u(t), \quad \omega_f > 0.$$

Show that the amplitude modulator system is linear and causal, but not time-invariant.

3.51 A *sample-and-hold* produces the output

$$y(t) = u(kT), \quad kT \leq t < (k+1)T, \quad k \in \mathbb{N},$$

in response to a continuous input, $u(t)$. Show that the sample-and-hold system is a linear system. Show that it is not time-invariant. Why the name *sample-and-hold*? Can you think of an application for such a system?

3.52 Show that the impulse response of the sample-and-hold system, P3.51, is the function

$$g_\tau(t) = \delta(kT - \tau), \quad kT \leq t < (k+1)T, \quad k \in \mathbb{N},$$

by verifying that

$$y(t) = \int_0^t g_\tau(t)u(\tau)d\tau.$$

Hint: Use P3.12.

In the following questions, assume zero initial conditions unless otherwise noted.

3.53 You have shown in P2.4 that the ordinary differential equation

$$m\dot{v} + bv = mg$$

is a simplified description of the motion of an object of mass m dropping vertically under constant gravitational acceleration, g, and linear air resistance, $-bv$. Solve this differential equation using the Laplace transform. Treating the gravitational force, $u = mg$, as an input, calculate the transfer-function from the input, u, to the velocity, v, then calculate the transfer-function from the input, u, to the object's position, $x = \int_0^t v(\tau)d\tau$. Assume that all constants are positive. Are these transfer-functions asymptotically stable?

3.54 You have shown in P2.10 and P2.12 that the ordinary differential equation

$$\left(J_1 r_2^2 + J_2 r_1^2\right) \dot{\omega}_1 + \left(b_1 r_2^2 + b_2 r_1^2\right) \omega_1 = r_2^2 \tau, \qquad \omega_2 = (r_1/r_2)\omega_1$$

is a simplified description of the motion of a rotating machine driven by a belt without slip as in Fig. 2.18(a), where ω_1 is the angular velocity of the driving shaft and ω_2 is the machine's angular velocity. Calculate the transfer-function from the input torque, τ, to the machine's angular velocity, ω_2, then calculate the transfer-function from the torque, τ, to the machine's angular position, $\theta_2 = \int_0^t \omega_2(\tau)d\tau$. Assume that all constants are positive. Are these transfer-functions asymptotically stable?

3.55 Consider the simplified model of the rotating machine described in P3.54. Calculate the machine's angular velocity, $\omega_2(t)$, obtained in response to a constant-torque input, $\tau(t) = \tilde{\tau}$, $t \geq 0$. Identify the transient and the steady-state components of the response.

3.56 Repeat P3.55 for the sinusoidal torque input $\tau(t) = \tilde{\tau} \cos(\omega t)$, $t \geq 0$.

3.57 Whenever possible, calculate the steady-state responses in P3.55 and P3.56 using the frequency-response method. Compare answers.

3.58 Consider the simplified model of the rotating machine described in P3.54. Calculate the machine's angular position, $\theta_2(t)$, obtained in response to a constant-torque input, $\tau(t) = \tilde{\tau}$, $t \geq 0$. Identify the transient and the steady-state components of the response.

3.59 Repeat P3.58 for the sinusoidal torque input $\tau(t) = \tilde{\tau} \cos(\omega t)$, $t \geq 0$.

3.60 Whenever possible, calculate the steady-state responses in P3.58 and P3.59 using the frequency-response method. Compare your answers.

3.61 Let $\tilde{\tau} = 1$ N m, $r_1 = 25$ mm, $r_2 = 500$ mm, $b_1 = 0.01$ kg m^2/s, $b_2 = 0.1$ kg m^2/s, $J_1 = 0.0031$ kg m^2, and $J_2 = 25$ kg m^2. Sketch or use MATLAB to plot the responses you obtained in P3.55, P3.56, P3.58, and P3.59.

3.62 You have shown in P2.18 that the ordinary differential equation

$$\left(J_1 + J_2 + r^2(m_1 + m_2)\right)\dot{\omega} + (b_1 + b_2)\omega = \tau + gr(m_1 - m_2), \qquad v_1 = r\omega,$$

is a simplified description of the motion of the elevator in Fig. 2.18(b), where ω is the angular velocity of the driving shaft and v_1 is the elevator's load linear velocity. Treating the gravitational torque, $w = gr(m_1 - m_2)$, as an input, calculate the transfer-function, G_w, from the gravitational torque, w, to the elevator's load linear velocity, v_1, and the transfer-function, G_τ, from the motor torque, τ, to the elevator's load linear velocity, v_1, and show that

$$V_1(s) = G_\tau(s) T(s) + G_w(s) W(s).$$

Assume that all constants are positive. Are these transfer-functions asymptotically stable?

3.63 Consider the simplified model of the elevator described in P3.62. Calculate the elevator's linear velocity, $v_1(t)$, obtained in response to a constant-torque input, $\tau(t) = \tilde{\tau}$, $t \geq 0$, and a constant gravitational torque $w(t) = gr(m_1 - m_2)$, $t \geq 0$. Identify the transient and the steady-state components of the response.

3.64 Repeat P3.63 for the sinusoidal torque input $\tau(t) = \tilde{\tau}\cos(\omega t)$, $t \geq 0$.

3.65 Whenever possible, calculate the steady-state responses in P3.63 and P3.64 using the frequency-response method. Compare answers.

3.66 Consider the simplified model of the elevator described in P3.62. Calculate the elevator's linear position, $x_1(t)$, obtained in response to a constant-torque input, $\tau(t) = \tilde{\tau}$, $t \geq 0$, and a constant gravitational torque $w(t) = gr(m_1 - m_2)$, $t \geq 0$. Identify the transient and the steady-state components of the response.

3.67 Repeat P3.66 for the sinusoidal torque input $\tau(t) = \tilde{\tau}\cos(\omega t)$, $t \geq 0$.

3.68 Whenever possible, calculate the steady-state responses in P3.66 and P3.67 using the frequency-response method. Compare answers.

3.69 Let $g = 10$ m/s², $r = 1$ m, $m_1 = m_2 = 1000$ kg, $b_1 = b_2 = 120$ kg m²/s, $J_1 = J_2 = 20$ kg m², and $\tilde{\tau} = 480$ N m. Sketch or use MATLAB to plot the responses you obtained in P3.63, P3.64, P3.66, and P3.67.

3.70 Repeat P3.69 with $m_2 = 800$ kg.

3.71 You have shown in P2.27 that the ordinary differential equation

$$m\ddot{x} + b\dot{x} + kx = f$$

is a simplified description of the motion of the mass–spring–damper system in Fig. 2.19(a), where x is the position of the mass, m, and f is a force applied on the mass. Calculate the transfer-function from the force f to the mass position x. Assume that all constants are positive. Is this transfer-function asymptotically stable?

3.72 Consider the simplified model of the mass–spring–damper system described in P3.71. Calculate the mass position, $x(t)$, obtained in response to a constant-force input, $f(t) = \tilde{f}$, $t \geq 0$. Identify the transient and the steady-state components of the response. How do the parameters m, k, and b affect the response?

3.73 Repeat P3.72 for the sinusoidal input $f(t) = \tilde{f}\cos(\omega t)$, $t \geq 0$.

3.74 Calculate the steady-state responses in P3.72 and P3.73 using the frequency-response method. Compare answers.

3.75 Let $m = 1$ kg, $b = 0.1$ kg/s, $k = 1$ N/m, and $\tilde{f} = 1$ N. Sketch or use MATLAB to plot the responses you obtained in P3.72 and P3.73.

3.76 You have shown in P2.32 that the ordinary differential equations

$$m_1\ddot{x}_1 + (b_1 + b_2)\dot{x}_1 + (k_1 + k_2)x_1 - b_2\dot{x}_2 - k_2x_2 = 0,$$
$$m_2\ddot{x}_2 + b_2(\dot{x}_2 - \dot{x}_1) + k_2(x_2 - x_1) = f_2$$

are a simplified description of the motion of the mass–spring–damper system in Fig. 2.20(b), where x_1 and x_2 are the positions of the masses m_1 and m_2 and f_2 is a force applied on mass m_2. Calculate the transfer-function from the force f_2 to the position x_1.

3.77 Consider the simplified model of the mass–spring–damper system described in P3.76. Assume that the parameters are such that the system is asymptotically stable and use the system's frequency response to calculate the steady-state component of the response of the position of the mass m_1, x_1, to a constant-force input $f_2(t) = \tilde{f}_2, t \geq 0$. Interpret the answer using what you know from physics.

3.78 Repeat P3.77 for the sinusoidal input $f_2(t) = \tilde{f}_2 \cos(\omega t), t \geq 0$.

3.79 Let $m_1 = m_2 = 1$ kg, $b_1 = b_2 = 0.1$ kg/s, $k_1 = 1$ N/m, $k_2 = 2$ N/m, and $\tilde{f}_2 = 1$ N. Sketch or use MATLAB to plot the responses to the inputs in P3.77 and P3.78.

3.80 You have shown in P2.34 that the ordinary differential equation

$$RC\dot{v}_C + v_C = v$$

is an approximate model for the RC electric circuit in Fig. 2.22(a). Calculate the transfer-function from the input voltage, v, to the capacitor voltage, v_C. Assume that all constants are positive. Is this transfer-function asymptotically stable?

3.81 Consider the model of the circuit described in P3.80. Calculate the capacitor voltage, $v_C(t)$, obtained in response to a constant input voltage, $v(t) = \tilde{v}, t \geq 0$. Identify the transient and the steady-state components of the response.

3.82 Repeat P3.81 for the sinusoidal voltage input $v(t) = \tilde{v} \cos(\omega t), t \geq 0$.

3.83 Calculate the steady-state responses in P3.81 and P3.82 using the frequency-response method. Compare answers.

3.84 Let $R = 1$ MΩ, $C = 10$ μF, and $\tilde{v} = 1$ V. Sketch or use MATLAB to plot the responses you obtained in P3.81 and P3.82.

3.85 You have shown in P2.36 that the ordinary differential equation

$$LC\ddot{v}_C + RC\dot{v}_C + v_C = v$$

is an approximate model for the RLC electric circuit in Fig. 2.22(b). Calculate the transfer-function from the input voltage, v, to the capacitor voltage, v_C. Assume that all constants are positive. Is this transfer-function asymptotically stable?

3.86 Consider the model of the circuit described in P3.85. Calculate the capacitor voltage, $v_C(t)$, obtained in response to a constant input voltage, $v(t) = \tilde{v}, t \geq 0$. Identify the transient and the steady-state components of the response. How do the parameters R, L, and C affect the response?

3.87 Repeat P3.84 for the sinusoidal input $v(t) = \tilde{v}\cos(\omega t), t \geq 0$.

3.88 Calculate the steady-state responses in P3.86 and P3.87 using the frequency-response method. Compare answers.

3.89 Let $L = 1$ H, $C = 100$ µF, and $R = 200\,\Omega$. Sketch or use MATLAB to plot the responses you obtained in P3.86 and P3.87.

3.90 You have shown in P2.38 that the ordinary differential equation

$$RC_2\ddot{v}_o + RC_1\dot{v} + v = 0$$

is an approximate model for the electric circuit in Fig. 2.23. Calculate the transfer-function from the input voltage, v, to the output voltage, v_o. Assume that all constants are positive. Is this transfer-function asymptotically stable?

3.91 Consider the model of the circuit described in P3.90. Calculate the output voltage, $v_o(t)$, obtained in response to a constant input voltage, $v(t) = \tilde{v}, t \geq 0$. Identify the transient and the steady-state components of the response.

3.92 Repeat P3.91 for the sinusoidal voltage input $v(t) = \tilde{v}\cos(\omega t), t \geq 0$.

3.93 Calculate the steady-state responses in P3.91 and P3.92 using the frequency-response method. Compare answers.

3.94 Let $R = 1$ MΩ, $C_1 = 100$ µF, $C_2 = 10$ µF, and $\tilde{v} = 1$ V. Sketch or use MATLAB to plot the responses you obtained in P3.91 and P3.92.

3.95 You have shown in P2.41 that the ordinary differential equation

$$J\dot{\omega} + \left(b + \frac{K_t K_e}{R_a}\right)\omega = \frac{K_t}{R_a}v_a$$

is a simplified description of the motion of the rotor of the DC motor in Fig. 2.24, where ω is the rotor angular velocity. Calculate the transfer-function from the armature voltage, v_a, to the rotor's angular velocity, ω, then calculate the transfer-function from the voltage, v_a, to the motor's angular position, $\theta = \int_0^t \omega(\tau)d\tau$. Assume that all constants are positive. Are these transfer-functions asymptotically stable?

3.96 Consider the simplified model of the DC motor described in P3.95. Calculate the motor's angular velocity, $\omega(t)$, obtained in response to a constant-armature-voltage input, $v_a(t) = \tilde{v}_a, t \geq 0$. Identify the transient and the steady-state components of the response.

3.97 Repeat P3.96 for the sinusoidal voltage input $v_a(t) = \tilde{v}_a\cos(\omega t), t \geq 0$.

3.98 Calculate the steady-state responses in P3.96 and P3.97 using the frequency-response method. Compare answers.

3.99 Let $J = 227 \times 10^{-6}$ kg m², $K_t = 0.02$ N m/A, $K_e = 0.02$ V s/rad, $b = 289.4 \times 10^{-6}$ kg m²/s, $R_a = 7\,\Omega$, and $\tilde{v}_a = 12$ V. Sketch or use MATLAB to plot the responses you obtained in P3.96 and P3.97.

3.100 You have shown in P2.49 that the temperature of a substance, T (in K or in °C), flowing in and out of a container kept at the ambient temperature, T_o, with an inflow temperature, T_i, and a heat source, q (in W), can be approximated by the differential equation

$$m c \dot{T} = q + w c (T_i - T) + \frac{1}{R}(T_o - T),$$

where m and c are the substance's mass and specific heat, and R is the overall system's thermal resistance. The input and output flow mass rates are assumed to be constant and equal to w (in kg/s). Calculate the transfer-functions from the inflow temperature, T_i, the ambient temperature, T_o, and the heat source, q, to the substance's temperature, T. Assume that all constants are positive. Are these transfer-functions asymptotically stable?

3.101 Consider the simplified model of the thermodynamic system described in P3.100. Calculate the substance's temperature, $T(t)$, obtained in response to a constant ambient temperature, $T_o(t) = \tilde{T}_o, t \geq 0$, a constant inflow temperature, $T_i(t) = \tilde{T}_i, t \geq 0$, and a constant heat source $q(t) = \tilde{q}, t \geq 0$. Identify the transient and the steady-state components of the response.

3.102 Repeat P3.101 for the sinusoidal heat source input $q(t) = \tilde{q} \cos(\omega t), t \geq 0$.

3.103 Calculate the steady-state responses in P3.101 and P3.102 using the frequency-response method. Compare your answers.

3.104 Assume that water's density and specific heat are $\rho = 997.1$ kg/m³ and $c = 4186$ J/kg K and consider a 50 gal (≈ 0.19 m³) tank with $R = 0.27$ K/W, $T_i = T_o = 77\,°F$ ($\approx 25\,°C$), and $\tilde{q} = 40{,}000$ BTU/h (≈ 12 kW). Sketch or use MATLAB to plot the responses you obtained in P3.101 and P3.102.

4 Feedback Analysis

You have already learned how to use the relationship (3.14), $Y(s) = G(s)U(s)$, to compute the Laplace transform of the output of a linear time-invariant system, $Y(s)$, from its transfer-function, $G(s)$, and the Laplace transform of the input, $U(s)$. The ability to relate the Laplace transform of the input, $U(s)$, to the Laplace transform of the output, $Y(s)$, by a simple product of complex functions has far broader implications, which we will study throughout the rest of the book.

Using the Laplace transform we can operate with dynamic linear systems and their interconnections as if they were static systems, as was done in Chapter 1. Consider for instance the series connection of systems depicted in Fig. 4.1. We apply the Laplace transform to the signals $u(t)$, $z(t)$, and $y(t)$:

$$Y(s) = G_2(s)Z(s), \qquad Z(s) = G_1(s)U_1(s),$$

where $G_1(s)$ and $G_2(s)$ are the transfer-functions for the systems in blocks G_1 and G_2. These are the same expressions that would have been obtained if G_1 and G_2 were static linear gains. Eliminating $Z(s)$ we obtain

$$Y(s) = G(s)U(s), \qquad G(s) = G_2(s)G_1(s),$$

where $G(s)$ is the transfer-function from the transform of the input, $U(s)$, to the transform of the output, $Y(s)$.

Virtually all formulas computed so far that describe the interconnection of static systems hold verbatim for the interconnection of dynamic systems if static gains are replaced by transfer-functions. This is true also for closed-loop feedback configurations such as the one in Fig. 4.2, which is a reproduction of the block-diagram originally depicted in Fig. 1.8. In this feedback loop

$$Y(s) = G(s)U(s), \qquad U(s) = K(s)E(s), \qquad E(s) = \bar{Y}(s) - Y(s).$$

Eliminating $E(s)$ and $U(s)$

$$[1 + G(s)K(s)]Y(s) = G(s)K(s)\bar{Y}(s).$$

We assume that $G(s)K(s) \neq -1$ to compute

$$Y(s) = H(s)\bar{Y}(s), \qquad H(s) = \frac{G(s)K(s)}{1 + G(s)K(s)},$$

Feedback Analysis

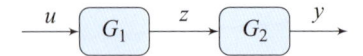

Figure 4.1 Two systems in series.

where $H(s)$ is the closed-loop transfer-function from the transform of the reference input, $\bar{Y}(s)$, to the transform of the output, $Y(s)$. For example, in Chapter 2 we derived the ordinary differential equation (2.3) to model the dynamics of the car used in the cruise control problem for which we later calculated, (3.16) in Section 3.3, the associated transfer-function:

$$G(s) = \frac{p/m}{s + b/m}.$$

Closing the loop as in Fig. 4.2 with a static proportional controller $K(s) = K$, we calculate the closed-loop transfer-function:

$$H(s) = \frac{G(s)K(s)}{1 + G(s)K(s)} = \frac{\frac{(p/m)K}{s + b/m}}{1 + \frac{(p/m)K}{s + b/m}} = \frac{(p/m)K}{s + b/m + (p/m)K}. \quad (4.1)$$

This result agrees with (3.35) and (3.36), which were computed earlier in Section 3.5 from the closed-loop ordinary differential equation (2.12).

The simplicity that frequency-domain methods bring to the analysis of interconnected dynamic linear time-invariant systems is remarkable: the formulas are virtually the same, with dynamic system transfer-functions replacing static gains. It is for this reason that diagrams and most expressions in this book have simply systems, e.g. G, K, etc., as opposed to transfer-functions, e.g. $G(s)$, $K(s)$, etc. Whenever no confusion is possible we drop the dependence of time-domain signals on t and frequency-domain functions on s. At times we also used mixed capitalization of letters when describing the flow of block-diagrams, with signals in lower case and systems or transfer-functions in upper case. For example, the flow in the series diagram in Fig. 4.1 is denoted simply by

$$y = G_2 z, \quad z = G_1 u.$$

The dynamic formulas should be interpreted in the light of the Laplace transform formalism discussed in Chapter 3. Take, for example, the condition $G(s)K(s) \neq -1$, which was assumed to hold in order to compute $H(s)$. What it really means is that s is taken to be in $\text{Re}(s) > \alpha_c$, where α_c is the abscissa of convergence of the function

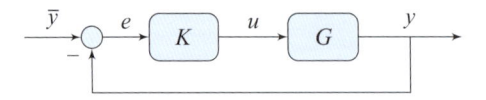

Figure 4.2 Closed-loop feedback configuration for tracking.

$(1 + G(s)K(s))^{-1}$. For all such s, the function $(1 + G(s)K(s))^{-1}$ converges and therefore is bounded, hence $G(s)K(s) \neq -1$. Note that there might exist particular values of s for which $G(s)K(s) = -1$, but those have to be outside of the region of convergence. In fact, we will spend much time later in Chapter 7 investigating the behavior of the function $G(s)K(s)$ near this somewhat special point "-1." When several interconnections are involved, the algebraic manipulations are assumed to have been performed in an appropriate region of convergence. If all of the functions involved are rational then such a region of convergence generally exists and is well defined.

4.1 Tracking, Sensitivity, and Integral Control

The notion of tracking embodies the idea of *following* a given reference. In Section 1.5, we proposed a feedback solution based on the block-diagram in Fig. 4.2. A key signal is the *tracking error*:

$$e = \bar{y} - y. \tag{4.2}$$

We expect the tracking error, e, to be small if the closed-loop system is to be able to *track* the reference signal, $\bar{y}$. Because the system, G, and the controller, K, can be dynamic, it is not hard to be convinced that one usually cannot achieve *perfect* tracking, that is, $e(t) = 0$ for all $t \geq 0$. What if $e(0) \neq 0$ when we *turn the controller on*? Alternatively, we shall consider a less ambitious objective, say, to have the tracking error, e, become small in some sense.

A basic requirement is that implied by stability. At a minimum we shall expect that the closed-loop system will be well behaved. For example, in response to a bounded reference signal, $\bar{y}$, we expect a bounded tracking error, e. With that goal in mind we compute

$$e = \bar{y} - y = \bar{y} - H\bar{y} = S\bar{y},$$

where

$$S = 1 - H = 1 - \frac{GK}{1 + GK} = \frac{1}{1 + GK}$$

is the *sensitivity*[1] transfer-function. As discussed in Section 3.6, it is necessary that S be asymptotically stable in order to obtain a bounded tracking error in response to a bounded arbitrary reference.

For example, the closed-loop transfer-function, $H(s)$, of the car model under proportional feedback control computed in (4.1) is associated with the sensitivity transfer-function:

$$S(s) = 1 - H(s) = 1 - \frac{(p/m)K}{s + b/m + (p/m)K} = \frac{s + b/m}{s + b/m + (p/m)K}. \tag{4.3}$$

---
[1] Compare S with the closed-loop sensitivity function computed earlier in Section 1.6 to understand the reason behind the name.

The functions S and H have the same denominator, the roots of which are the poles of S and H, i.e. the roots of the characteristic equation

$$s + \frac{b}{m} + \frac{p}{m}K = 0. \tag{4.4}$$

The roots of (4.4) are the same as the zeros of the equation

$$1 + G(s)K(s) = 1 + \frac{(p/m)K}{s + b/m} = \frac{s + b/m + (p/m)K}{s + b/m} = 0.$$

For asymptotic stability, the single root of the characteristic equation (4.4) must satisfy

$$-\left(\frac{b}{m} + \frac{p}{m}K\right) < 0 \qquad \Longrightarrow \qquad K > -\frac{b}{p}.$$

Because b, m, and p are all positive, S has its pole located on the left-hand side of the complex plane and the closed-loop is stable for all positive values of the proportional controller gain, K.

A measure of performance for a tracking controller is the steady-state tracking error. When the reference input $\bar{y}$ is the sinusoidal function

$$\bar{y}(t) = \bar{y}\cos(\omega t + \phi), \quad t \geq 0, \tag{4.5}$$

where $\bar{y}$, ω, and ϕ are constant, a general formula for the calculation of the steady-state tracking error can be obtained using the frequency-response method introduced in Section 3.8. Compare the above waveform with (3.44). Assuming that S is linear time-invariant and asymptotically stable, the steady-state closed-loop response to the above sinusoidal input is given by the frequency-response formula (3.45):

$$e_{ss}(t) = \bar{y}|S(j\omega)|\cos(\omega t + \phi + \angle S(j\omega)),$$

from where

$$|e_{ss}(t)| \leq |S(j\omega)||\bar{y}| \tag{4.6}$$

measures the closed-loop tracking performance. When $|S(j\omega)|$ is small, we expect small tracking errors. In the particular case of a constant reference, i.e. a step or constant function of amplitude $\bar{y}$, this formula reduces to

$$|e_{ss}(t)| \leq |S(0)||\bar{y}|.$$

For example, the car cruise control with proportional feedback responds to a reference input in the form of a step of amplitude $\bar{y}$ with steady-state error

$$|e_{ss}(t)| \leq |S(0)||\bar{y}| = \frac{\bar{y}\,b/m}{b/m + (p/m)K} = \frac{\bar{y}}{1 + (p/b)K}.$$

This is (2.14) obtained in Section 2.5. The higher K the smaller the steady-state error, as seen in Fig. 2.8.

A more ambitious objective is *asymptotic tracking*, that is,

$$\lim_{t \to \infty} e(t) = 0. \tag{4.7}$$

As discussed above, it is not enough that S be asymptotically stable for the tracking error to converge to zero. Indeed, it is necessary that the product

$$E = S\bar{Y}$$

be asymptotically stable as well. To make matters worse, interesting reference signals often have Laplace transforms with poles on the imaginary axis or the right-hand side of the complex plane, for instance, steps, ramps, cosines, etc. For this reason, in order for the tracking error to converge to zero, pole–zero cancellations in the product $S\bar{Y}$ must occur. In this case, our only hope is that zeros of S cancel out all the poles of the Laplace transform of the reference input, $\bar{Y}$, that are imaginary or lie on the right-hand side of the complex plane. When the reference is the constant or sinusoidal function (4.5) then formula (4.6) reveals the key to asymptotic tracking: $S(j\omega)$ must be zero. Because the zeros of $S = (1 + GK)^{-1}$ are the poles of GK, we can state the following result about asymptotic tracking.

LEMMA 4.1 (Asymptotic tracking) *Let $S = (1 + GK)^{-1}$ be the asymptotically stable transfer-function from the reference input, $\bar{y}$, to the tracking error, $e = \bar{y} - y$, in the closed-loop diagram in Fig. 4.2. Given constants $\bar{y}$, ϕ, and ω, if $\bar{y}(t) = \bar{y}\cos(\omega t + \phi)$, $t \geq 0$, and GK has a pole at $s = j\omega$, then $\lim_{t \to \infty} e(t) = 0$.*

This lemma provides an alternative explanation for *integral action*, discussed in Section 2.8. When the system to be controlled does not already contain an integrator, adding an integrator to the controller ensures that the product GK has a pole at the origin and that S has a zero at the origin, ensuring asymptotic tracking of constant reference inputs of *any* amplitude! The ability to achieve asymptotic tracking is a remarkable feature of feedback control. When G already contains an integrator, asymptotic tracking is achieved for any controller that does not have a zero at the origin. This was what happened with the toilet water tank discussed in Section 2.8, which we revisit in the frequency domain for illustration. The toilet water tank has open-loop transfer-function

$$y = Gu, \qquad G(s) = \frac{1}{sA}. \qquad (4.8)$$

The tank is in feedback with the static valve (proportional controller),

$$u = K(\bar{y} - y).$$

The corresponding closed-loop sensitivity transfer-function is

$$S(s) = \frac{1}{1 + G(s)K(s)} = \frac{1}{1 + K/(sA)} = \frac{s}{s + K/A}, \qquad (4.9)$$

confirming the presence of the zero at $s = 0$, the single pole of G, which appears in the numerator of the sensitivity transfer-function. Because of this zero, $S(0) = 0$ and the toilet water tank achieves asymptotic zero tracking error in response to a constant reference fill line, $\bar{y}$, regardless of the area, A, the valve gain, K, and the position of the fill line, $\bar{y}$, which can be adjusted to meet the needs of the user's installation site.

Feedback Analysis

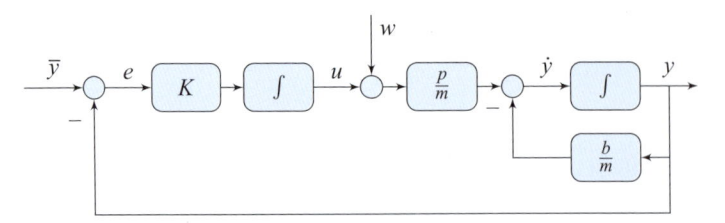

Figure 4.3 Closed-loop connection of the car with integral controller.

Returning to the car cruise control with Lemma 4.1 in mind, let us add an integrator to the controller as shown in the closed-loop diagram in Fig. 4.3. In this case

$$u = Ke, \qquad K(s) = \frac{K_\text{i}}{s}, \qquad (4.10)$$

and the corresponding closed-loop sensitivity transfer-function is

$$S(s) = \frac{1}{1 + G(s)K(s)} = \frac{1}{1 + \frac{K_\text{i}}{s} \times \frac{p/m}{s + b/m}} = \frac{s(s + b/m)}{s^2 + (b/m)s + (p/m)K_\text{i}}. \qquad (4.11)$$

As expected, S has a zero at the origin, $S(0) = 0$, and the closed-loop system achieves asymptotic tracking of constant references as long as K_i is chosen so as to keep the closed-loop system asymptotically stable. As we will see in Section 4.2, any $K_\text{i} > 0$ will stabilize the closed-loop. The closed-loop response of the car with the *integral controller* (I controller) is shown in Fig. 4.4 for various values of the gain. All solutions converge to the desired target velocity, but the plot now shows oscillations. Oscillations are present despite the fact that the gains seem to be one order of magnitude lower than the proportional gains used to plot Fig. 2.8. Oddly enough, high gains seem to

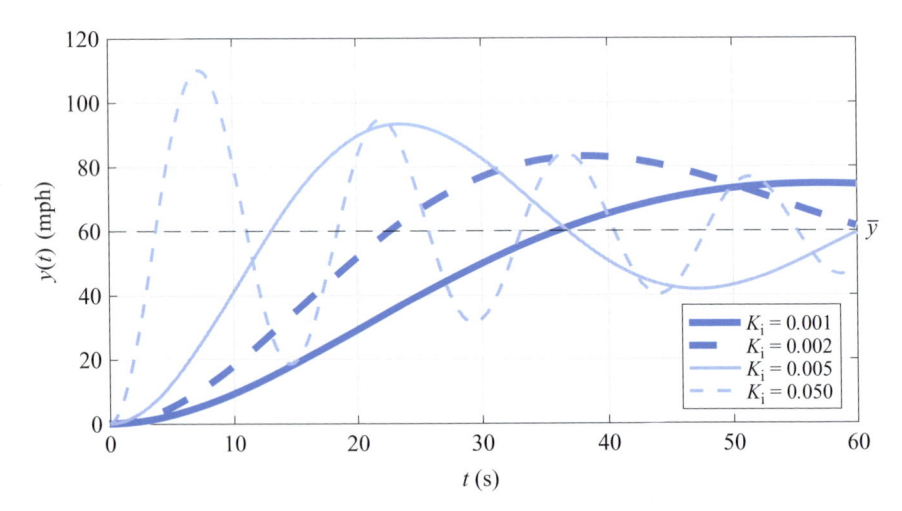

Figure 4.4 Closed-loop dynamic response, $y(t)$, calculated for $b/m = 0.05$ and $p/b = 73.3$ and a constant target output of $\bar{y} = 60$ mph with integral control (4.10) for various values of K_i.

de-stabilize the system, leading to larger oscillations. The addition of the integrator solved one problem, steady-state asymptotic tracking, but compromised the closed-loop transient response. We will *fix* this problem in the next section.

Before we get to the transient response, it is important to note that the idea that high gains drive down tracking error is still present in this section. The new information provided by the dynamic analysis of Lemma 4.1 is that high gains are not needed at all frequencies, but just at those that match the frequencies of the reference being tracked. An integrator, $K(s) = K_i s^{-1}$, effectively achieves a very high gain. It is indeed true that $\lim_{\omega \to 0} |K(j\omega)| \to \infty$, but only at the constant zero frequency $\omega \to 0$. The trade-off imposed by the dynamic solution is that the results of such high gains are achieved only asymptotically, in steady-state as time grows. We will return to this issue in a broader context in Chapter 7.

The results in this section can be extended to handle more complex reference inputs. The key property is that the zeros of S, i.e. the poles of GK, lead to asymptotic tracking when they cancel out the poles of $\bar{Y}$. For instance, if $\bar{y}(t)$ is a ramp of slope μ, that is $\bar{Y}(s) = \mu s^{-2}$, then asymptotic tracking happens if GK has at least two poles at $s = 0$, in other words, if S has two zeros at $s = 0$.

4.2 Stability and Transient Response

The sensitivity transfer-function, S, plays a key role in asymptotic tracking of reference inputs. In order to analyze the transient behavior of closed-loop systems we first establish some basic facts, most of which you already know:

(a) the zeros of $1 + GK$ are the poles of S;
(b) the poles of the product GK are the zeros of S;
(c) when G and K are rational the order of S is the order of the product GK.

Fact (c) is new information. It is a consequence of the fact that no additional pole–zero cancellations can happen when computing S from G and K.[2] This means that, when the order of the system to be controlled, G, is n_G and the order of the controller, K, is n_K, and no pole–zero cancellations occur when forming the product GK, then the order of S, that is, the order of the closed-loop system, is $n_S = n_G + n_K$. A consequence of this fact is

---

[2] It is not possible for a pole–zero cancellation to take place when computing values of S that have not already happened when computing the product GK. Let $GK = N/D$, where the polynomials N and D have no common zeros, that is, GK does not have a pole–zero cancellation. Verify that

$$S = \frac{N_S}{D_S}, \qquad N_S = D, \qquad D_S = D + N.$$

For a pole–zero cancellation to occur, the polynomials N_S and D_S must have a common factor, i.e.

$$N_S = (s - p)\tilde{N}_S, \qquad D_S = (s - p)\tilde{D}_S.$$

But for this to happen the terms of D_S also have to have a common factor,

$$D = (s - p)\tilde{D}, \qquad N = (s - p)\tilde{N},$$

which is possible only if the product GK has a pole–zero cancellation ...

that higher-order controllers will necessarily lead to higher-order closed-loop systems. As seen in our examples, higher-order often means more complex dynamic behavior. A fundamental question is therefore that of how to design K to achieve asymptotic tracking or other long-term control objectives without compromising the short-term transient response.

Let us start by revisiting the closed-loop cruise control with proportional control of Fig. 2.7. The car, G, has order 1, and the controller, K, has order 0, so the closed-loop system has order 1. The sensitivity transfer-function, S, has been computed in (4.3) and its single pole is the root of the closed-loop first-order characteristic equation, that is

$$s + \frac{b}{m} + \frac{p}{m}K = 0.$$

Asymptotic closed-loop stability is attained for all values of K such that

$$\frac{b}{m} + \frac{p}{m}K > 0.$$

When b, p, and m are positive, any nonnegative K will lead to stability. The choice of K affects the *speed* of the loop as it is directly associated with the closed-loop time-constant and rise-time. For instance, from (2.9),

$$t_r \approx \frac{2.2}{b/m + (p/m)K}.$$

For $b/m = 0.05$ and $p/b = 73.3$, rise-times faster than 10 s are possible if

$$K > 0.05.$$

Compare this with the closed-loop responses plotted in Fig. 2.8.

Now consider the closed-loop poles in the case of the cruise control with integral control of Fig. 4.3. The car model, G, has order 1, and the integral controller, K, has order 1, so the closed-loop system has order 2. The poles of the sensitivity transfer-function, S, computed in (4.11), are the zeros of the second-order characteristic equation

$$s^2 + \frac{b}{m}s + \frac{p}{m}K_i = 0. \tag{4.12}$$

It is not difficult to study the roots of this second-order polynomial equation in terms of its coefficient. For a more general analysis of second-order systems see Section 6.1. The roots of (4.12) are

$$s = -\frac{1}{2}\frac{b}{m} \pm \frac{1}{2}\sqrt{\left(\frac{b}{m}\right)^2 - 4\frac{p}{m}K_i}.$$

With b, p, and m positive, and $K_i < 0$, we have

$$\sqrt{\left(\frac{b}{m}\right)^2 - 4\frac{p}{m}K_i} \geq \frac{b}{m},$$

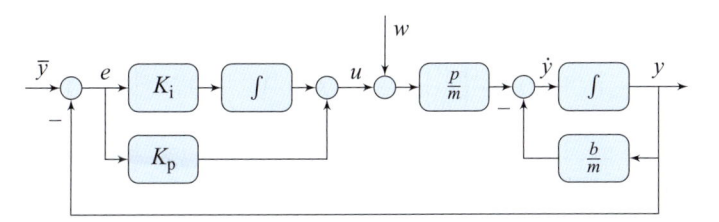

Figure 4.5 Closed-loop connection of the car model with proportional–integral controller (PI controller).

so at least one root has a positive real part, leading to an unstable closed-loop system. When $K_i = 0$ the closed-loop system has one root at the origin. Finally, when $K_i > 0$, the closed-loop is asymptotically stable. Furthermore, if the discriminant of the second-order polynomial (4.12) is positive, that is

$$\left(\frac{b}{m}\right)^2 - 4\frac{p}{m}K_i > 0,$$

then the resulting closed-loop poles are real and negative. For $p/m = 3.7$ and $p/b = 73.3$ this occurs in the interval

$$0 < K_i \leq 0.00017. \tag{4.13}$$

If $K_i > 0.00017$ then the discriminant is negative and the closed-loop poles become complex-conjugates with negative real part. This explains why all of the responses shown in Fig. 4.4, which were obtained for $K_i > 0.001$, are oscillatory. Clearly, it is not possible to improve tracking without making the closed-loop response of the car cruise control more oscillatory with integral-only control.

A possible solution is to increase the complexity of the controller. As seen earlier, proportional control provided good transient-response (see Fig. 2.8) while integral control provided good asymptotic tracking (see Fig. 4.3). A natural choice of controller is therefore one that combines the positive features of proportional and integral controllers. Consider the connection of the car with the *proportional–integral controller* (PI controller) in Fig. 4.5. The transfer-function of the controller is

$$u = Ke, \qquad K(s) = K_p + \frac{K_i}{s}, \tag{4.14}$$

and the constant gains K_p and K_i can be selected independently. The difference between the PI controller and a pure integral controller is the presence of a zero, which becomes evident if we rewrite (4.14) as

$$K(s) = \frac{K_p s + K_i}{s} = K_p \frac{s + z_i}{s}, \qquad z_i = K_i/K_p. \tag{4.15}$$

In closed-loop one obtains

$$S(s) = \frac{1}{1 + G(s)K(s)} = \frac{s(s + b/m)}{s^2 + \left(b/m + (p/m)K_p\right)s + (p/m)K_i}. \tag{4.16}$$

The closed-loop characteristic equation is

$$s^2 + \left(\frac{b}{m} + \frac{p}{m}K_p\right)s + \frac{p}{m}K_i = 0. \qquad (4.17)$$

It is now possible to coordinate the two gains K_p and K_i to *place* the closed-loop poles anywhere in the complex plane by manipulating the two coefficients of the closed-loop characteristic equation. For instance, let us revisit the requirement that the poles be real and negative. In this case a positive discriminant leads to

$$0 < 4\frac{p}{m}K_i \leq \left(\frac{b}{m} + \frac{p}{m}K_p\right)^2.$$

One way to satisfy this inequality is to choose

$$K_i = \frac{b}{m}K_p, \qquad (4.18)$$

with which the closed-loop characteristic equation (4.17) becomes

$$s^2 + \left(\frac{b}{m} + \frac{p}{m}K_p\right)s + \frac{p}{m}\frac{b}{m}K_p = \left(s + \frac{b}{m}\right)\left(s + \frac{p}{m}K_p\right) = 0. \qquad (4.19)$$

The special structure of the characteristic equation results in a *pole–zero cancellation* when forming the product

$$G(s)K(s) = \frac{p/m}{s + b/m} \times K_p \frac{s + b/m}{s} = \frac{(p/m)K_p}{s},$$

which leads to cancellations in the closed-loop transfer-functions

$$S(s) = \frac{s}{s + (p/m)K_p}, \qquad H(s) = \frac{(p/m)K_p}{s + (p/m)K_p}. \qquad (4.20)$$

The canceled pole is $s = -b/m$. The pole at $s = 0$, which is responsible for asymptotic tracking of constant reference inputs, is still present as a zero of S. We will present a more detailed discussion of possible issues with pole–zero cancellations in Section 4.7. As a side-effect of the pole–zero cancellation, the closed-loop transient analysis has been simplified down to first order.

Having set K_i as a function of $K_p > 0$ in a way that guarantees closed-loop stability and asymptotic tracking, we can now select K_p to achieve other goals. For example, we can select K_p so that the control input remains within the admissible bounds, $u(t) \in [0, 3]$, or to achieve a certain time-constant or rise-time (Section 2.3). Let us first take a look at the control input issue. Examining the loop in Fig. 4.5 we find that the transfer-function from the reference input, $\bar{y}$, to the control input, u, is

$$u = Ke, \qquad e = S\bar{y} \qquad \Longrightarrow \qquad u = Q\bar{y}, \qquad Q = KS.$$

With the particular choice of K_i in (4.18) this is

$$Q(s) = K(s)S(s) = \frac{s(K_p s + K_i)}{s(s + (p/m)K_p)} = K_p \frac{s + b/m}{s + (p/m)K_p}. \qquad (4.21)$$

4.2 Stability and Transient Response

Table 4.1 Maximum value of K_p for which the response of the closed-loop connection of the linear car velocity model (2.12), $b/m = 0.05$, $p/b = 73.3$, with the proportional–integral controller (4.14), $K_i/K_p = b/m$, to a constant target output, $\bar{y}$, is such that the throttle never saturates, that is, $u(t) \leq \bar{u} = 3, t \geq 0$. The last three columns are the corresponding integral gain, K_i, the closed-loop time-constant, τ, and the closed-loop rise-time, t_r.

$\bar{y}$ (mph)	K_p (in/mph)	$10^3 K_i$ (in/miles)	τ (s)	t_r (s)
6	0.50	2.05	0.5	1.2
30	0.10	5.00	2.7	6.0
60	0.05	2.50	5.4	12.0
90	0.03	1.50	9.1	20.0

With $K_p > 0$, you will show in P4.1 that the largest value of the control input in response to a step reference $\bar{y}(t) = \bar{y}\, 1(t)$ will be achieved at $t = 0$. We use the initial value property in Table 3.2 to compute

$$\lim_{t \to 0^+} u(t) = \lim_{s \to \infty} s U(s) = \lim_{s \to \infty} s Q(s) \frac{\bar{y}}{s} = \bar{y} \lim_{s \to \infty} Q(s) = K_p \bar{y}. \quad (4.22)$$

For $u(t) \leq \bar{u}$ we need to have

$$K_p \leq \bar{u}/\bar{y},$$

which we compute in Table 4.1 for various values of $\bar{y}$ and $\bar{u} = 3$ in. As with proportional-only control, the trend is that bigger gains lead to a faster response. For example, the closed-loop time-constant is $\tau = (K_p p/m)^{-1}$, with which one can calculate the time-constants and rise-times shown in Table 4.1. Those should be compared with the time-responses plotted in Fig. 4.6 in response to a constant reference input $\bar{y} = 60$ mph. Note the extraordinary rise-time $t_r \approx 1.2$ s obtained with $K_p = 0.5$.

For perspective, the magnitude of the frequency response of the sensitivity transfer-function, S, is plotted in Fig. 4.7. Note the zero at $s = 0$ both for the integral (I) and for the proportional–integral (PI) controllers, and the accentuated response of the integral

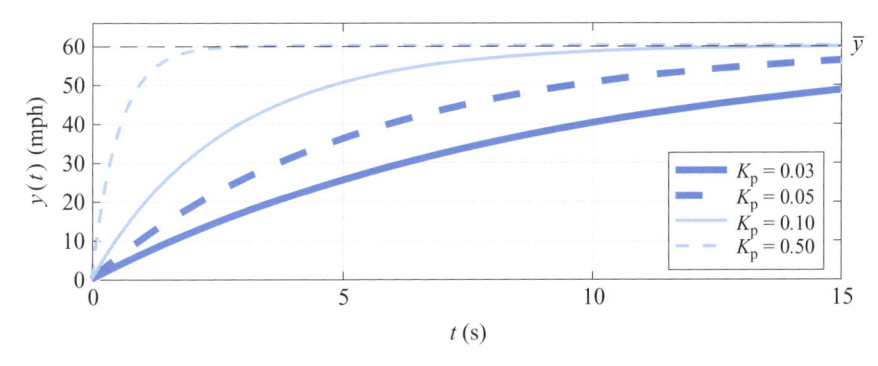

Figure 4.6 Closed-loop dynamic response, $y(t)$, for the linear car velocity model (2.12), $b/m = 0.05$, $p/b = 73.3$, to a constant target output $\bar{y} = 60$ mph with proportional–integral control (4.14), $K_i/K_p = b/m$, for various values of the proportional gain, K_p.

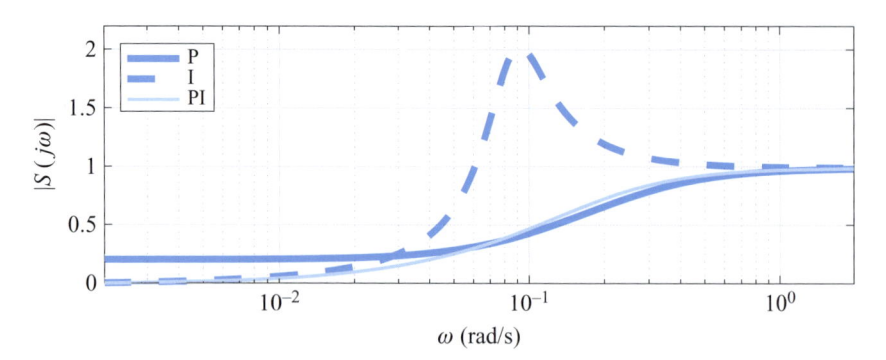

Figure 4.7 Magnitude of the frequency response of the closed-loop sensitivity transfer-function, $|S(j\omega)|$, for the linear car velocity model (2.12) calculated for $b/m = 0.05$ and $p/b = 73.3$ with the following controllers: P, proportional, Fig. 2.7, $K = 0.5$; I, integral, Fig. 4.3, $K_i = 0.002$; and PI, proportional–integral, Fig. 4.5, with $K_p = 0.05$ and $K_i/K_p = b/m$.

controller (I) near $\omega \approx 0.1$ rad/s. As we will explore in detail in Chapter 7, this peak in the frequency response explains the oscillatory behavior in Fig. 4.4. The PI design preserves the same overall frequency response of the proportional (P) design with the addition of the zero at $s = 0$ to ensure asymptotic tracking.

4.3 Integrator Wind-up

By now you should have learned not to trust responses which look too good to be true. Because the input produced by the proportional–integral controller is guaranteed to be less than $\bar{u} = 3$ in for $K_p = 0.5$ only if $\bar{y} \leq 6$ mph, the throttle input must have saturated in response to a reference input $\bar{y} = 60$ mph in Fig. 4.6, just as it did with proportional control. Simulation of the nonlinear model introduced in Section 2.6 in closed-loop with the PI controller (4.14), with K_i chosen as (4.18), reveals saturation for $K_p = 0.1$ and $K_p = 0.5$ of the system input in Fig. 4.8, as predicted by (4.22).

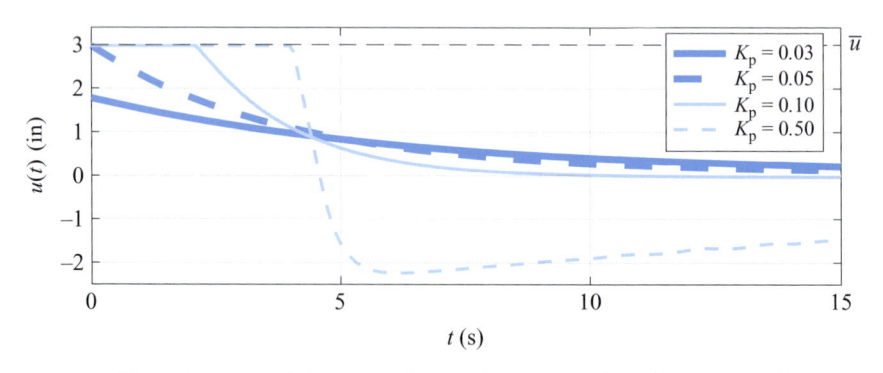

Figure 4.8 Closed-loop control input, pedal excursion, $u(t)$, produced by the car velocity nonlinear model (2.17) under proportional–integral control (2.11).

4.3 Integrator Wind-up

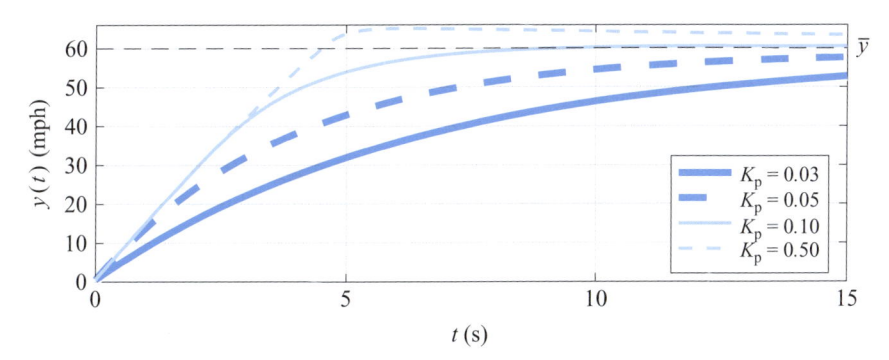

Figure 4.9 Closed-loop dynamic response, $y(t)$, for the nonlinear car velocity model (2.17) to a constant target output $\bar{y} = 60$ mph with proportional–integral control (4.14), $K_i/K_p = b/m$, for various values of proportional gain, K_p. Compare with the linear response in Fig. 4.6.

The plots in Fig. 4.9 show the corresponding output time-responses, from which we can see that the saturation of the throttle imposes a limit on how fast the car accelerates, limiting the slope of the initial response and significantly increasing the actual time-constant and rise-time. Note that the initial slope coincides for all values of K_p that saturate the throttle input, namely $K_p = 0.1$ and $K_p = 0.5$, and that, as seen in Fig. 4.9, the output responses split as soon as the input is no longer saturated. Compared with the nonlinear responses in the case of proportional control, Fig. 2.10, something new happens in the case of proportional–integral control: the response obtained with the largest gain, $K_p = 0.5$, *overshoots* the target velocity, forcing the system input to take negative values. The car must now brake after its velocity has exceeded the target. This is something that could never be predicted by the linear closed-loop model, which is of order one.

No less disturbing is the fact that it seems to take a very long time for the response to converge back to the target *after* it overshoots. This phenomenon happens frequently with controllers that have integral action and systems with saturation nonlinearities, and is known as *integrator wind-up*. The reason for the name wind-up can be understood

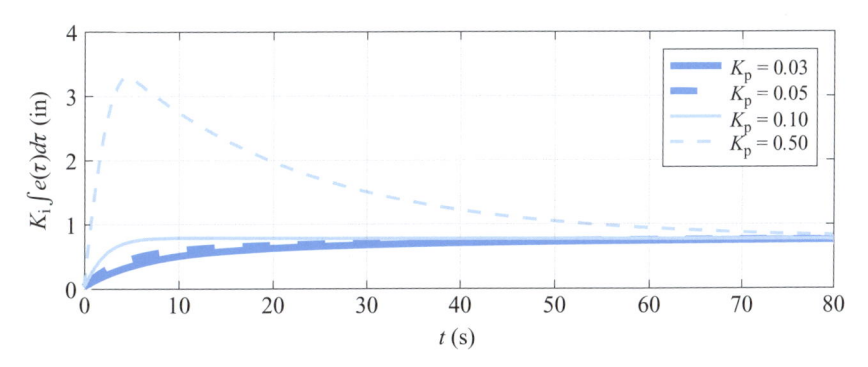

Figure 4.10 Integral component of the closed-loop dynamic response, $u_i(t)$, for the nonlinear car velocity model (2.17) to a constant target output $\bar{y} = 60$ mph with proportional–integral control (4.14), $K_i/K_p = b/m$, for various values of the proportional gain, K_p.

by looking at the plots in Fig. 4.10. No matter which form of control law is used, if a closed-loop system achieves zero tracking error with respect to a nonzero step input, an additional nonzero constant term must be present at the input of the system, u, for the tracking error, e, to be zero. We will discuss this in some detail in Section 5.8. This nonzero constant is precisely the steady-state solution of the integral component of the controller.[3] In the case of the proportional–integral controller (4.14)

$$u_i(t) = K_i \int_0^t e(\tau) d\tau, \qquad (4.23)$$

which we plot in Fig. 4.10. Beware that the time scale in this plot is different than the time scale in previous plots.

Note how the integral components, $u_i(t)$ from (4.23), converge to a common steady-state value when zero tracking error is achieved, irrespective of the values of the particular gains, K_p and K_i. Comparing these plots with the ones in Fig. 4.8 we see that the saturation of the system input, u, causes the integrated error component, u_i, to grow faster than the linear model predicted. The reason for this behavior is that the system produces less output than expected, due to the saturated input, generating larger tracking errors. Even as the error is reduced, the component of the control due to the integrator, u_i from (4.23), remains large: the integrator *wound-up*. As a result, the system input, $u = K_p e + u_i$, is not small by the time the output gets close to its target, causing the system to overshoot.

Integrator wind-up can lead to instability and oscillation. With integrators in the loop, saturation of a component, which might not be relevant when controllers without integral action are used, may reveal itself in unexpected ways. One example is *any* physical system in which tracking of position is asserted by integral action. No matter how sophisticated your controller may be, in the end, as your system gets close to its target position, it will have to stop. But it is right there, when you thought you had everything "under control," that the devil is hidden. If any amount of dry friction is present and the controller reduces the system input as it approaches the target, as in any linear[4] control law, the system will end up stopping *before* reaching the target! The controller is left operating in closed-loop with a nonzero tracking error which might be small at first, but gets integrated by the controller until it produces an output large enough to move your system and, most likely, miss the target again, this time by overshooting. Before overcoming the dry friction, the system is effectively saturated, the input has a *dead-zone*. The closed-loop system with integral action ends up oscillating around the target without ever hitting it.

It is because of these difficulties that position control is rarely implemented with "pure" integral action. In some cases, certain types of special devices can help. For instance, it is common to use *stepper motors* for position control. A stepper motor is essentially a DC motor that is built and operated in such a way as to allow rotation only in set increments. It allows precise movement to one of the set positions but makes

---

[3] Or the system if integral action is provided by the presence of a pole at the origin in the system.
[4] In fact any control law not necessarily linear that is continuous at the origin!

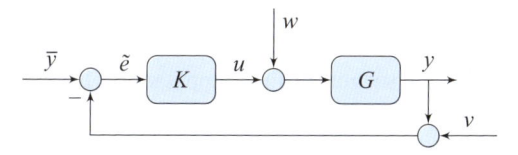

Figure 4.11 Closed-loop feedback configuration with reference, $\bar{y}$, input disturbance, w, and measurement noise, v.

it impossible for the system to be located anywhere between the increments, therefore limiting the overall accuracy. An alternative is to use *motor drivers* which are not linear. It is often the case that the voltage applied to regular DC motors, even if produced by a linear controller, is first converted into a discontinuous signal. This will increase the available torque and help overcome dry friction. The conversion is done by using the continuous signal to *modulate* a discontinuous signal, which is done in some periodic fashion. A popular choice is *pulse-width modulation* (PWM), in which the instantaneous amplitude of a continous signal is sampled and periodically converted into the width of a pulse of fixed amplitude before being applied to the motor.

4.4 Feedback with Disturbances

In this section we study the impact of disturbances on the closed-loop performance. Look at the feedback diagram in Fig. 4.11. Compared with Fig. 4.2, this diagram has two additional *disturbance* signals: the input disturbance, w, and the output disturbance, v. The input disturbance, w, was introduced earlier in Section 2.7 to model the effects of a slope in the car cruise control problem. The output disturbance, v, is often used to model imperfect measurement, that is, measurement noise.

Start by writing all relations in the closed-loop system of Fig. 4.11:

$$y = G(u + w), \qquad u = K\tilde{e}, \qquad \tilde{e} = \bar{y} - (y + v).$$

Treating $\bar{y}$ and the disturbances w and v as inputs, we write

$$\begin{aligned} y &= H\bar{y} - Hv + Dw, \\ u &= Q\bar{y} - Qv - Hw, \end{aligned} \qquad (4.24)$$

which relate the inputs $\bar{y}$, w, and v to the outputs y and u. We also compute the tracking error,

$$e = \bar{y} - y = S\bar{y} + Hv - Dw. \qquad (4.25)$$

Note that the tracking error signal, e, is different from the controller error signal, $\tilde{e}$, which is corrupted by the measurement noise v. Remarkably, these nine transfer-functions can be written in terms of only four transfer-functions:

$$\begin{aligned} S &= \frac{1}{1 + GK}, & D &= \frac{G}{1 + GK}, \\ H &= \frac{GK}{1 + GK}, & Q &= \frac{K}{1 + GK}. \end{aligned} \qquad (4.26)$$

We have already encountered all of these four transfer-functions before. By now you should be more than familiar with H and S and their role in tracking. The transfer-function Q appeared in the previous section in the computation of the amplitude of the control input signal, u, in response to the reference input, $\bar{y}$. Indeed, $u = Q\bar{y}$ when $w = v = 0$. The transfer-function D will be investigated in detail in Section 4.5, but it has already made a discrete appearance in Section 2.7. Before proceeding, note that

$$H = GKS, \qquad D = GS, \qquad Q = KS.$$

Because S has as zeros the poles of the product GK, if no pole–zero cancellations occur when forming the product GK, the transfer-functions H, D, and Q cannot have any poles that have not already appeared in S. The conclusion is that closed-loop stability analysis can be performed entirely on the basis of S as was done in Section 4.2, even in the presence of the disturbances w and v. Differences will result mostly from the zeros of H, D, and Q, as we will see in the coming sections.

4.5 Input-Disturbance Rejection

Input disturbances are often used to model variations in the system's operating conditions, such as the road slope in the cruise control problem (Section 2.7). After setting $v = 0$ we obtain

$$e = S\bar{y} - Dw, \qquad D = GS = \frac{G}{1 + GK},$$

from (4.25) and (4.26). As seen earlier, we need to have S small near the poles of the reference input $\bar{Y}$ for good tracking. Similarly we will need to have D small in the range of frequencies where the input disturbances are most prominent. A natural question is that of whether we can make S and D simultaneously small at the frequencies of interest. Say $\bar{y}$ and w are constants. Can we make $|S(0)|$ and $|D(0)|$ both small? Because the zeros of S are the poles of the product GK, the zeros of D are only the poles of K. The poles of G cancel out with the product $D = GS$. Therefore, for both S and D to be small at $s = 0$, the DC gain[5] of the controller, $|K(0)|$, has to be large.

More generally, for asymptotic tracking *and* asymptotic input-disturbance rejection of sinusoidal inputs with a given frequency, ω, it is necessary that the transfer-function of the controller, K, have a pole at $s = j\omega$. That is because poles of the disturbance, w, that need to be canceled by the feedback loop will have to be canceled only by the controller, K. This leads to the following modified version of Lemma 4.1.

LEMMA 4.2 (Asymptotic tracking and input-disturbance rejection) *Let* $S = (1 + GK)^{-1}$ *be the asymptotically stable transfer-function from the reference input, $\bar{y}$, to the tracking error, $e = \bar{y} - y$, and let $D = GS$ be the transfer-function from the input disturbance, w, to the tracking error, e, in the closed-loop diagram in Fig. 4.11. Given*

---

[5] See Section 7.1.

4.5 Input-Disturbance Rejection

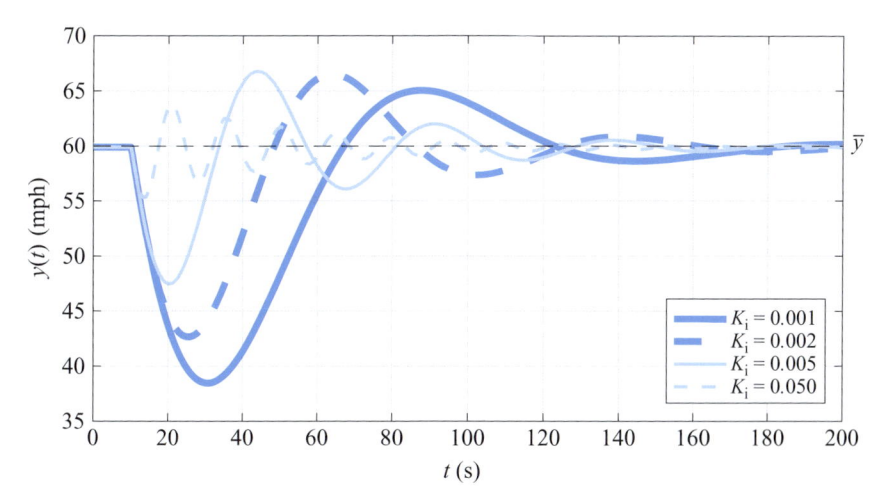

Figure 4.12 Closed-loop response of the velocity of the car cruise control with integral control (Fig. 4.3) to a change in road slope at $t = 10$ s, from flat to 10 % grade, for $b/m = 0.05$ and $p/b = 73.3$ and various values of the control gain, K_i.

constants $\bar{y}$, ϕ, $\bar{w}$, ψ, and ω, if $v(t) = 0$, $\bar{y}(t) = \bar{y}\cos(\omega t + \phi)$, $w(t) = \bar{w}\cos(\omega t + \psi)$, $t \geq 0$, and K has a pole at $s = j\omega$, then $\lim_{t\to\infty} e(t) = 0$.

A special case is when $\bar{y}(t) = 0$ and one is interested only in disturbance rejection, a control problem known as *regulation*. Since $e(t) = y(t)$, asymptotic tracking here means simply that the output is asymptotically *regulated*, that is $\lim_{t\to\infty} y(t) = 0$.

Consider the example of the car cruise control with proportional control (P controller) in Fig. 2.7. With G from (3.16) and $K(s) = K$, we compute

$$S(s) = \frac{s + b/m}{s + b/m + (p/m)K}, \quad D(s) = G(s)S(s) = \frac{p/m}{s + b/m + (p/m)K}.$$

Evaluation at $s = 0$ coincides with (2.14) and (2.21). Because there is no value of ω for which either $S(j\omega)$ or $D(j\omega)$ is zero, in this example, proportional control does not achieve asymptotic tracking and asymptotic input-disturbance rejection. As seen in Fig. 2.14, a change in road slope will imply a change in the regulated speed.

We compute S and D for the car cruise control with integral control (I controller), $K(s) = K_i s^{-1}$, Fig. 4.3, to obtain

$$S(s) = \frac{s(s + b/m)}{s^2 + (b/m)s + (p/m)K_i}, \quad D(s) = G(s)S(s) = \frac{(p/m)s}{s^2 + (b/m)s + (p/m)K_i}. \quad (4.27)$$

This time, both S and D have a zero at $s = 0$, hence the closed-loop achieves asymptotic tracking and asymptotic disturbance rejection of step input references and disturbances. This is illustrated in Fig. 4.12, where a step change in the slope from flat to a 10% grade slope ($\theta \approx 5.7°$) happens at $t = 10$ s. The car eventually recovers the originally set speed of 60 mph without error, that is, asymptotic input-disturbance rejection is achieved. The

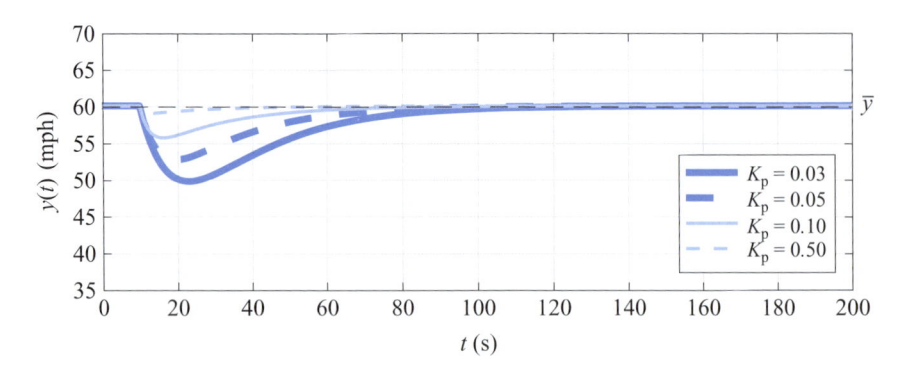

Figure 4.13 Closed-loop response of the velocity of the car cruise control with proportional–integral control (Fig. 4.5) with $K_i/K_p = b/m$ to a change in road slope at $t = 10$ s, from flat to 10 % grade, for $b/m = 0.05$ and $p/b = 73.3$ and various values of the proportional control gain, K_p.

transient performance is, however, adversely affected by the integral controller. This should be no surprise since the transient behavior in response to an input disturbance is also dictated by the poles of S.

At this point you should know what is coming: the cruise control with proportional–integral control (PI controller), $K(s) = K_p + K_i s^{-1}$, Fig. 4.5, will display asymptotic tracking and input-disturbance rejection with better transient performance. Indeed, we compute S and D with the choice of K_i in (4.18). Recall that this leads to a pole–zero cancellation in GK so that

$$S(s) = \frac{s}{s + (p/m)K_p}, \quad D(s) = G(s)S(s) = \frac{(p/m)s}{(s + b/m)(s + (p/m)K_p)}. \quad (4.28)$$

As with pure integral control, both S and D have a zero at $s = 0$, therefore the closed-loop system achieves asymptotic tracking and asymptotic disturbance rejection of step inputs. Compare the responses of the integral controller, Fig. 4.12, and of the proportional–integral controller, Fig. 4.13, to a step change in the slope from flat to a 10% grade slope ($\theta \approx 5.7°$) at $t = 10$ s. The response of the PI controller is not oscillatory (the poles of D are real) and is much faster than that of the integral controller. A surprising observation is that pole–zero cancellations take place in S and H but not in D. The canceled pole, $s = -b/m$, still appears in D, which is second-order! We will comment on this and other features of pole–zero cancellations later in Section 4.7.

Differences among the various control schemes can be visualized in the plot of the magnitude of the frequency response of the transfer-function D shown in Fig. 4.14. The high peak in the frequency response of the sensitivity function, S, of the I controller seen in Fig. 4.7 appears even more accentuated in the frequency response of D in Fig. 4.14. For both controllers with integral action, I and PI, the frequencies around $\omega = 0.1$ rad/s are the ones which are more sensitive to input disturbances. This was already evident for the I controller in Fig. 4.7. However, it became clear for the PI controller only in Fig. 4.14. In this example, the relative weakness at $\omega = 0.1$ rad/s should be evaluated

4.5 Input-Disturbance Rejection

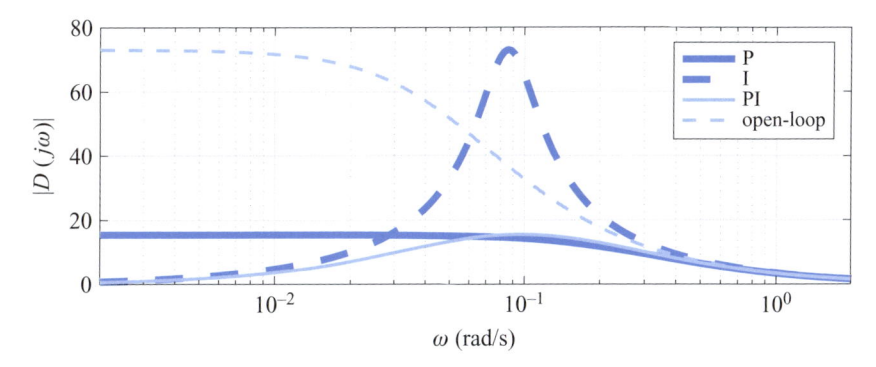

Figure 4.14 Magnitude of the frequency response of the closed-loop transfer-function, $|D(j\omega)|$, for the linear car velocity model (2.12) calculated for $b/m = 0.05$ and $p/b = 73.3$ with the following controllers: (P) proportional, Fig. 2.7, $K = 0.5$; (I) integral, Fig. 4.3, $K_i = 0.002$; and (PI) proportional–integral, Fig. 4.5, with $K_p = 0.05$ and $K_i/K_p = b/m$.

with respect to possible road conditions that can excite this frequency. An analysis is possible with our current tools.

Consider a road on rolling hills with sinusoidal elevation profile:

$$h(x) = \frac{\bar{h}}{2}[1 + \cos(2\pi\lambda^{-1}x)],$$

where x denotes the horizontal distance traveled and h the elevation of the road. The constant λ is the characteristic length of the hill oscillation. See Fig. 4.15. Starting at an elevation of $h = \bar{h}$, the car will reach the lowest elevation $h = 0$ at $x = \lambda/2$ and will reach $h = \bar{h}$ after traversing λ units of horizontal length. A car traversing this road with a constant horizontal speed $\dot{x} = v$ will experience a slope profile

$$h(t) = h(vt) = \frac{\bar{h}}{2}[1 + \cos(\omega t)], \qquad \omega = 2\pi\lambda^{-1}v.$$

For example, with $v = 60$ mph and $\lambda = 1$ mile,

$$\omega = 2\pi\lambda^{-1}v = \frac{2\pi \times 60}{1 \times 3600} \approx 0.1 \text{ rad/s},$$

and the designed cruise control system will experience the worst possible input disturbance. Such a road profile is not so far-fetched!

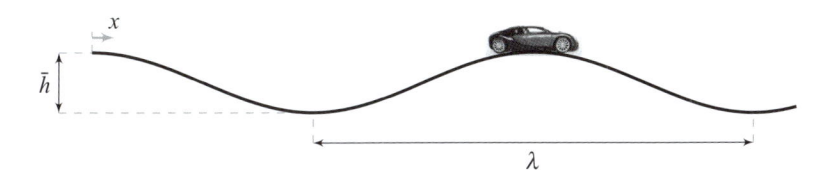

Figure 4.15 Car traveling on a rolling hill.

If the difference between the peak and the valley on the road is that of a 1% grade road, that is,

$$\bar{h} = 0.01 \frac{\lambda}{4} = 0.0025 \text{ miles} \approx 13 \text{ feet},$$

then the slope of the road profile will be

$$\theta(t) = \tan^{-1}\left(\frac{\omega \bar{h}}{2v} \sin(\omega t)\right) = \tan^{-1}\left(\frac{\bar{h}\pi}{\lambda} \sin(\omega t)\right).$$

For $b/m = 0.05$ and $p/b = 73.3$ and (2.18), the input disturbance seen by the car would be approximately

$$w(t) = -\frac{mg}{p} \sin\left(\tan^{-1}\left(\frac{\bar{h}\pi}{\lambda} \sin(\omega t)\right)\right) \approx \frac{mg}{p} \frac{\bar{h}\pi}{\lambda} \sin(\omega t) \approx 0.04 \sin(\omega t).$$

With $\omega = 0.1$ rad/s, pure integral (I) control produces a steady-state velocity error of

$$|\Delta y(t)| \leq 0.04|D(j0.1)| \approx 3.0557 \text{ mph},$$

where D is from (4.27). The proportional–integral (PI) controller produces the much smaller error

$$|\Delta y(t)| \leq 0.04|D(j0.1)| \approx 0.7324 \text{ mph},$$

where D is from (4.28). The I controller responds with oscillations of amplitude 6 mph, which might be noticeable, whereas the PI controller produces oscillations of amplitude four times smaller.

We could not conclude without going back to the toilet. For the toilet water tank example

$$G(s) = \frac{1}{sA}, \qquad S(s) = \frac{s}{s + K/A}, \qquad D(s) = \frac{1/A}{s + K/A},$$

after (4.8) and (4.9). As expected, the sensitivity function, S, contains a zero at $s = 0$ but D does not, since the integrator is located on the system, G, and not on the controller, K. The water tank system achieves asymptotic tracking but not asymptotic input-disturbance rejection of step inputs. For example, if a leak is present that offsets the input flow, the valve may end up permanently open as the water never reaches the fill line. In an actual toilet with a real valve, this generates oscillations such as the valve opening periodically even if no one flushed the toilet, which is the number one sign you should check the rubber gasket and flapper valve of your toilet water tank.

4.6 Measurement Noise

Let us now analyze the effect of the measurement noise disturbance, v, on the measured output, y. We use Fig. 4.11 and Equation (4.25) to calculate

$$e = S\bar{y} + Hv,$$

after setting $w = 0$. At this point, a fundamental limitation of feedback systems is apparent: for better tracking we need S to be small. However, from (4.26), we always have

$$S + H = 1. \tag{4.29}$$

Because of the above relation, the transfer-function H is also known as the *complementary sensitivity* transfer-function. Consequently, having S small means that H should be close to one or, in other words, that measurement noise, v, will likely increase the tracking error, e. If measurement errors are significant, the only way to mitigate this problem is to reduce the magnitude of H, which will reduce the loop's ability to track. In a static scenario, there is no way around the inevitable trade-off imposed by (4.29). Dynamically, however, there may be a way out.

The key is that (4.29) has to hold for every frequency, that is,

$$S(j\omega) + H(j\omega) = 1,$$

so it may be possible to make $|S(j\omega)|$ small for ω close to the frequencies present in the reference input while letting $|S(j\omega)|$ be large elsewhere. Typically, measurement noise is rich in high frequencies while reference inputs are often of low frequency, e.g. a step or a low-frequency sinusoidal function.

For instance, the cruise control closed-loop sensitivity functions shown in Fig. 4.7 are all such that $|S(j\omega)|$ is small for frequencies close to zero while $|S(j\omega)|$ is close to one for $\omega \gg 1$ rad/s. This means that they will perform adequately if the measurement noise is confined to frequencies higher than, say, 1 rad/s. A complementary picture is Fig. 4.16, which shows the magnitude of the frequency response of the transfer-function $H(s) = 1 - S(s)$. The peak for the integral controller around $\omega \approx 0.1$ rad/s is likely to cause trouble in closed-loop if measurement noise is significant at those frequencies.

Note that (4.29) does not imply that $|S(j\omega)| + |H(j\omega)| = 1$. Therefore S and H can be simultaneously large at a given frequency as long as they have opposite phases. Indeed, the large peaks in $|S(j\omega)|$ and $|H(j\omega)|$ in the case of the integral controller both happen very close to $\omega \approx 0.1$ rad/s. What is true, however, is that when $|S(j\omega)|$ is small then $|H(j\omega)|$ should be close to one, and vice versa. We will have much more to say about this in Chapter 7.

4.7 Pole–Zero Cancellations and Stability

The proportional–integral controller design proposed in Section 4.2 performs a *pole–zero cancellation*: a pole (or a zero) of the controller, K, cancels a zero (or a pole) of

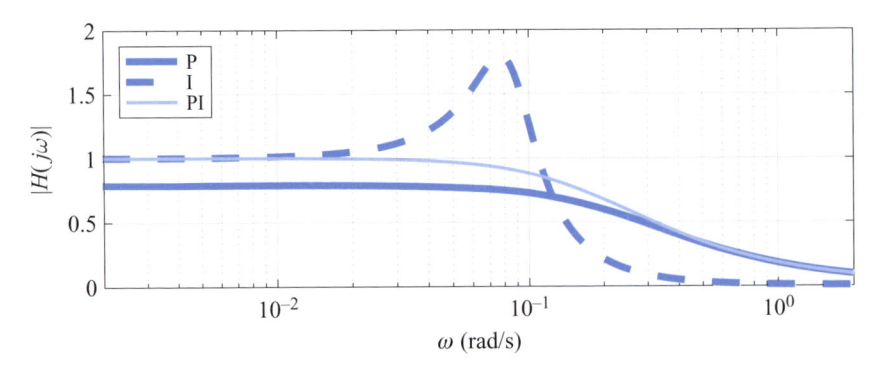

Figure 4.16 Magnitude of the frequency response of the closed-loop complementary sensitivity transfer-function, $|H(j\omega)|$, for the linear car velocity model (2.12) calculated for $b/m = 0.05$ and $p/b = 73.3$ with the following controllers: (P) proportional, Fig. 2.7, $K = 0.5$; (I) integral, Fig. 4.3, $K_i = 0.002$; and (PI) proportional–integral, Fig. 4.5, with $K_p = 0.05$ and $K_i/K_p = b/m$.

the system, G. Before we discuss the potential pitfalls of pole–zero cancellations, let us first revisit the open-loop solution proposed in Section 1.4.

The basic open-loop diagram of Fig. 1.7 is reproduced in Fig. 4.17. In a static context, the open-loop solution to the problem of tracking the reference input, $\bar{y}$, consisted simply of *inverting* the system, G. In a dynamic context, inverting G means performing a pole–zero cancellation. Indeed, if G is rational,

$$G = \frac{N_G}{D_G},$$

and a rational controller K is chosen so that

$$K = G^{-1} = \frac{D_G}{N_G},$$

where N_G and D_G are polynomials, then

$$y = GK\bar{y} = \frac{N_G}{D_G} \times \frac{D_G}{N_G} \bar{y} = \bar{y}.$$

The validity of this solution in a dynamic scenario is at best suspicious. There are at least two obstacles. (a) Is the connection of K and G stable? (b) If it is stable, can one construct a physical realization of the controller? We will consider the issue of stability in the rest of this section. The next chapter will address the problem of constructing a physical realization of the controller in detail. For now, be aware that it might not be possible to implement open-loop controllers that attempt to invert the system. When the system, G, is rational and strictly proper, its inverse is not proper. That is, the degree of

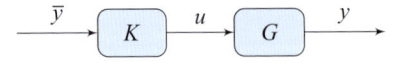

Figure 4.17 Open-loop configuration.

4.7 Pole–Zero Cancellations and Stability

the numerator is greater than the degree of the denominator. For example,

$$G(s) = \frac{1}{s+1}, \qquad K(s) = G(s)^{-1} = s+1.$$

As we will see in Chapter 5, it is not possible to physically realize transfer-functions which are not proper. See also Section 8.6.

The transfer-function from $\bar{y}$ to y, which is equal to one, is clearly asymptotically stable. However, if G has zeros with positive real part, K will have poles with positive real part and therefore will not be asymptotically stable. For example,

$$G(s) = \frac{s-1}{s+1}, \qquad K(s) = G(s)^{-1} = \frac{s+1}{s-1}, \qquad (4.30)$$

and K has a pole with positive real part, hence it is not asymptotically stable.

At this point one must ask why is stability relevant if the unstable pole or zero is being canceled? It is not uncommon to find explanations based on the following argument: it is impossible to perfectly cancel a pole with a zero, and any imperfection in the cancellation will reveal itself in the form of instabilities. Take (4.30), and suppose the actual zero was not at 1 but at $1 - \epsilon$. In this case

$$Y(s) = G(s)K(s)\bar{Y}(s) = \frac{s-1+\epsilon}{s+1} \times \frac{s+1}{s-1} \times \bar{Y}(s) = \frac{s-1+\epsilon}{s-1}\bar{Y}(s).$$

Therefore, from Section 3.5, the signal $y(t) = \mathscr{L}^{-1}\{Y(s)\}$ has a term[6]

$$y(t) = k_1 e^t + \cdots, \quad t \geq 0 \qquad k_1 = \lim_{s \to 1}(s-1+\epsilon)Y(s) = \epsilon \bar{Y}(1).$$

If ϵ is small but not zero, this term would eventually grow out of bounds, hence revealing the instability in the connection.

A less publicized but much better reasoning is that canceling a pole or a zero with positive real part is a really bad idea *even if perfect cancellation were possible*! If we modify Fig. 4.17 to include an input disturbance as in Fig. 1.12, then, with perfect cancellation,

$$y = GK\bar{y} + Gw = \bar{y} + Gw, \qquad u = K\bar{y},$$

so that if either G or K were not asymptotically stable then the corresponding signal y or u would grow unbounded. A more detailed argument will be provided next as a particular case of the closed-loop analysis.

To analyze the effects of disturbances in closed-loop we first introduce the slightly more complicated closed-loop diagram of Fig. 4.18. This diagram has an additional feedback filter element, F. Figure 4.11 is a particular case of Fig. 4.18 in which the feedback filter F is unitary, i.e. $F = 1$. Interestingly, the open-loop diagram in Fig. 4.17 is also a particular case of Fig. 4.18 in which the feedback filter F is zero, i.e. $F = 0$. We analyze Fig. 4.18 by writing the relations

$$y = G(u+w), \qquad u = K(\bar{y}-z), \qquad z = F(y+v).$$

---

[6] Unless $Y(s)$ has a zero exactly at "1," in which case a similar argument about the *exactness* of the location of the zero would apply.

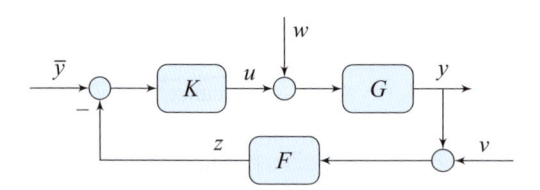

Figure 4.18 Closed-loop feedback configuration with reference, $\bar{y}$, input disturbance, w, and measurement noise, v, and feedback filter F.

Identifying the reference, $\bar{y}$, and the disturbances, w and v, as inputs, and y, u, and z as outputs, we write

$$y = SGK\bar{y} - SGKFv + SGw,$$
$$u = SK\bar{y} - SKFv - SGKFw,$$
$$z = SGKF\bar{y} + SFv - SGFw,$$

where S is the sensitivity transfer-function:

$$S = \frac{1}{1 + FGK}.$$

Note that in open-loop, that is when $F = 0$, the sensitivity reduces to $S = 1$. In closed-loop, that is when $F \neq 0$, we also calculate the tracking error,

$$e = \bar{y} - y = S\bar{y} + SGKFv - SGw.$$

We say that the closed-loop connection in Fig. 4.18 is *internally stable* if the eight transfer-functions appearing above, namely

$$S, \quad SG, \quad SF, \quad SK, \quad SGK, \quad SGF, \quad SKF, \quad SGKF,$$

are asymptotically stable. The rationale is that all output signals in the loop will be bounded if their inputs are bounded (see Section 3.6).

In open-loop,[7] $F = 0$ and $S = 1$. Therefore, internal stability means asymptotic stability of the transfer-functions

$$G, \quad K, \quad GK.$$

Note the important fact that asymptotic stability of the product GK is not enough! Indeed, as shown earlier, in the presence of a nonzero input disturbance, w, we shall have

$$y = GK\bar{y} + Gw, \qquad u = K\bar{y},$$

which shows that the signals y and u will have unbounded components if any one of G, K, and GK is not asymptotically stable. It was necessary to introduce the disturbance, w, but also to look *inside the loop*, at the internal signal u for traces of instability, hence the name *internal stability*. Requiring that G, K, and GK be asymptotically stable has

---

[7] Compare this with Section 1.6!

important implications in the case of open-loop solutions: (a) it means that open-loop solutions cannot be used to control unstable systems (G must be asymptotically stable); (b) it rules out choices of K that perform a pole–zero cancellation with positive real part (K must be asymptotically stable); (c) it reveals that open-loop solutions do not possess any capability to reject input disturbances other than those already rejected by the original system G, since $D = GS = G$.

Now consider internal stability in the case of the unit-feedback loop in Fig. 4.11, that is, Fig. 4.18 with $F = 1$. In this case, internal stability is equivalent to asymptotic stability of the four transfer-functions

$$S, \quad SG, \quad SK, \quad SGK,$$

which are nothing but the four transfer-functions S, D, Q, and H presented earlier in (4.26). Because $H = 1 - S$, this is really stability of the three transfer-functions

$$S, \quad SG, \quad SK.$$

As discussed in Section 4.2, no pole–zero cancellation occurs in S that has not already occurred when forming the product GK. Furthermore, the zeros of S are the poles of GK. If no pole–zero cancellations occur in GK, then all we need to check is stability of the poles of S. On the other hand, if there are pole–zero cancellations in the product GK, then these cancellations have to be of stable poles. This is better illustrated with an example. Suppose that

$$G = \frac{(s-z)}{(s-p)}\tilde{G} = \frac{(s-z)}{(s-p)}\frac{N_{\tilde{G}}}{D_{\tilde{G}}}, \qquad K = \frac{(s-p)}{(s-z)}\tilde{K} = \frac{(s-p)}{(s-z)}\frac{N_{\tilde{K}}}{D_{\tilde{K}}}$$

perform a pole–zero cancellation of the zero z and the pole p. Assume that there are no further cancellations when forming the product

$$GK = \tilde{G}\tilde{K} = \frac{N_{\tilde{G}} N_{\tilde{K}}}{D_{\tilde{G}} D_{\tilde{K}}};$$

that is, $N_{\tilde{G}} N_{\tilde{K}}$ and $D_{\tilde{G}} D_{\tilde{K}}$ are polynomials without a common factor. If the roots of the polynomial

$$N_{\tilde{G}} N_{\tilde{K}} + D_{\tilde{G}} D_{\tilde{K}} = 0$$

all have negative real part, then

$$S = \frac{1}{1+GK} = \frac{D_{\tilde{G}} D_{\tilde{K}}}{N_{\tilde{G}} N_{\tilde{K}} + D_{\tilde{G}} D_{\tilde{K}}}$$

and $H = 1 - S$ are asymptotically stable. However,

$$SG = \frac{(s-z)}{(s-p)} \frac{N_{\tilde{G}} D_{\tilde{K}}}{N_{\tilde{G}} N_{\tilde{K}} + D_{\tilde{G}} D_{\tilde{K}}} \quad \text{and} \quad SK = \frac{(s-p)}{(s-z)} \frac{N_{\tilde{K}} D_{\tilde{G}}}{N_{\tilde{G}} N_{\tilde{K}} + D_{\tilde{G}} D_{\tilde{K}}}$$

will be stable only if p and z are both negative. This argument can be extended without difficulty to complex-conjugate poles and zeros. The discussion is summarized in the next lemma.

Feedback Analysis

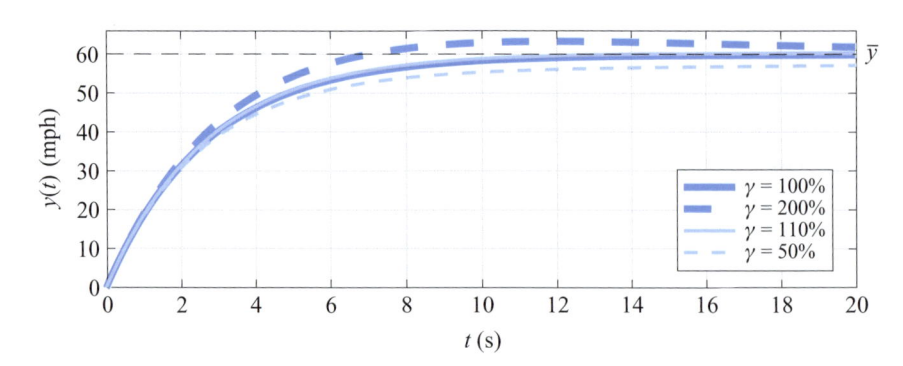

Figure 4.19 Closed-loop dynamic response, $y(t)$, for the linear car velocity model (2.12) calculated for $b/m = 0.05$ and $p/b = 73.3$ and a constant target output of $\bar{y} = 60$ mph with proportional–integral control where $K_i/K_p = \gamma \, b/m$ and $K_p = 0.005$. When $\gamma = 1$ (100 %), the controller performs an exact pole–zero cancellation.

LEMMA 4.3 (Internal stability) *Consider the closed-loop diagram in Fig. 4.11. The closed-loop system is internally stable if and only if S, the transfer-function from the reference input, $\bar{y}$, to the tracking error, $e = \bar{y} - y$, is asymptotically stable and any pole–zero cancellations performed when forming the product GK are of poles and zeros with negative real part.*

Recall that we designed the proportional–integral controller (PI controller) for the cruise controller with a choice of K_i in (4.18) that performed a pole–zero cancellation. In that case we had

$$G(s) = \frac{p/m}{s + b/m}, \qquad K(s) = \frac{K_p \left(s + K_i/K_p\right)}{s}, \qquad \frac{K_i}{K_p} = \frac{b}{m}.$$

You will verify in P4.2 that

$$S(s) = \frac{s}{s + (p/m)K_p}, \qquad H(s) = \frac{(p/m)K_p}{s + (p/m)K_p},$$

$$Q(s) = \frac{K_p(s + b/m)}{s + (p/m)K_p}, \qquad D(s) = \frac{(p/m)s}{(s + b/m)(s + (p/m)K_p)},$$

(4.31)

which are all stable when b, p, m, and K_p are positive. If one chooses to perform a pole–zero cancellation in the product GK, this does not mean that the canceled pole or zero will simply *disappear* from the loop. As illustrated above, the canceled pole or zero will still appear as a pole or a zero in one of the loop transfer-functions and therefore should have negative real part for internal stability.

Interestingly, even when the cancellation of a pole or zero with negative real part is not perfect, the resulting transfer-function should not be significantly affected by a small mismatch in the value of the canceled zero or pole. How large the mismatch can be is, of course, dependent on the problem data. Too much mismatch might even lead to closed-loop instability. For the car cruise control with proportional–integral control (PI controller) we evaluate this possibility in Fig. 4.19 after recalculating the step responses for the choice of integral gain:

$$K_i = \gamma \frac{b}{m} K_p.$$

The resulting controller performs a perfect pole–zero cancellation only when $\gamma = 1$. Note that the response is not much affected in the case of a small 10% variation, $\gamma = 1.1$, but becomes somewhat slower when the mismatch is by about 50%, $\gamma = 0.5$, or gets a bit faster with a slight overshoot when the mismatch is by 200%, $\gamma = 2$. We will understand the reason for this behavior in Sections 6.1 and 6.4.

Problems

4.1 Compute the response of a system with transfer-function $u = Q\bar{y}$, where Q is given by (4.21), to a constant reference input $\bar{y}(t) = \bar{y}$, $t \geq 0$. Use your answer to show that the largest value of $u(t)$ is at $u(0)$.

4.2 Calculate the transfer-functions S, H, Q, and D in (4.31) for $G(s)$ and $K(s)$ corresponding to the PI control of the car cruise controller with the choice of K_i from (4.18).

4.3 Calculate

$$S = \frac{1}{1+GK}, \qquad D = \frac{G}{1+GK},$$

$$H = \frac{GK}{1+GK}, \qquad Q = \frac{K}{1+GK},$$

for the system and controller transfer-functions

$$G(s) = \frac{a}{s+a}, \qquad K(s) = K_p.$$

What is the order of each transfer-function?

4.4 Repeat P4.3 for the following system and controller transfer-functions:

(a) $G = \dfrac{a}{s+a}$, $\qquad K = \dfrac{K_i}{s}$;

(b) $G = \dfrac{a}{s+a}$, $\qquad K = K_p + \dfrac{K_i}{s}$;

(c) $G = \dfrac{a}{s+a}$, $\qquad K = K_l \dfrac{s+c}{s+b}$;

(d) $G = \dfrac{a_2}{s^2 + a_1 s + a_2}$, $\qquad K = K_p$;

(e) $G = \dfrac{a_2}{s^2 + a_1 s + a_2}$, $\qquad K = K_p + K_d s$;

(f) $G = \dfrac{a_2}{s^2 + a_1 s + a_2}$, $\qquad K = K_p + \dfrac{K_i}{s}$,

in which K_i, K_p, K_d, and K_l are scalar gains.

Feedback Analysis

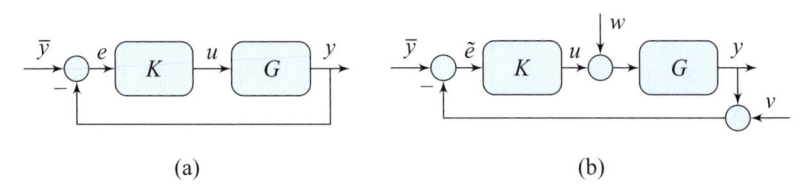

(a) (b)

Figure 4.20 Diagrams for P4.5 and P4.7.

4.5 Consider the standard feedback connection in Fig. 4.20(a) in which the system and controller transfer-functions are

$$G(s) = \frac{1}{s}, \qquad K(s) = 1.$$

Is the closed-loop system internally stable? Does the closed-loop system achieve asymptotic tracking of a constant input $y(t) = \bar{y}, t \geq 0$?

4.6 Repeat P4.5 for the following combinations of system and controller transfer-functions:

(a) $G = \dfrac{1}{s}$, $\qquad K = \dfrac{1}{s+1}$;

(b) $G = \dfrac{1}{s}$, $\qquad K = \dfrac{1}{s}$;

(c) $G = \dfrac{1}{s+1}$, $\qquad K = 1$;

(d) $G = \dfrac{1}{s+1}$, $\qquad K = \dfrac{1}{s}$;

(e) $G = \dfrac{s}{s+1}$, $\qquad K = \dfrac{1}{s}$;

(f) $G = \dfrac{1}{s^2+s+1}$, $\qquad K = 1$;

(g) $G = \dfrac{1}{s^2+s+1}$, $\qquad K = \dfrac{1}{s}$;

(h) $G = \dfrac{1}{s^2+s+1}$, $\qquad K = \dfrac{s+1}{s}$;

(i) $G = \dfrac{1}{s^2+s+1}$, $\qquad K = \dfrac{1}{s+1}$;

(j) $G = \dfrac{1}{s^2+s+1}$, $\qquad K = \dfrac{1}{s(s+1)}$;

(k) $G = \dfrac{1}{s+1}$, $\qquad K = \dfrac{s+1}{s+2}$;

(l) $G = \dfrac{1}{s-1}$, $\qquad K = \dfrac{s-1}{s+1}$;

(m) $G = \dfrac{1}{s(s+1)}$, $\qquad K = \dfrac{s}{s+1}$;

(n) $G = \dfrac{s-1}{s+1}$, $\qquad K = \dfrac{1}{s-1}$.

Problems

4.7 Consider the closed-loop connection in Fig. 4.20(b) with $v = 0$ in which the system and controller transfer-functions are

$$G(s) = \frac{1}{s}, \qquad K(s) = 1.$$

Does the closed-loop system achieve asymptotic rejection of a constant disturbance $w(t) = \bar{w}, t \geq 0$?

4.8 Repeat P4.7 for the combination of system and controller transfer-functions in P4.6.

4.9 Consider the closed-loop connection in Fig. 4.20(b) with $v = 0$ and the system and controller transfer-functions

$$G(s) = \frac{1}{s}, \qquad K(s) = \frac{(s+1)^2}{s^2+1}.$$

Show that the closed-loop system asymptotically tracks a constant reference input $y(t) = \bar{y}, t \geq 0$, and asymptotically rejects an input disturbance $w(t) = \bar{w}\cos(t), t \geq 0$.

4.10 Consider the closed-loop connection in Fig. 4.20(b) with $w = 0$ and the system and controller transfer-functions

$$G(s) = \frac{1}{s+1}, \qquad K(s) = \frac{1}{s}.$$

Calculate the steady-state component of the output, y, and the tracking error, $e = \bar{y} - y$, in response to

$$\bar{y}(t) = \bar{y}, \qquad v(t) = \bar{v} + \cos(\omega t), \qquad t \geq 0.$$

Does the closed-loop achieve asymptotic tracking of the reference input, $\bar{y}(t)$? Does the closed-loop achieve asymptotic rejection of the measurement noise input, $v(t)$? What happens if ω is very large and $\bar{v}$ is zero?

4.11 The scheme in the diagram in Fig. 4.21 is used to control a time-invariant system with a time-delay $\tau \geq 0$, represented by the transfer-function $e^{-s\tau}G$. Show that $u = \hat{K}(\bar{y} - y)$, where

$$\hat{K} = \frac{K}{1 + (1 - e^{-s\tau})GK}.$$

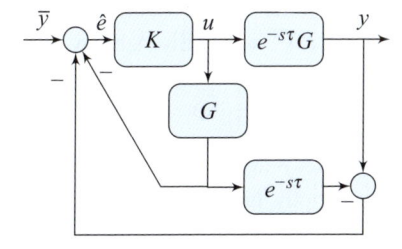

Figure 4.21 Diagram for P4.11.

The controller $\hat{K}$ is known as a *Smith predictor*. Rearrange the closed-loop diagram in Fig. 4.21 so as to reveal the controller $\hat{K}$.

4.12 Show that if G and K are rational transfer-functions and there are no pole–zero cancellations when forming the product GK then the zeros of $\hat{K}$, from P4.11, are the poles of G.

4.13 Explain why the controller in Fig. 4.21 can be used only if G is asymptotically stable. *Hint: P4.12 and internal stability.*

4.14 Show that the transfer-function from $\bar{y}$ to $\hat{e}$ is $\hat{S} = (1 + GK)^{-1}$ and that from $\bar{y}$ to $e = \bar{y} - y$ is $S = \left(1 + (1 - e^{-s\tau})GK\right)(1 + GK)^{-1}$. Assuming that G is asymptotically stable, explain how to select K so that the closed-loop connection in Fig. 4.21 is internally stable.

4.15 Show that the block-diagrams in Fig. 4.22 have the same transfer-function from $\bar{y}$ to e:

$$S = \frac{1}{1 + \tilde{K}(G + D)},$$

and that

$$K = \frac{\tilde{K}}{1 + \tilde{K}D}$$

is the transfer-function from $\tilde{e}$ to u in Fig. 4.22(b).

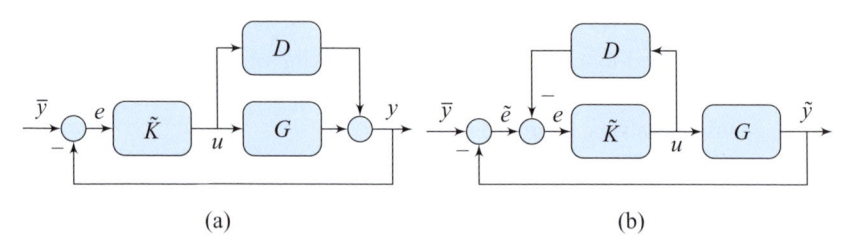

Figure 4.22 Diagrams for P4.15.

4.16 Show that if $\tilde{K}$ is such that an internally stable controller $K = \tilde{K}(1 + \tilde{K}D)^{-1}$ internally stabilizes the system G in the standard feedback diagram from Fig. 4.20(a) then $\tilde{K}$ internally stabilizes both feedback diagrams in Figs. 4.22(a) and (b). Conversely, if an internally stable controller $\tilde{K} = K(1 - KD)^{-1}$ internally stabilizes any of the feedback diagrams in Figs. 4.22(a) and (b) then K internally stabilizes the system G in the standard diagram of Fig. 4.20(a). *Hint: Use P4.15.*

4.17 Show that there exists a proper controller $\tilde{K}(s)$ that stabilizes the proper transfer-function $\tilde{G}(s) = \alpha + G(s)$, where $G(s)$ is strictly proper if and only if there exists a proper controller $K(s)$ that stabilizes the strictly proper transfer-function $G(s)$. Calculate the relationship between the two controllers. *Hint: Use P4.15 and P4.16.*

Problems

4.18 What is wrong with the feedback diagram in Fig. 4.23(a)?

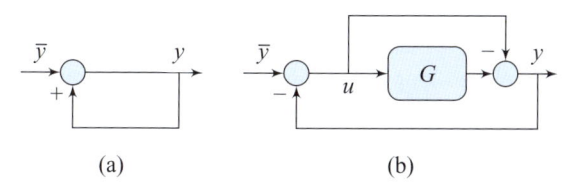

Figure 4.23 Diagrams for P4.18 and P4.19.

4.19 What is wrong with the feedback diagram in Fig. 4.23(b)? *Hint: Use P4.15 and P4.18.*

4.20 You have shown in P2.10 and P2.12 that the ordinary differential equation

$$(J_1 r_2^2 + J_2 r_1^2)\dot{\omega}_1 + (b_1 r_2^2 + b_2 r_1^2)\omega_1 = r_2^2 \tau, \qquad \omega_2 = (r_1/r_2)\omega_1$$

is a simplified description of the motion of a rotating machine driven by a belt without slip as in Fig. 2.18(a), where ω_1 is the angular velocity of the driving shaft and ω_2 is the machine's angular velocity. Let $r_1 = 25$ mm, $r_2 = 500$ mm, $b_1 = 0.01$ kg m²/s, $b_2 = 0.1$ kg m²/s, $J_1 = 0.0031$ kg m², and $J_2 = 25$ kg m². Design a feedback controller

$$\tau = K(\bar{\omega}_2 - \omega_2),$$

and select K such that the closed-loop system is internally stable. Can the closed-loop system asymptotically track a constant reference $\bar{\omega}_2(t) = \bar{\omega}_2$, $t \geq 0$? Assuming zero as the initial condition, sketch or use MATLAB to plot the closed-loop response when $\bar{\omega}_2 = 4.5$ rad/s.

4.21 Repeat P4.20 with a controller

$$\tau = K_i(\bar{\theta}_2 - \theta_2),$$

where

$$\theta_2(t) = \theta_2(0) + \int_0^t \omega_2(\sigma) d\sigma,$$

and the angular reference $\bar{\theta}_2(t) = \bar{\theta}_2 = \pi, t \geq 0$.

4.22 In P4.21, does it matter whether the angle $\theta_2(t)$ used by the controller is coming from an angular position sensor or from integrating the output of an angular velocity sensor?

4.23 You have shown in P2.18 that the ordinary differential equation

$$(J_1 + J_2 + r^2(m_1 + m_2))\dot{\omega} + (b_1 + b_2)\omega = \tau + gr(m_1 - m_2), \qquad v_1 = r\omega,$$

is a simplified description of the motion of the elevator in Fig. 2.18(b), where ω is the angular velocity of the driving shaft and v_1 is the elevator's load linear velocity.

Feedback Analysis

Let $g = 10\,\text{m/s}^2$, $r = 1\,\text{m}$, $m_1 = m_2 = 1000\,\text{kg}$, $b_1 = b_2 = 120\,\text{kg m}^2/\text{s}$, and $J_1 = J_2 = 20\,\text{kg m}^2$. Design a feedback controller

$$\tau = K(\bar{v}_1 - v_1),$$

and select K such that the closed-loop system is internally stable. Can the closed-loop system asymptotically track a constant velocity reference $v_1(t) = \bar{v}_1$, $t \geq 0$? Assuming zero as the initial condition, sketch or use MATLAB to plot the closed-loop response when $\bar{v}_1 = 3\,\text{m/s}$.

4.24 Repeat P4.23 with $m_2 = 800\,\text{kg}$.

4.25 Repeat P4.23 with a controller

$$\tau = K_\text{i}(\bar{x}_1 - x_1),$$

where

$$x_1(t) = x_1(0) + \int_0^t v_1(\tau)d\tau,$$

and the position reference $\bar{x}_1(t) = \bar{x}_1 = 10\,\text{m}$, $t \geq 0$.

4.26 In P4.25, does it matter whether the position $x_1(t)$ used by the controller is coming from a position sensor or from integrating the output of a velocity sensor?

4.27 Repeat P4.25 with $m_2 = 800\,\text{kg}$.

4.28 Modify the feedback controller in P4.25 so that the closed-loop elevator system can asymptotically track a constant reference $\bar{x}_1(t) = \bar{x}_1 = 10\,\text{m}$, $t \geq 0$, when $m_2 = 800\,\text{kg}$.

4.29 You have shown in P2.41 that the ordinary differential equation

$$J\dot{\omega} + \left(b + \frac{K_\text{e}K_\text{t}}{R_\text{a}}\right)\omega = \frac{K_\text{t}}{R_\text{a}}v_\text{a}$$

is a simplified description of the motion of the rotor of the DC motor in Fig. 2.24, where ω is the rotor angular velocity. Let $J = 227 \times 10^{-6}\,\text{kg m}^2$, $K_\text{t} = 0.02\,\text{N m/A}$, $K_\text{e} = 0.02\,\text{V s/rad}$, $b = 289.4 \times 10^{-6}\,\text{kg m}^2/\text{s}$, and $R_\text{a} = 7\,\Omega$. Design a feedback controller

$$v_\text{a} = K(\bar{\omega} - \omega),$$

and select K such that the closed-loop system is internally stable. Can the closed-loop system asymptotically track a constant-angular-velocity reference $\bar{\omega}(t) = \bar{\omega}$, $t \geq 0$? Assuming zero as the initial condition, sketch or use MATLAB to plot the closed-loop response when $\bar{\omega} = 900\,\text{RPM}$.

4.30 Repeat P4.29 with a controller

$$v_\text{a}(t) = K_\text{i}\int_0^t e(\tau)d\tau, \qquad e = \bar{\omega} - \omega.$$

4.31 Repeat P4.29 with a controller

$$v_a = K_i(\bar{\theta} - \theta),$$

where

$$\theta(t) = \theta(0) + \int_0^t \omega(\tau) d\tau,$$

and the angular reference $\bar{\theta}(t) = \bar{\theta} = \pi, t \geq 0$.

4.32 It seems that the controllers in P4.30 and P4.31 are the same if $\bar{\omega} = \bar{\theta} = 0$. Explain their differences.

4.33 Why would you want to run the controller from problem P4.29 with a zero velocity reference $\bar{\omega} = 0$? What is the role of the control gain K in this case?

4.34 Recall from P2.41 that the rotor torque is

$$\tau = K_t i_a.$$

The armature current, i_a, is related to the armature voltage, v_a, and the rotor angular velocity, ω, through

$$v_a = R_a i_a + K_e \omega.$$

Show that

$$\frac{T(s)}{V_a(s)} = \frac{K_t}{R_a} \frac{(s + b/J)}{(s + b/J + K_e K_t/(R_a J))}$$

is the transfer-function from v_a to τ.

4.35 Consider the DC motor in P4.34 with the same physical parameters as in P4.29. Design a feedback controller

$$v_a = K(\bar{\tau} - \tau),$$

and select K such that the closed-loop system is internally stable. Can the closed-loop system asymptotically track a constant-torque reference $\bar{\tau}(t) = \bar{\tau}, t \geq 0$? Assuming zeros as the initial conditions, sketch or use MATLAB to plot the closed-loop response when $\bar{\tau} = 0.5$ N m.

4.36 Repeat P4.35 with a controller

$$v_a(t) = K_i \int_0^t e(\tau) d\tau, \qquad e = \bar{\tau} - \tau.$$

4.37 Contrast the similarities and differences between the solutions to P4.30 and P4.35.

4.38 You have shown in P2.49 that the temperature of a substance, T (in K or in °C), flowing in and out of a container kept at the ambient temperature, T_o, with an inflow temperature, T_i, and a heat source, q (in W), can be approximated by the differential equation

$$mc\dot{T} = q + wc(T_i - T) + \frac{1}{R}(T_o - T),$$

where m and c are the substance's mass and specific heat, and R is the overall system's thermal resistance. The input and output flow mass rates are assumed to be equal to w (in kg/s). Assume that water's density and specific heat are $\rho = 997.1\,\text{kg/m}^3$ and $c = 4186\,\text{J/kg K}$. Design a feedback controller

$$q = K(\bar{T} - T)$$

for a 50 gal ($\approx 0.19\,\text{m}^3$) water heater rated at $\bar{q} = 40{,}000\,\text{BTU/h}$ ($\approx 12\,\text{kW}$) and thermal resistance $R = 0.27\,\text{K/W}$ at ambient temperature $T_o = 77\,°\text{F}$ ($\approx 25\,°\text{C}$). Select K such that the closed-loop system is asymptotically stable and so that a heater initially at ambient temperature never exceeds its maximum power without any in/out flow, $w = 0$, and $\bar{T} = 140\,°\text{F}$ ($\approx 60\,°\text{C}$). Calculate the resulting closed-loop time-constant in hours and compare your answer with the open-loop time-constant. Is the closed-loop system capable of asymptotically tracking a constant reference temperature $\bar{T}(t) = \bar{T}, t \geq 0$? Use MATLAB to plot the temperature of the water during 3 days. Compute the average water temperature over the entire period and over the last 2 days. Compare your answer with that from P2.52.

4.39 Repeat P4.38 with a constant in/out flow of 20 gal/h ($\approx 21 \times 10^{-6}\,\text{m}^3/\text{s}$) at ambient temperature. Is the closed-loop system capable of asymptotically rejecting this constant in/out flow perturbation?

4.40 Repeat P4.38 with a controller

$$q(t) = K_p\, e(t) + K_i \int_0^t e(\tau)\, d\tau,$$

where $e = \bar{T} - T$.

4.41 Repeat P4.40 with a constant in/out flow of 20 gal/h ($\approx 21 \times 10^{-6}\,\text{m}^3/\text{s}$) at ambient temperature. Is the closed-loop system capable of asymptotically rejecting this constant in/out flow perturbation?

4.42 Repeat P4.38 with a sinusoidal in/out flow perturbation

$$w(t) = \frac{\bar{w}}{2}(1 + \cos(\omega t)),$$

where $\bar{w} = 20\,\text{gal/h}$ ($\approx 21 \times 10^{-6}\,\text{m}^3/\text{s}$) at ambient temperature and $\omega = 2\pi/24\,\text{h}^{-1}$. What can this in/out flow represent? Use the approximation $w(t)(T_i - T(t)) \approx \bar{w}(T_i - \bar{T})$. When is this approximation reasonable? Can you solve the problem without approximating? Is the closed-loop system capable of asymptotically rejecting this in/out flow perturbation? If not, what form would the controller need to have in order to reject this disturbance?

4.43 Repeat P4.42 with a controller

$$q(t) = K_p\, e(t) + K_i \int_0^t e(\tau)d\tau,$$

where $e = \bar{T} - T$.

5 State-Space Models and Linearization

In this chapter we introduce ideas that can be used to implement controllers on physical hardware. The resulting block-diagrams and equations also serve as the basis for simulation of dynamic systems in computers, a topic that we use to motivate the introduction of *state-space models*. The state-space formalism provides a framework for computing linearized models from nonlinear differential equations, and sometimes relates the stability of the linearized model to the stability of a more complex nonlinear model. We finish with a discussion about possible issues that can arise when a linear controller is used in feedback with a nonlinear system.

5.1 Realization of Dynamic Systems

The simplest dynamic system for which we can envision a construction is the *integrator*. Any device that is capable of storing mass, charge, or energy in some form is basically an integrator. Indeed, we have already met one such device in Section 2.8: the *toilet water tank*. In Fig. 2.16, the water level, y, is the results of integration of the water input flow, u,

$$y(t) = \frac{1}{A} \int_0^t u(\tau)d\tau,$$

where A is the constant cross-section area of the tank. The voltage across the terminals of a *capacitor*, v, is the integral of the current, i,

$$v(t) = \frac{1}{C} \int_0^t i(\tau)d\tau,$$

where C is the capacitor's capacitance. A *fly-wheel* with negligible friction integrates the input torque, f, to produce an angular velocity, ω,

$$\omega(t) = \frac{1}{J} \int_0^t f(\tau)d\tau,$$

where J is the wheel's moment of inertia. It is of significance in modern control and signal processing algorithms that an integrator can be implemented via an approximate

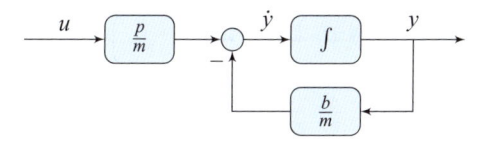

Figure 5.1 First-order differential equation, $m\dot{y} + by = pu$, $m > 0$.

integration rule, such as the *trapezoidal integration rule*:

$$y(kT) - y(kT - T) = \int_{kT-T}^{kT} u(t)dt \approx \frac{T}{2}(u(kT) - u(kT - T)),$$

which can be implemented in the form of the recursive algorithm

$$y(kT) = y(kT - T) + \frac{T}{2}(u(kT) - u(kT - T)),$$

indexed by the integer k and where T is a small enough *sampling period* at which periodic samples of the continuous input $u(t)$ are obtained.

Two difficulties are common to all physical realizations of integrators: (a) providing enough storage capacity, be it in the form of mass, volume, charge, energy, or numbers in a computer; and (b) controlling the *losses* or *leaks*. The storage capacity should be sized to fit the application in hand, and losses must be kept under control with appropriate materials and engineering. In the following discussion we assume that these issues have been worked out and proceed to utilize integrators to realize more complex dynamic systems.

We have already used integrators in Chapter 2 to represent simple differential equations. Here we expand on the basic idea to cover differential equations of higher order that might also involve the derivative of the input. As seen in Section 3.3, transfer-functions and linear ordinary differential equations are closely related, and the techniques and diagrams obtained from differential equations can be readily used to implement the corresponding transfer-function.

Let us start by revisiting the diagram in Fig. 2.3 which represents the linear ordinary differential equation (2.3). This diagram is reproduced in Fig. 5.1. The trick used to represent the differential equation (2.3) in a block-diagram with integrators was to isolate the highest derivative:

$$\dot{y} = \frac{p}{m}u - \frac{b}{m}y.$$

The highest derivative is then integrated in a series of cascade integrators from which all lower-order derivatives become available. For example, in order to represent the second-order differential equation

$$\ddot{y} + a_1\dot{y} + a_2 y = b_2 u$$

in a block-diagram with integrators we first isolate the second derivative

$$\ddot{y} = b_2 u - a_1 \dot{y} - a_2 y,$$

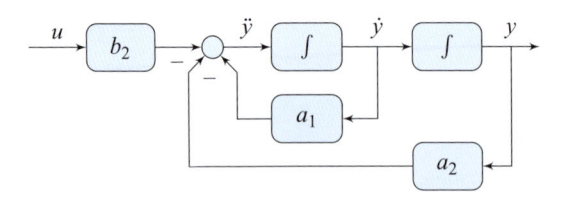

Figure 5.2 Second-order differential equation, $\ddot{y} + a_1\dot{y} + a_2 y = b_2 u$.

which is then integrated twice. The input signal u and the signals y and $\dot{y}$ are run through amplifiers (gains) and a summer is used to reconstruct $\ddot{y}$ as shown in Fig. 5.2. Of course, one could use the exact same scheme to implement the associated transfer-function:

$$\frac{Y(s)}{U(s)} = \frac{b_2}{s^2 + a_1 s + a_2}.$$

In this case, isolate the term with the highest power of s:

$$s^2 Y(s) = b_2 U(s) - s a_1 Y(s) - a_2 Y(s)$$

and replace integrators by "s^{-1}." The result is Fig. 5.3, which is virtually the same as Fig. 5.2. Initial conditions for the differential equation, $y(0)$ and $\dot{y}(0)$ in this case, are implicitly incorporated in the block-diagrams as initial conditions for the integrators. For example, the second integrator in Fig. 5.2 implements the definite integral

$$y(t) = y(0) + \int_0^t \dot{y}(\tau) d\tau.$$

Physically, $y(0)$ is the amount of the quantity being integrated, water, current, torque, etc., which is present in the integrator at $t = 0$. There will be more about that in Section 5.2.

A slightly more challenging task is to represent the differential equation

$$\ddot{y} + a_1 \dot{y} + a_2 y = b_0 \ddot{u} + b_1 \dot{u} + b_2 u$$

using only integrators. The main difficulty is the derivatives of the input signal u. One idea is to use linearity. First solve the differential equation

$$\ddot{z} + a_1 \dot{z} + a_2 z = u.$$

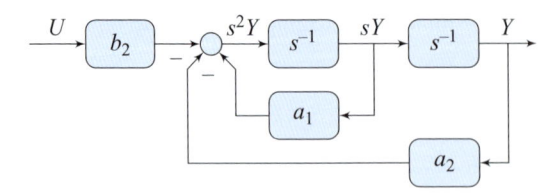

Figure 5.3 Second-order transfer-function, $G(s) = b_2/(s^2 + a_1 s + a_2)$.

5.1 Realization of Dynamic Systems

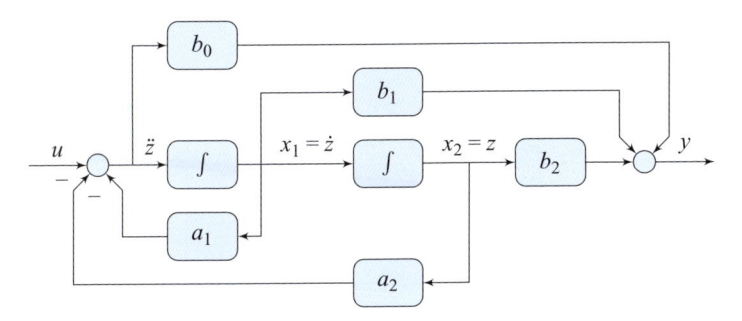

Figure 5.4 Second-order differential equation, $\ddot{y} + a_1\dot{y} + a_2 y = b_0\ddot{u} + b_1\dot{u} + b_2 u$.

One can use the diagram in Fig. 5.2 with $b_2 = 1$ for that. A solution to

$$\ddot{y} + a_1\dot{y} + a_2 y = u_0 + u_1 + u_2, \qquad u_0 = b_0\ddot{u}, \qquad u_1 = b_1\dot{u}, \qquad u_2 = b_2 u$$

can be calculated from z, $\dot{z}$, and $\ddot{z}$ using superposition,

$$y = b_0\ddot{z} + b_1\dot{z} + b_2 z.$$

This idea is implemented in the diagram of Fig. 5.4. See also P2.38.

Here is an alternative: isolate the highest derivative and collect the right-hand terms with the same degree, that is,

$$\ddot{y} = b_0\ddot{u} + (b_1\dot{u} - a_1\dot{y}) + (b_2 u - a_2 y).$$

Now integrate[1] twice to obtain

$$y(t) = b_0 u(t) + \int_0^t \left[b_1 u(\tau) - a_1 y(\tau) + \int_0^\tau (b_2 u(\sigma) - a_2 y(\sigma))d\sigma \right] d\tau.$$

These operations are represented in the diagram in Fig. 5.5. Because Figs. 5.4 and 5.5 represent the same system, the realization of a differential equation, or its corresponding transfer-function, is not unique. Other forms are possible, which we do not have room to discuss here. They can be found, for instance, in the excellent text [Kai80].

Besides the issue of uniqueness, a natural question is the following: can we apply these ideas to *any* differential equation and obtain a block-diagram using only integrators? The answer is *no*. Consider for example

$$y = \dot{u},$$

where u is an input and y is the output. It is not possible to represent this equation in a diagram where y is obtained as a function of u without introducing a derivative block. In general, by a straightforward generalization of the techniques discussed above, it should be possible to represent a general linear ordinary differential equation of the form (3.17):

$$y^{(n)}(t) + a_1 y^{(n-1)}(t) + \cdots + a_n y(t) = b_0 u^{(m)}(t) + b_1 u^{(m-1)}(t) + \cdots + b_m u(t),$$

---

[1] Assuming zero initial conditions $y(0) = \dot{y}(0) = 0$ for convenience.

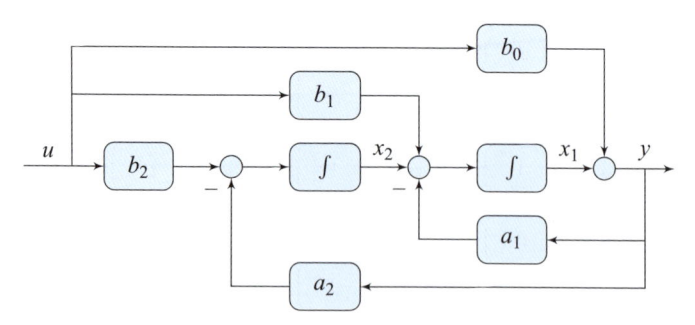

Figure 5.5 Second-order differential equation, $\ddot{y} + a_1\dot{y} + a_2 y = b_0\ddot{u} + b_1\dot{u} + b_2 u$.

using only integrators if the highest derivative of the input signal, u, appearing in (3.17) is not higher than the highest derivative of the output signal, y. That is, if $m \leq n$. In terms of the associated rational transfer-function, G in (3.18), it means that the degree of the numerator, m, is no higher than the degree of the denominator, n, that is, G is proper. As we will soon argue, it is not possible to physically implement a differentiator, hence one should not ordinarily find transfer-function models of physical systems which are not proper. Nor should one expect to be able to implement a controller obtained from a transfer-function which is not proper.

One obstruction for implementing differentiators is the required amount of amplification: differentiators need to deliver an ever-increasing gain to high-frequency signals. Indeed, the transfer-function of a differentiator, $G(s) = s$, produces in the presence of the sinusoidal input, $u(t) = \cos(\omega t)$, with frequency ω and unit amplitude, a steady-state response

$$y_{ss}(t) = |j\omega|\cos(\omega t + \angle j\omega) = -|\omega|\sin(\omega t).$$

See Section 3.8. As ω grows, the amount of amplification required to produce $y_{ss}(t)$ grows as well, becoming unbounded as $\omega \to \infty$. Clearly no physical system can have this capability. Even if one is willing to construct such a system in the hope that the input signals have low enough frequency content, the system will likely face problems if it has to deal with discontinuities or large derivatives of the input signal. Using the Laplace transform formalism, one might find it useful to think of the generalized derivative of a signal at a discontinuity as producing an impulse, e.g. $(d/dt)1(t) = \delta(t)$. A physical system that is modeled as a differentiator will not be able to produce an impulse, as discussed in Chapter 3.

That is, however, not to say that *components* of a system cannot be modeled as differentiators or have improper transfer functions. Take for example the electric circuit in Fig. 5.6. The relationship between the voltage and current of an *ideal* capacitor is the differentiator:

$$i_C(t) = C\dot{v}_C(t).$$

The transfer-function of the ideal capacitor component is therefore

$$\frac{I_C(s)}{V_C(s)} = sC,$$

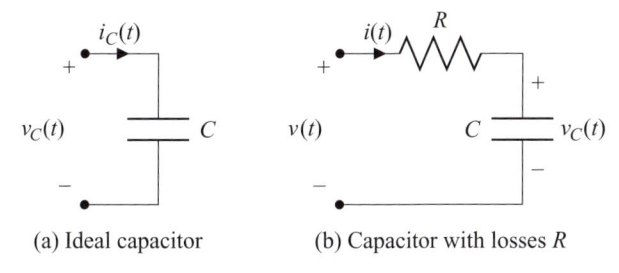

Figure 5.6 Capacitor circuit models.

which is not proper. A *real* capacitor will, however, have losses, which can be represented by some small nonzero resistance R appearing in series with the capacitor in the circuit of Fig. 5.6. The complete circuit relations are

$$Ri(t) + v_C(t) = v(t), \qquad i(t) = C\dot{v}_C(t),$$

from which we obtain the transfer-function

$$G(s) = \frac{I(s)}{V(s)} = \frac{sC}{1 + sRC},$$

after eliminating v_C. See P2.34 and P3.80. The overall circuit has a proper transfer-function, $G(s)$. The smaller the losses, that is, R, the more the circuit behaves as an ideal capacitor, and hence as a differentiator. Indeed,

$$\lim_{R \to 0} G(s) = sC,$$

which is not proper. It was nevertheless very useful to work with the *ideal capacitor* model and its improper component transfer-function to understand the overall circuit behavior and its limitations. It is in this spirit that one should approach improper transfer-functions.

Even if the issue of gain could be addressed, there are still potential problems with losses in practical differentiators. For example, the unit step response of the ideal capacitor with model sC is the impulse $C\delta(t)$. The unit step response of the capacitor with losses is

$$\mathcal{L}^{-1}\left\{\frac{sC}{1+sRC} \times \frac{1}{s}\right\} = \frac{1}{R}e^{-\frac{1}{RC}t}, \quad t \geq 0,$$

which approaches an impulse of amplitude C as $R \to 0$. A capacitor with losses would therefore produce large *spikes* of current for small amounts of time every time a 1 V source is connected to its terminals, as plotted[2] in Fig. 5.7. Such intense but short-lived currents and voltages will cause other problems for the circuit and the physical materials it is made of; one of them is that the materials will stop responding linearly.

---

[2] The response plotted in Fig. 5.7 is still idealized as it assumes a voltage source that has a step discontinuity. This model could be improved to account for that as well if wanted.

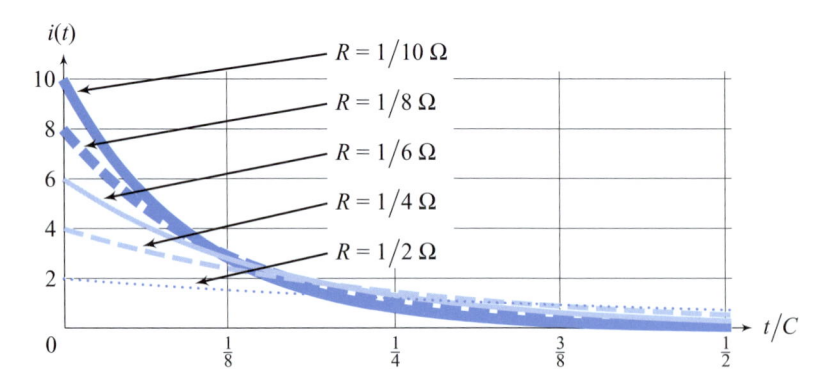

Figure 5.7 Normalized response of a capacitor with losses to a 1V step input.

On the issue of high-frequency gains, the steady-state response of the capacitor with losses to a sinusoidal input of frequency $\omega > 0$ and unit amplitude, $u(t) = \cos(\omega t)$, is

$$y_{ss}(t) = |G(j\omega)|\cos(\omega t + \angle G(j\omega)).$$

At high frequencies the gain is approximately

$$|G(j\omega)| \approx \frac{1}{R},$$

which shows that the amount of amplification in this circuit is limited by the losses represented by the resistor R. Note also that

$$\lim_{R \to 0} |G(j\omega)| = |\omega|C,$$

revealing the high-frequency amplification of a differentiator when R is small.

We conclude our discussion with another electric circuit. The circuit in Fig. 5.8 contains a resistor, R_1, two capacitors, C_1 and C_2, and an (operational) amplifier with very high gain, the triangular circuit element. You have shown in problems P2.38 and P3.90

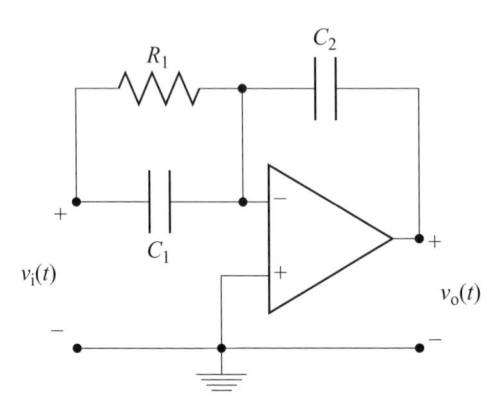

Figure 5.8 Circuit for PI control.

Figure 5.9 Analog computer used by NASA in the mid twentieth century. The device was located in the Engine Research Building at the Lewis Flight Propulsion Laboratory, now John H. Glenn Research Center, Cleveland, Ohio.

that the transfer-function from the voltage v_i to the voltage v_o is

$$\frac{V_o(s)}{V_i(s)} = -K\frac{s+z}{s}, \qquad K = \frac{C_1}{C_2}, \qquad z = \frac{1}{R_1 C_1}.$$

Therefore, by adjusting the ratio between the two capacitors, C_1 and C_2, and the resistor, R_1, it is possible to set the gain K and the zero z to be exactly the ones needed to implement the PI controller (4.15). The particular case when the capacitor C_1 is removed from the circuit, i.e. $C_1 = 0$, is important. In this case the circuit transfer-function reduces to

$$\frac{V_o(s)}{V_i(s)} = -\frac{K}{s}, \qquad K = \frac{1}{R_1 C_2},$$

which is a pure integrator with gain K (see P2.40). This circuit can be combined with amplifiers to build a physical realization for the diagrams in Figs. 5.1–5.5. Items of electronic hardware with specialized circuitry implementing blocks such as the one in Fig. 5.8 were manufactured and used for analysis and simulation of dynamic systems in the second half of the twentieth century under the name *analog computers*. An analog computer used by NASA is shown in Fig. 5.9. These computers have all but been replaced by *digital computers*, in which dynamic systems are simulated via numerical integration.

5.2 State-Space Models

After learning how to convert differential equations and transfer-functions into block-diagrams using integrators we will now introduce a formalism to transform differential equations of arbitrary order into a set of *vector* first-order differential equations. The key is to look at the integrators in the block-diagrams. Start by defining a new variable

State-Space Models and Linearization

for each variable being integrated. For example, in the diagram in Fig. 5.4, define two variables,

$$x_1 = \dot{z}, \quad x_2 = z,$$

one per output of each integrator. Next write two first-order differential equations at the input of each integrator:

$$\dot{x}_1 = u - a_1 x_1 - a_2 x_2,$$
$$\dot{x}_2 = x_1.$$

Then rewrite the output, y, in terms of x_1 and x_2 after eliminating derivatives:

$$y = b_0 \dot{x}_1 + b_1 x_1 + b_2 x_2$$
$$= b_0(u - a_1 x_1 - a_2 x_2) + b_1 x_1 + b_2 x_2$$
$$= (b_1 - a_1 b_0) x_1 + (b_2 - a_2 b_0) x_2 + b_0 u.$$

With the help of vectors and matrices rearrange:

$$\begin{pmatrix} \dot{x}_1 \\ \dot{x}_2 \end{pmatrix} = \begin{bmatrix} -a_1 & -a_2 \\ 1 & 0 \end{bmatrix} \begin{pmatrix} x_1 \\ x_2 \end{pmatrix} + \begin{bmatrix} 1 \\ 0 \end{bmatrix} u,$$

$$y = \begin{bmatrix} b_1 - a_1 b_0 & b_2 - a_2 b_0 \end{bmatrix} \begin{pmatrix} x_1 \\ x_2 \end{pmatrix} + \begin{bmatrix} b_0 \end{bmatrix} u. \tag{5.1}$$

A similar procedure works for the diagram in Fig. 5.5. With x_1 and x_2 representing the output of each integrator, we write the differential equations

$$\dot{x}_1 = b_1 u - a_1 y + x_2,$$
$$\dot{x}_2 = b_2 u - a_2 y.$$

Because these equations depend on the output, y, substitute

$$y = b_0 u + x_1$$

to obtain

$$\dot{x}_1 = -a_1 x_1 + x_2 + (b_1 - a_1 b_0) u,$$
$$\dot{x}_2 = -a_2 x_1 + (b_2 - a_2 b_0) u.$$

Rearranging using vectors and matrices:

$$\begin{pmatrix} \dot{x}_1 \\ \dot{x}_2 \end{pmatrix} = \begin{bmatrix} -a_1 & 1 \\ -a_2 & 0 \end{bmatrix} \begin{pmatrix} x_1 \\ x_2 \end{pmatrix} + \begin{bmatrix} b_1 - a_1 b_0 \\ b_2 - a_2 b_0 \end{bmatrix} u,$$

$$y = \begin{bmatrix} 1 & 0 \end{bmatrix} \begin{pmatrix} x_1 \\ x_2 \end{pmatrix} + \begin{bmatrix} b_0 \end{bmatrix} u. \tag{5.2}$$

Equations (5.1) and (5.2) are in a special form called *state-space* equations. Differential equations for linear time-invariant systems are in state-space form when they match

5.2 State-Space Models

the template

$$\dot{x} = Ax + Bu,$$
$$y = Cx + Du, \qquad (5.3)$$

for some appropriate quadruple of constant matrices (A, B, C, D). As before, u and y denote the input and output signals. The vector x is known as the *state vector*. The terminology *state* alludes to the fact that knowledge of the state (of the system), x, and the input, u, implies that all signals, including the output, y, can be determined. Furthermore, knowledge of $x(t)$ at a given instant of time, say at $t = 0$, and knowledge of the inputs $u(t)$, $t \geq 0$, is enough to predict or reconstruct any signal at time $t \geq 0$.

In (5.3), not only the state, x, but also the input, u, and output, y, can be vectors. This means that state-space is capable of providing a uniform representation for single-input–single-output (SISO) as well as multiple-input–multiple-output (MIMO) systems. Furthermore, the matrix and linear algebra formalism enables the use of compact and powerful notation. For instance, applying the Laplace transform[3] to the state-space equations (5.3), we obtain

$$sX(s) - x(0^-) = AX(s) + BU(s),$$
$$Y(s) = CX(s) + DU(s).$$

We can *easily* solve for $X(s)$ in the first equation,

$$sX(s) - AX(s) = (sI - A)X(s) = BU(s) + x(0^-)$$
$$\implies X(s) = (sI - A)^{-1}BU(s) + (sI - A)^{-1}x(0^-),$$

and compute

$$Y(s) = G(s)U(s) + F(s)x(0^-),$$

where

$$G(s) = C(sI - A)^{-1}B + D, \qquad (5.4)$$

which is the transfer function, and

$$F(s) = C(sI - A)^{-1} \qquad (5.5)$$

parametrize the response to the *initial state*, $x(0^-)$, which plays the role of the initial conditions. The simplicity of these formulas hides the complexity of the underlying calculations. For example, with

$$A = \begin{bmatrix} -a_1 & -a_2 \\ 1 & 0 \end{bmatrix}, \quad B = \begin{bmatrix} 1 \\ 0 \end{bmatrix}, \quad C = \begin{bmatrix} b_1 - a_1 b_0 & b_2 - a_2 b_0 \end{bmatrix}, \quad D = \begin{bmatrix} b_0 \end{bmatrix},$$

---

[3] The Laplace transform of a matrix or vector is to be interpreted entry-wise.

we compute
$$sI - A = \begin{bmatrix} s + a_1 & a_2 \\ -1 & s \end{bmatrix},$$

from which
$$(sI - A)^{-1} = \frac{\text{Adj}(sI - A)}{|sI - A|} = \frac{1}{s^2 + a_1 s + a_2} \begin{bmatrix} s & -a_2 \\ 1 & s + a_1 \end{bmatrix}, \quad (5.6)$$

and
$$\begin{aligned} G(s) &= C(sI - A)^{-1} B + D \\ &= \frac{1}{s^2 + a_1 s + a_2} \begin{bmatrix} b_1 - a_1 b_0 & b_2 - a_2 b_0 \end{bmatrix} \begin{bmatrix} s & -a_2 \\ 1 & s + a_1 \end{bmatrix} \begin{bmatrix} 1 \\ 0 \end{bmatrix} + [b_0] \\ &= \frac{b_1 s - a_1 b_0 s + b_2 - a_2 b_0}{s^2 + a_1 s + a_2} + b_0 \\ &= \frac{b_0 s^2 + b_1 s + b_2}{s^2 + a_1 s + a_2}. \end{aligned}$$

Because $\text{Adj}(sI - A)$ is a polynomial matrix and B, C, and D are real matrices, the associated transfer-function, $G(s)$, is always *rational*. The denominator of $G(s)$ is equal to the determinant $|sI - A|$, a polynomial of degree n, where n is the dimension of the state vector. The determinant equation
$$|sI - A| = 0$$

is the *characteristic equation*. If s_0 is a root of the characteristic equation then it is an *eigenvalue* of the square matrix A and satisfies
$$A x_0 = s_0 x_0, \quad x_0 \neq 0.$$

The vector x_0 corresponding to an eigenvalue s_0 is an *eigenvector* of A. Eigenvalues of A are poles of the transfer-function $G(s)$. A matrix that has all eigenvalues with negative real part is called a *Hurwitz matrix*, and internal stability[4] of the state-space system (A, B, C, D) is equivalent to A being a Hurwitz matrix. Another consequence of formula (5.4) is that
$$\lim_{|s| \to \infty} (sI - A)^{-1} = 0 \quad \Longrightarrow \quad \lim_{|s| \to \infty} G(s) = D.$$

Consequently, the transfer-function of a linear time-invariant system that can be put in state-space form always satisfies (3.25). In other words, $G(s)$ is always rational and proper. Furthermore, if $D = 0$ then (3.23) holds and $G(s)$ is strictly proper.

Another deceptively simple formula is the impulse response:
$$g(t) = \mathscr{L}^{-1}\{G(s)\} = \mathscr{L}^{-1}\{C(sI - A)^{-1} B + D\} = C e^{At} B + D \delta(t), \quad t \geq 0.$$

---

[4] See Section 4.7.

Of course the trick is to be able to pull out the *rabbit*:

$$\mathscr{L}^{-1}\{(sI - A)^{-1}\} = e^{At}, \quad t \geq 0,$$

from one's empty hat. The exponential[5] function of a matrix hides the complexities which will ultimately be able to correctly compute the response of linear systems even in the most complicated cases,[6] e.g. for systems with multiple roots or complex-conjugate roots (see Chapter 3). Likewise, in response to a nonzero initial condition we have that

$$f(t) = \mathscr{L}^{-1}\{F(s)\} = \mathscr{L}^{-1}\{C(sI - A)^{-1}\} = Ce^{At}, \quad t \geq 0, \quad (5.7)$$

and[7]

$$\|f(t)x(0^-)\|_2^2 = \int_{0-}^{\infty} x(0^-)^{\mathrm{T}} f(t)^{\mathrm{T}} f(t) x(0^-) dt$$
$$= x(0^-)^{\mathrm{T}} P x(0^-) \leq \max_j \lambda_j(P) \|x(0^-)\|_2^2, \quad (5.8)$$

where $\lambda_j(P)$ denotes the jth (real) eigenvalue of the $n \times n$ symmetric matrix

$$P = \int_{0-}^{\infty} e^{A^{\mathrm{T}} t} C^{\mathrm{T}} C e^{At} dt,$$

which is known as the *observability Gramian*. It is possible to show that when A is Hurwitz then the observability Gramian is finite and positive-semidefinite, $P \succeq 0$, which implies that all $\lambda_j(P)$ are real, finite, and bounded, and therefore $\|f(t)x(0^-)\|_2$ is bounded by the 2-norm of the initial state, $x(0^-)$ [Kai80, Theorem 2.6-1 and Corollary 2.6-2]. Note that because A is Hurwitz

$$\lim_{t \to \infty} e^{At} x(0^-) = 0 \quad \Longrightarrow \quad \lim_{t \to \infty} f(t) x(0^-) = 0.$$

This is in fact the very definition of *asymptotic stability*.

---

[5] There are many ways to make peace with the notion of a square matrix exponential. One is through the power-series expansion

$$e^A = \sum_{i=0}^{\infty} \frac{1}{i!} A^i = I + A + \frac{1}{2} A^2 + \frac{1}{6} A^3 + \cdots,$$

which is a direct extension of the standard power series of the scalar exponential function. For the most part, the exponential function of a matrix operates like, and has properties similar to those of, the regular exponential, e.g. $(d/dt)e^{At} = Ae^{At} = e^{At}A$. Beware that some properties hold only if the matrices involved commute, e.g. $e^{A+B} \neq e^A e^B$ unless $AB = BA$. Other properties require that A be nonsingular, e.g. $\int e^{A\tau} d\tau = A^{-1} e^{At}$. Note that A, A^{-1}, and e^A all commute, that is $Ae^A = e^A A$, and $AA^{-1} = A^{-1}A = I$.

[6] For example, for

$$A = \begin{bmatrix} -a_1 & -a_2 \\ 1 & 0 \end{bmatrix}$$

the matrix exponential e^A is given by

$$e^A = \frac{e^{-\frac{a_1}{2}}}{\Delta} \begin{bmatrix} \Delta \cosh\left(\frac{1}{2}\Delta\right) - a_1 \sinh\left(\frac{1}{2}\Delta\right) & -2a_2 \sinh\left(\frac{1}{2}\Delta\right) \\ 2 \sinh\left(\frac{1}{2}\Delta\right) & \Delta \cosh\left(\frac{1}{2}\Delta\right) + a_1 \sinh\left(\frac{1}{2}\Delta\right) \end{bmatrix},$$

where $\Delta = \sqrt{a_1^2 - 4a_2}$.

[7] The next lines may be a bit too advanced for some readers. The important idea is that $\|f(t)x(0^-)\|_2$ is bounded by the norm of the initial state.

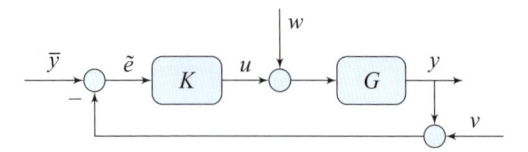

Figure 5.10 Closed-loop feedback configuration with reference, $\bar{y}$, input disturbance, w, and measurement noise, v.

As mentioned before, state-space provides a unified treatment of SISO and MIMO systems. To see how this can work, consider the block-diagram in Fig. 4.11 reproduced for convenience in Fig. 5.10. Assume that G is strictly proper[8] and let the system, G, and the controller, K, have the state-space representations

$$\dot{x}_g = A_g x_g + B_g u_g, \qquad \dot{x}_k = A_k x_k + B_k u_k,$$
$$y_g = C_g x_g, \qquad y_k = C_k x_k + D_k u_k.$$

The closed-loop diagram in Fig. 5.10 implies the following connections:

$$u_g = w + u, \qquad u = y_k, \qquad u_k = \tilde{e} = \bar{y} - y - v, \qquad y = y_g.$$

After some algebra, we obtain first-order vector equations for the system state,

$$\begin{aligned}\dot{x}_g &= A_g x_g + B_g(w + u) \\ &= A_g x_g + B_g C_k x_k + B_g w + B_g D_k (\bar{y} - y - v) \\ &= (A_g - B_g D_k C_g) x_g + B_g C_k x_k + B_g w + B_g D_k \bar{y} - B_g D_k v,\end{aligned}$$

the controller state,

$$\begin{aligned}\dot{x}_k &= A_k x_k + B_k (\bar{y} - y - v) \\ &= -B_k C_g x_g + A_k x_k + B_k \bar{y} - B_k v,\end{aligned}$$

and the *outputs*,

$$y = y_g = C_g x_g,$$
$$u = y_k = C_k x_k.$$

These equations are put in the form (5.3) after collecting all inputs, outputs, and closed-loop state components into the vectors

$$\mathbf{u} = \begin{pmatrix} \bar{y} \\ w \\ v \end{pmatrix}, \qquad \mathbf{y} = \begin{pmatrix} y \\ u \end{pmatrix}, \qquad \mathbf{x} = \begin{pmatrix} x_g \\ x_k \end{pmatrix},$$

---

[8] Messier formulas are available in the case of proper systems. See P5.2.

and rearranging:

$$\dot{x} = Ax + Bu,$$
$$y = Cx + Du,$$

where

$$A = \begin{bmatrix} A_g - B_g D_k C_g & B_g C_k \\ -B_k C_g & A_k \end{bmatrix}, \qquad C = \begin{bmatrix} C_g & 0 \\ 0 & C_k \end{bmatrix},$$
$$B = \begin{bmatrix} B_g D_k & B_g & -B_g D_k \\ B_k & 0 & -B_k \end{bmatrix}, \qquad D = 0.$$

Closed-loop internal stability (Lemma 4.3) can be shown to be equivalent to matrix A being Hurwitz.

5.3 Minimal State-Space Realizations

The discussion in Sections 5.1 and 5.2 may leave one with the impression that the passage from transfer-function to state-space model and vice versa is a relatively straightforward process. Other than the issue of non-unicity of state-space realizations, state-space formulas look and feel simple. However, we would like to draw the reader's attention to an important issue we have overlooked so far, which we motivate with two simple examples.

Consider for instance the state-space realization in the form (5.3) with

$$A = \begin{bmatrix} 0 & 1 \\ 0 & 0 \end{bmatrix}, \qquad B = \begin{bmatrix} 1 \\ \beta \end{bmatrix}, \qquad C = \begin{bmatrix} \alpha & 1 \end{bmatrix}, \qquad D = \gamma. \qquad (5.9)$$

The formulas in Section 5.2 produce the associated second-order transfer-function

$$G(s) = C(sI - A)^{-1}B + D = \frac{\alpha + \beta}{s} + \frac{\alpha\beta}{s^2} + \gamma.$$

What happens if $\alpha = 0$ or $\beta = 0$? Well, nothing seems to have changed in the state-space representation: some entries in matrices B and C assume particular values. However, much has changed in the transfer-function, $G(s)$, as its order drops to one as either α or β becomes zero. In the extreme, if $\alpha = \beta = 0$ then the order drops by two and $G(s) = \gamma$. What you are witnessing here is pole–zero cancellations happening in state-space. When α or β is zero, we say that the realization is not minimal, and the order of the associated transfer-function no longer matches the dimension of the state-space vector. Minimality in state-space realizations (and much more) is associated with the notions of *controllability* and *observability*, which we will not have much time to discuss in the present book. A detailed presentation of such issues can be found in [Kai80]. As a teaser, a realization is said to be observable if there exist no vector x and complex number λ such that

$$Ax = \lambda x, \qquad Cx = 0, \qquad x \neq 0. \qquad (5.10)$$

You will verify in P5.4 that, for the matrices in (5.9) and $\alpha = 0$,

$$x = \begin{pmatrix} 1 \\ 0 \end{pmatrix}, \qquad \lambda = 0$$

satisfy (5.10), which shows that this state-space realization is not observable if $\alpha = 0$.

An interpretation for the loss of observability is that not all coordinates of the state vector can be *estimated* by taking measurements of the output and its derivatives alone [Kai80]. A similar statement, this time involving the matrices A and B, can be used to test for controllability (see P5.3). Lack of controllability can be interpreted as the impossibility of computing a control input that can steer all coordinates of the state vector to a desired location [Kai80].

In the case of multiple-input–multiple-output (MIMO) systems, the situation is much more complex, a fact that puzzled many and motivated intense research in the 1950s and 1960s. For example, the state-space realization with matrices

$$A = -\begin{bmatrix} 1 & 0 \\ 0 & 1 \end{bmatrix}, \quad B = \begin{bmatrix} 2 & 0 \\ 0 & 2 \end{bmatrix}, \quad C = -\begin{bmatrix} 0 & 1/2 \\ 1 & 1/2 \end{bmatrix}, \quad D = \begin{bmatrix} 1 & 1 \\ 1 & 1 \end{bmatrix} \quad (5.11)$$

is minimal (see P5.5). Yet, it is associated with the transfer-function

$$G(s) = \begin{bmatrix} 1 & \dfrac{s}{s+1} \\ \dfrac{s-1}{s+1} & \dfrac{s}{s+1} \end{bmatrix},$$

which appears to be of first order. Also $s = 0$ is a zero for this transfer-function but $s = 1$ is not.

In many cases, lack of controllability and observability is indicative of structural issues that may have deeper roots. For instance, consider two masses traveling in one dimension:

$$m_1 \dot{v}_1 = u_1, \qquad m_2 \dot{v}_2 = u_2,$$

where v_1 and v_2 are their velocities. If they are to be controlled using a force f which is *internal*, that is,

$$u_1 = f, \qquad u_2 = -f,$$

then it is easy to verify that

$$\dot{z} = 0, \qquad z = m_1 v_1 + m_2 v_2,$$

no matter what f is. That is, the quantity z, which is the sum of the particles' linear momentum, is conserved. This is of course a restatement of Newton's third law that internal forces cannot change a system's total linear momentum. No control action through internal forces can change that!

At a more philosophical level, one might argue that controllability and observability are not to be verified for any physical system because, at some level, there will always be variables that cannot be controlled or that cannot be estimated without explicit measurement. Can we control or observe the exact position of atoms in any object subject to automatic control? Controllability and observability, and hence minimality, are properties of one's mathematical model, never of the underlying physical system.

5.4 Nonlinear Systems and Linearization

Nonlinear dynamic systems can also be represented in state-space form:

$$\dot{x}(t) = f(x(t), u(t)),$$
$$y(t) = g(x(t), u(t)). \tag{5.12}$$

When f and g are continuous and differentiable functions and the state vector, x, and the input, u, are in the neighborhood of a point $(\bar{x}, \bar{u})$ or trajectory $(\bar{x}(t), \bar{u}(t))$, it is natural to expect that the behavior of the nonlinear system can be approximated by that of a properly defined linear system. The procedure used to compute such a linear system is known as *linearization*, and the resulting system is a *linearized* approximation.

Assume that $(\bar{x}, \bar{u})$ is a point at which both f and g are continuous and differentiable. Expanding f and g in Taylor series around $(\bar{x}, \bar{u})$ we obtain

$$f(x, u) \approx f(\bar{x}, \bar{u}) + A(x - \bar{x}) + B(u - \bar{u}),$$
$$g(x, u) \approx g(\bar{x}, \bar{u}) + C(x - \bar{x}) + D(u - \bar{u}),$$

where

$$A = \left.\frac{\partial f}{\partial x}\right|_{\substack{x=\bar{x}\\u=\bar{u}}}, \quad B = \left.\frac{\partial f}{\partial u}\right|_{\substack{x=\bar{x}\\u=\bar{u}}}, \quad C = \left.\frac{\partial g}{\partial x}\right|_{\substack{x=\bar{x}\\u=\bar{u}}}, \quad D = \left.\frac{\partial g}{\partial u}\right|_{\substack{x=\bar{x}\\u=\bar{u}}}. \tag{5.13}$$

For reasons which will become clear soon, one is often interested in special points $(\bar{x}, \bar{u})$ satisfying the nonlinear equation

$$f(\bar{x}, \bar{u}) = 0. \tag{5.14}$$

Such points are called *equilibrium points*. Indeed, at $(\bar{x}, \bar{u})$ we have $\dot{x} = f(\bar{x}, \bar{u}) = 0$; hence, *in the absence of perturbations*, trajectories of a dynamic system starting at an equilibrium point $(\bar{x}, \bar{u})$ will simply stay at $(\bar{x}, \bar{u})$. Around an equilibrium point we define deviations

$$\tilde{x}(t) = x(t) - \bar{x}, \quad \tilde{u}(t) = u(t) - \bar{u}, \quad \tilde{y}(t) = y(t) - g(\bar{x}, \bar{u}), \tag{5.15}$$

to obtain the *linearized system*

$$\dot{\tilde{x}}(t) = A\tilde{x}(t) + B\tilde{u}(t),$$
$$\tilde{y}(t) = C\tilde{x}(t) + D\tilde{u}(t), \tag{5.16}$$

in the standard state-space form (5.3). The next lemma, due to Lyapunov and presented without a proof,[9] links asymptotic stability of the linearized system with asymptotic stability of the nonlinear system around the equilibrium point.

---

[9] See for instance [Kha96, Theorem 3.7].

LEMMA 5.1 (Lyapunov) *Consider the nonlinear dynamic system in state-space form defined in* (5.12). *Let* $(\bar{x}, \bar{u})$ *be an equilibrium point satisfying* (5.14) *and consider the linearized system* (5.16) *for which the quadruple* (A, B, C, D) *is given in* (5.13).

If A is Hurwitz then there is $\epsilon > 0$ *for which any trajectory with initial condition in* $\|x(0) - \bar{x}\| < \epsilon$ *and input* $u(t) = \bar{u}$, $t \geq 0$, *converges asymptotically to the equilibrium point* $(\bar{x}, \bar{u})$, *that is,* $\lim_{t \to \infty} x(t) \to \bar{x}$.

On the other hand, if A has at least one eigenvalue with positive real part then for any $\epsilon > 0$ *there exists at least one trajectory with initial condition in* $\|x(0) - \bar{x}\| < \epsilon$ *and input* $u(t) = \bar{u}$, $t \geq 0$, *that diverges from the equilibrium point* $(\bar{x}, \bar{u})$, *that is, there exists* $T > 0$ *for which* $\|x(t) - \bar{x}\| > 0$ *for all* $t > T$.

This lemma is of major significance for control systems. The first statement means that it suffices to check whether a linearized version of a nonlinear system around an equilibrium point is asymptotically stable in order to ensure convergence to that equilibrium point. The lemma's main weakness is that it does not tell us anything about the size of the neighborhood of the equilibrium point in which convergence to equilibrium happens, that is, the size of ϵ. The second statement says that instability of the linearized system implies instability of the original nonlinear system. Lemma 5.1 is inconclusive when A has eigenvalues on the imaginary axis.

For some systems, linearizing around an equilibrium point might be too restrictive. Take, for example, an airplane or spacecraft in orbit which cannot be in equilibrium with zero velocities. Another example is a bicycle. In such cases it is useful to linearize around a time-dependent *equilibrium trajectory* $(\bar{x}(t), \bar{u}(t))$ satisfying

$$\dot{\bar{x}}(t) = f(\bar{x}(t), \bar{u}(t)).$$

As before, we can define deviations from the equilibrium trajectory

$$\tilde{x}(t) = x(t) - \bar{x}(t), \qquad \tilde{u}(t) = u(t) - \bar{u}(t), \qquad \tilde{y}(t) = y(t) - g(\bar{x}(t), \bar{u}(t)),$$

and the *time-varying* linearized system

$$\begin{aligned}\dot{\tilde{x}}(t) &= A(t)\tilde{x}(t) + B(t)\tilde{u}(t), \\ \tilde{y}(t) &= C(t)\tilde{x}(t) + D(t)\tilde{u}(t),\end{aligned} \qquad (5.17)$$

where $(A(t), B(t), C(t), D(t))$ are computed as in (5.13). Note the potential dependence on time due to the evaluation at $(\bar{x}(t), \bar{u}(t))$. Stability of time-varying linear systems is a much more complicated subject, which is beyond the scope of this book. See the excellent [Bro15] and the brief discussion in Section 3.6.

A familiar example in which equilibrium trajectories, as opposed to equilibrium points, arise naturally is a simple mass subject to Newton's *second* law, $m\ddot{y} = u$, where u is an external force, which we write in state-space form:

$$\dot{x} = f(x, u), \qquad x = \begin{pmatrix} x_1 \\ x_2 \end{pmatrix} = \begin{pmatrix} \dot{y} \\ y \end{pmatrix}, \qquad f(x, u) = \begin{pmatrix} u/m \\ x_1 \end{pmatrix}.$$

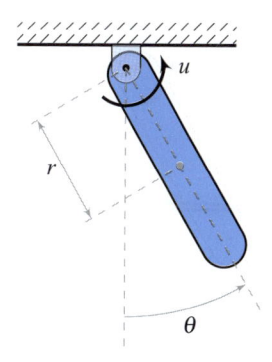

Figure 5.11 Simple pendulum.

Equilibrium trajectories in which the external force is zero, $\bar{u} = 0$, satisfy

$$\dot{\bar{x}}(t) = f(\bar{x}(t), 0) \quad \Longrightarrow$$
$$\dot{\bar{x}}_1 = 0, \quad \dot{\bar{x}}_2 = \bar{x}_1(t) \quad \Longrightarrow$$
$$\bar{x}_1(t) = v, \quad \bar{x}_2(t) = \bar{x}(0) + vt, \quad (5.18)$$

which is the familiar statement that a mass in equilibrium will be at rest or at constant velocity, that is, Newton's *first* law. See Section 5.7 for another example.

Beware that state-space realizations obtained via linearization might fail to be controllable and observable (see P5.7 for an example) and that, in some cases, one may be able to use more sophisticated tools to directly control or estimate the state of the nonlinear model.

In the next sections we will illustrate how to obtain linearized systems from nonlinear models through a series of simple examples.

5.5 Simple Pendulum

Consider the simple planar pendulum depicted in Fig. 5.11. The equation of motion of the simple pendulum obtained from Newton's law in terms of the pendulum's angle θ is the second-order nonlinear differential equation

$$J_r \ddot{\theta} + b\dot{\theta} + mgr \sin \theta = u,$$

where

$$J_r = J + mr^2 > 0, \qquad (5.19)$$

J is the pendulum's moment of inertia about its center of mass, m is the pendulum's mass, b is the (viscous) friction coefficient, and r is the distance to the pendulum's center of mass. For example, if the pendulum is a uniform cylindrical rod of length ℓ

State-Space Models and Linearization

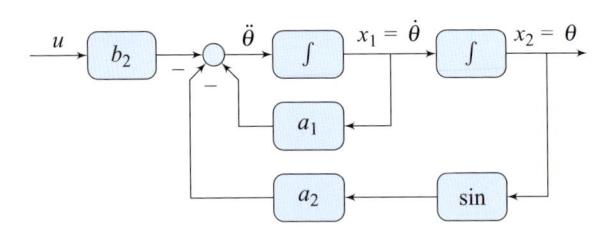

Figure 5.12 Block-diagram for a simple pendulum.

then $r = \ell/2$ and the moment of inertia about its center of mass[10] is $J = m\ell^2/12$. The input, u, is a torque, which is applied by a motor mounted on the axis of the pendulum. We assume that the motor is attached in such a way that the pendulum can rotate freely and possibly complete multiple turns around its axis. The motor will not be modeled.[11]

We isolate the highest derivative to write

$$\ddot{\theta} = b_2 u - a_1 \dot{\theta} - a_2 \sin\theta, \quad a_1 = \frac{b}{J_r} \geq 0, \quad a_2 = \frac{mgr}{J_r} > 0, \quad b_2 = \frac{1}{J_r} > 0.$$

This differential equation is represented as a block-diagram in Fig. 5.12, which is a version of Fig. 5.2 modified to accommodate the nonlinear block with the sine function. Following Section 5.2, we represent the pendulum's nonlinear equation of motion in the state-space form (5.12) after defining

$$x = \begin{pmatrix} x_1 \\ x_2 \end{pmatrix} = \begin{pmatrix} \dot{\theta} \\ \theta \end{pmatrix}, \quad f(x, u) = \begin{pmatrix} b_2 u - a_1 x_1 - a_2 \sin x_2 \\ x_1 \end{pmatrix}, \quad g(x, u) = x_2.$$

We set $\bar{u} = 0$ and look for equilibrium points by solving the system of equations

$$f(\bar{x}, \bar{u}) = \begin{pmatrix} -a_1 \bar{x}_1 - a_2 \sin \bar{x}_2 \\ \bar{x}_1 \end{pmatrix} = \begin{pmatrix} 0 \\ 0 \end{pmatrix}.$$

Solutions must satisfy $\bar{x}_1 = \sin \bar{x}_2 = 0$ or, in other words,

$$\bar{x}_1 = 0 \quad \text{and} \quad \bar{x}_2 = k\pi, \quad k \in \mathbb{Z}.$$

We compute the derivatives in (5.13):

$$\frac{\partial f}{\partial x} = \begin{bmatrix} -a_1 & -a_2 \cos x_2 \\ 1 & 0 \end{bmatrix}, \quad \frac{\partial f}{\partial u} = \begin{bmatrix} b_2 \\ 0 \end{bmatrix}, \quad \frac{\partial g}{\partial x} = \begin{bmatrix} 0 & 1 \end{bmatrix}, \quad \frac{\partial g}{\partial u} = \begin{bmatrix} 0 \end{bmatrix},$$

which we first evaluate at

$$(\bar{x}, \bar{u}) = \left(\begin{pmatrix} 0 \\ 0 \end{pmatrix}, 0 \right)$$

---

[10] In this case $J_r = J + mr^2 = m\ell^2/12 + m\ell^2/4 = m\ell^2/3$, which is the rod's moment of inertia about one of its ends.
[11] See P2.41 and P3.95 for a model of a DC motor.

to obtain the linearized system (5.16) with matrices

$$A_0 = \begin{bmatrix} -a_1 & -a_2 \\ 1 & 0 \end{bmatrix}, \qquad B_0 = \begin{bmatrix} b_2 \\ 0 \end{bmatrix}, \qquad C_0 = \begin{bmatrix} 0 & 1 \end{bmatrix}, \qquad D_0 = \begin{bmatrix} 0 \end{bmatrix}.$$

When $a_1 > 0$ and $a_2 > 0$ the matrix A is Hurwitz.[12] The transfer-function associated with the linearized system is

$$G_0(s) = C_0(sI - A_0)^{-1} B_0 + D_0$$

$$= \frac{1}{s^2 + a_1 s + a_2} \begin{bmatrix} 0 & 1 \end{bmatrix} \begin{bmatrix} s & -a_2 \\ 1 & s + a_1 \end{bmatrix} \begin{bmatrix} b_2 \\ 0 \end{bmatrix} = \frac{b_2}{s^2 + a_1 s + a_2},$$

where (5.6) was used in place of $(sI - A_0)^{-1}$. Because A_0 is Hurwitz, $G_0(s)$ is asymptotically stable and Lemma 5.1 guarantees that trajectories starting close enough to $\theta = 0$ and with small enough velocity will converge to 0, as would a physical pendulum. We expect that a physical pendulum will eventually converge to $\theta = 2k\pi$, $k \in \mathbb{Z}$, that is an integer multiple of 2π, no matter what the initial conditions are, since the energy available due to the initial conditions will eventually dissipate due to friction. This is also true in the simple pendulum model when $b > 0$. The idea of using energy or a related positive measure of the state of a system in order to assess stability is the main idea behind Lyapunov functions, which are widely used with linear and nonlinear systems. Lyapunov stability is discussed in detail in standard nonlinear systems and control references, e.g. [Kha96].

Linearization around the next equilibrium point

$$(\bar{x}, \bar{u}) = \left(\begin{pmatrix} 0 \\ \pi \end{pmatrix}, 0 \right)$$

produces the linearized system (5.16) with matrices

$$A_\pi = \begin{bmatrix} -a_1 & a_2 \\ 1 & 0 \end{bmatrix}, \qquad B_\pi = \begin{bmatrix} b_2 \\ 0 \end{bmatrix}, \qquad C_\pi = \begin{bmatrix} 0 & 1 \end{bmatrix}, \qquad D_\pi = \begin{bmatrix} 0 \end{bmatrix}.$$

This time, however, matrix A is never Hurwitz.[13] The transfer-function associated with the linearized system is

$$G_\pi(s) = C_\pi(sI - A_\pi)^{-1} B_\pi + D_\pi = \frac{b_2}{s^2 + a_1 s - a_2},$$

which is not asymptotically stable. According to Lemma 5.1, trajectories starting close enough to $\theta = \pi$ will diverge from π, again as we would expect from a physical pendulum. All other equilibria in which $\bar{x}_2$ is an integer multiple of π will lead to one of the above linearized systems.

---

[12] The eigenvalues of A are the roots of the characteristic equation $0 = |sI - A_0| = s^2 + a_1 s + a_2$, which is similar to Equation (4.12) studied in Section 4.2. See also Section 6.1.
[13] The characteristic equation $0 = |sI - A_\pi| = s^2 + a_1 s - a_2$ will always have a root with positive real part. See Section 6.1.

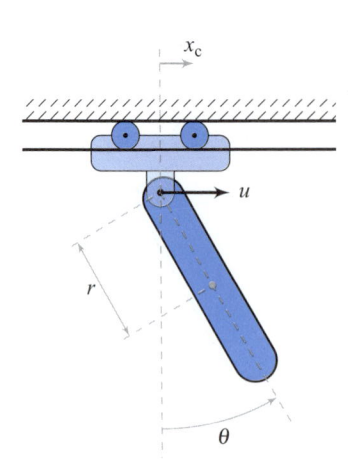

(a) Pendulum suspended in a rail

(b) Inverted pendulum in a cart

Figure 5.13 Pendula in carts.

5.6 Pendulum in a Cart

We now complicate the pendulum by attaching it to a cart that can move only in the x_c direction, as shown in Fig. 5.13(a). In Fig. 5.13(a) the cart is on a rail, which would be the case, for example, in a crane. A similar model can be used to describe the *inverted* pendulum in a cart shown in Fig. 5.13(b). Without delving into the details of the derivation, the equations of motion for the pendulum are the following pair of coupled nonlinear second-order differential equations:

$$(J_p + m_p r^2)\ddot{\theta} + m_p r \ddot{x}_c \cos\theta + b_p \dot{\theta} + m_p g r \sin\theta = 0,$$
$$m_p r \ddot{\theta} \cos\theta + (m_p + m_c)\ddot{x}_c + b_c \dot{x}_c - m_p r \dot{\theta}^2 \sin\theta = u, \tag{5.20}$$

where θ is the pendulum's angle and x_c is the cart's position, shown in the diagram in Fig. 5.13(a). The positive constants J_p, m_p, and b_p are the pendulum's moment of inertia, mass, and viscous friction coefficient, r is the distance to the pendulum's center of mass, and m_c and b_c are the cart's mass and viscous friction coefficient. An important difference is that the input, u, is a force applied to the cart, as opposed to a torque applied to the pendulum. The goal is to equilibrate the pendulum by moving the cart, as done in the Segway® Personal Transporter shown in Fig. 5.13(b).

When $\ddot{x}_c = 0$, the first equation in (5.20) reduces to the equation of motion of the simple pendulum developed in Section 5.5 but with zero input torque. Note that the term $-m_p r \ddot{x}_c \cos\theta$ is the torque applied to the pendulum by virtue of accelerating the cart in the x_c direction. As expected, a positive acceleration produces a negative torque. Check that in Fig. 5.13(a)!

As an intermediate step toward a state-space model, these equations can be cast into the second-order *vector* differential equation

$$M(q)\ddot{q} + F(q, \dot{q}) = Gu, \qquad q = \begin{pmatrix} \theta \\ x_c \end{pmatrix},$$

where q is the *configuration* vector and

$$M(q) = \begin{bmatrix} J_p + m_p r^2 & m_p r \cos q_1 \\ m_p r \cos q_1 & m_p + m_c \end{bmatrix},$$

$$F(q, \dot{q}) = \begin{pmatrix} b_p \dot{q}_1 + m_p gr \sin q_1 \\ b_c \dot{q}_2 - m_p r \dot{q}_1^2 \sin q_1 \end{pmatrix},$$

$$G = \begin{pmatrix} 0 \\ 1 \end{pmatrix}.$$

The matrix $M(q)$ is known as the *mass matrix*. A vector second-order system is the standard form for models of mechanical and electric systems derived from physical principles, such as the present model of the pendulum in a cart. For instance, a finite-element modeler of mechanical systems will produce models that conform to the above vector second-order form.

One can go from vector second-order to state-space form by defining

$$x = \begin{pmatrix} q \\ \dot{q} \end{pmatrix}, \quad f(x, u) = f(q, \dot{q}, u) = \begin{pmatrix} \dot{q} \\ M^{-1}(q)[Gu - F(q, \dot{q})] \end{pmatrix}.$$

When moving from vector second-order to state-space form one needs to *invert* the mass matrix, which might not always be a trivial task. Setting $\bar{u} = 0$ we calculate the equilibrium points:

$$\bar{x}_1 = \bar{q}_1 = k\pi, \quad k \in \mathbb{Z}, \quad \bar{x}_3 = \bar{x}_4 = \dot{\bar{q}}_1 = \dot{\bar{q}}_2 = 0.$$

Note that the constant $x_2 = q_2 = x_c$ is arbitrary, which means that equilibrium of the pendulum does not depend on a particular value of x_c. In the case of the inverted pendulum, this is due to the fact that the coordinate $q_2 = x_c$ does not appear directly in the equations of motion. This is analogous to what happens in the car model (2.1), and indicates the existence of equilibrium trajectories such as (5.18), in which the velocity $\dot{q}_2 = \dot{x}_c$ is constant as opposed to zero. Indeed, for the pendulum in a cart, $M(q) = M(q_1) = M(\theta)$ and $F(q, \dot{q}) = F(q_1, \dot{q}) = F(\theta, \dot{\theta}, \dot{x}_c)$ so that a reduced state-space model is possible, with

$$x = \begin{pmatrix} \theta \\ \dot{\theta} \\ \dot{x}_c \end{pmatrix}, \quad f(x, u) = \begin{pmatrix} \dot{x}_1 \\ M^{-1}(x_1)[Gu - F(x)] \end{pmatrix}. \quad (5.21)$$

The equilibrium points for (5.21) are

$$\bar{x}_1 = \bar{q}_1 = k\pi, \quad k \in \mathbb{Z}, \quad \bar{x}_2 = \bar{x}_3 = \dot{\bar{q}}_1 = \dot{\bar{q}}_2 = 0.$$

Linearized matrices (5.13) can be computed from (5.21) when $\bar{x}_1 = 0$:

$$A_0 = \frac{1}{J} \begin{bmatrix} 0 & J & 0 \\ -gm_r r & -b_p m_r/m_p & b_c r \\ gm_p r^2 & b_p r & -b_c J_r/m_p \end{bmatrix}, \quad B_0 = \frac{1}{J} \begin{bmatrix} 0 \\ -r \\ J_r/m_p \end{bmatrix},$$

or when $\bar{x}_1 = \pi$:

$$A_\pi = \frac{1}{J}\begin{bmatrix} 0 & J & 0 \\ gm_r r & -b_p m_r/m_p & -b_c r \\ gm_p r^2 & -b_p r & -b_c J_r/m_p \end{bmatrix}, \qquad B_\pi = \frac{1}{J}\begin{bmatrix} 0 \\ r \\ J_r/m_p \end{bmatrix},$$

where we defined and used the positive quantities

$$J_r = J_p + m_p r^2, \qquad m_r = m_p + m_t, \qquad J = J_r \frac{m_r}{m_p} - m_p r^2,$$

to simplify the entries in the matrices. The linearized matrices are very similar except for a couple of sign changes. However, these small changes are fundamental for understanding the behavior of the system around each equilibrium.

For example, consider that damping is zero, $b_c = b_t = 0$, which is simpler to analyze. In this case the eigenvalues of A_0 are

$$0, \qquad j\sqrt{g(m_t + m_t)r}, \qquad -j\sqrt{g(m_t + m_t)r},$$

where the imaginary eigenvalues are indicative of an oscillatory system. Indeed, when the damping coefficients b_c and b_p are positive, all eigenvalues of A_0 have negative real part (you should verify this) and, from Lemma 5.1, the equilibrium point $\theta = 0$ is asymptotically stable.

If $b_c = b_t = 0$ then the eigenvalues of A_π are real,

$$0, \qquad \sqrt{g(m_t + m_t)r}, \qquad -\sqrt{g(m_t + m_t)r},$$

and one of them will always have positive real part. When the damping coefficients b_c and b_p are positive, two of the eigenvalues of A_π have negative real part but one remains on the right-hand side of the complex plane. From Lemma 5.1, the equilibrium point $\theta = \pi$ is unstable.

5.7 Car Steering

Our third example is that of a simplified four-wheel vehicle traveling as depicted in Fig. 5.14. Without any slip, the wheels of the car remain tangent to circles centered at the virtual point c, as shown in the figure. A real car uses a more complicated steering mechanism to traverse the same geometry shown in Fig. 5.14: only the front wheels turn, as opposed to the front axle, and real tires allow some *slip* to occur. The front axle steering angle, ψ, is related to the radius of the circle that goes by the mid-point of the rear axle by the formula

$$\rho = \frac{\ell}{\tan \psi}.$$

We assume that the steering angle ψ is in the interval $(-\pi/2, \pi/2)$. If v is the rear axle's mid-point tangential velocity, then v is related to the car's angular velocity, $\dot{\theta}$, by

$$v = \rho \dot{\theta}.$$

5.7 Car Steering

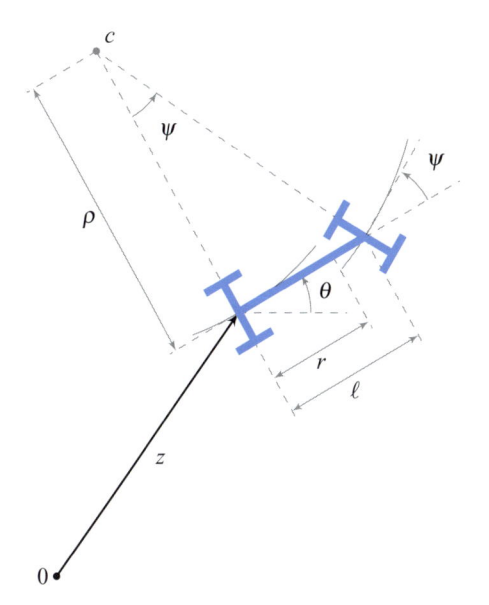

Figure 5.14 Schematic of a simplified car steering in the plane; the car is turning around a virtual circle of radius $\rho = \ell/\tan\psi$ centered at c.

If $z = (z_x, z_y)$ is the position of the mid-point of the rear axle then the velocity vector is

$$\dot{z} = (\dot{z}_x, \dot{z}_y), \qquad \dot{z}_x = v\cos\theta, \qquad \dot{z}_y = v\sin\theta.$$

We can put these equations together in state-space form (5.12) with

$$x = \begin{pmatrix} z_x \\ z_y \\ \theta \end{pmatrix}, \qquad u = \tan\psi, \qquad f(x, u) = \begin{pmatrix} v\cos\theta \\ v\sin\theta \\ uv/\ell \end{pmatrix}.$$

Note that we chose as input $u = \tan\psi$ rather than $u = \psi$. Therefore, by inverting u, we obtain $\psi = \tan^{-1} u \in (-\pi/2, \pi/2)$, which automatically enforces that ψ is always within its permissible range.

When $v \neq 0$, the car does not have any equilibrium points because there exists no $\bar{\theta}$ such that $\cos\bar{\theta} = \sin\bar{\theta} = 0$. For this reason we linearize around a moving trajectory. For example, a straight horizontal line

$$\bar{x}(t) = \begin{pmatrix} z_x(t) \\ z_y(t) \\ \theta(t) \end{pmatrix} = \begin{pmatrix} vt \\ 0 \\ 0 \end{pmatrix}, \qquad \bar{u}(t) = 0, \tag{5.22}$$

satisfies $\dot{\bar{x}}(t) = f(\bar{x}(t), \bar{u}(t))$ and hence is an equilibrium trajectory. We compute derivatives to obtain

$$\frac{\partial f}{\partial x} = \begin{bmatrix} 0 & 0 & -v\sin\theta \\ 0 & 0 & v\cos\theta \\ 0 & 0 & 0 \end{bmatrix}, \qquad \frac{\partial f}{\partial u} = \begin{bmatrix} 0 \\ 0 \\ v/\ell \end{bmatrix},$$

which are evaluated to compute the matrices of the linearized system (5.17):

$$A(t) = A = \begin{bmatrix} 0 & 0 & 0 \\ 0 & 0 & v \\ 0 & 0 & 0 \end{bmatrix}, \qquad B(t) = B = \begin{bmatrix} 0 \\ 0 \\ v/\ell \end{bmatrix}. \qquad (5.23)$$

In this case, the matrices $A(t)$ and $B(t)$ happen not to depend on the time t, and hence the linearized system is in fact time-invariant. See the footnote[14] for an example of a trajectory that leads to a *time-varying* linearized system. The model discussed in this section does not take into account the inertial properties of the car and, for this reason, it is known as a *kinematic* model. A full dynamic nonlinear model is obtained with the addition of the equation

$$\dot{v} = \frac{\ell^2 f - b\ell^2 v - J_r v u \dot{u}}{m\ell^2 + J_r u^2}, \qquad J_r = J + mr^2$$

where m and J are the car's mass and moment of inertia, r is the distance measured along the car's main axis from the mid-point of the rear axle to the car's center of mass, f is a tangential force applied at the rear axle, and b is a (viscous) damping coefficient. Note that when u is small or J_r is small then

$$\dot{v} \approx \frac{1}{m} f - \frac{b}{m} v,$$

which is the same equation as (2.1) used before to model a car moving in a straight line.

5.8 Linear Control of Nonlinear Systems

Suppose that one designs a linear controller based on a model, $\tilde{G}$, linearized around the equilibrium point, $(\bar{x}, \bar{u})$, of a certain nonlinear system, G. Two questions need to be

---

[14] Consider a trajectory in which $\bar{u}(t) = \bar{u} \neq 0$. Such an equilibrium trajectory satisfies

$$\dot{\bar{x}}(t) = \begin{pmatrix} \dot{\bar{z}}_x(t) \\ \dot{\bar{z}}_y(t) \\ \dot{\bar{\theta}}(t) \end{pmatrix} = \begin{pmatrix} v \cos \bar{\theta}(t) \\ v \sin \bar{\theta}(t) \\ \bar{u} v/\ell \end{pmatrix} = f(\bar{x}(t), \bar{u}) \implies \bar{\theta}(t) = \bar{\theta}(0) + \omega t, \quad \omega = \bar{u} v/\ell.$$

With $\bar{\theta}(0) = 0$,

$$\bar{x}(t) = \begin{pmatrix} \bar{z}_x(t) \\ \bar{z}_y(t) \\ \bar{\theta}(t) \end{pmatrix} = \begin{pmatrix} \bar{z}_x(0) + (v/\omega) \sin(\omega t) \\ \bar{z}_y(0) - (v/\omega) \cos(\omega t) \\ \omega t \end{pmatrix}, \qquad \bar{u}(t) = \bar{u},$$

which are circles centered at $(\bar{z}_x(0), \bar{z}_y(0))$ with radius $\rho = v/\omega = \ell/\bar{u}$. This time, however,

$$A(t) = \begin{bmatrix} 0 & 0 & -v \sin(\omega t) \\ 0 & 0 & v \cos(\omega t) \\ 0 & 0 & 0 \end{bmatrix}, \qquad B(t) = B = \begin{bmatrix} 0 \\ 0 \\ v/\ell \end{bmatrix},$$

and the linearized system is therefore time-varying.

5.8 Linear Control of Nonlinear Systems

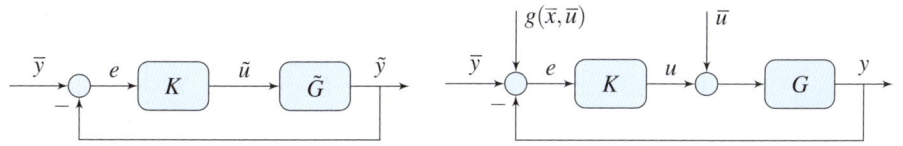

(a) Design of linear controller, K, based on the linearized model, $\tilde{G}$.

(b) Controlling the nonlinear model, G, with the linear controller, K.

Figure 5.15 Closed-loop feedback configuration for controlling a nonlinear system, G, with reference, $\bar{y}$, reference offset, $g(\bar{x}, \bar{u})$, and input offset, $\bar{u}$; $\tilde{G}$ is a linearized model obtained from the nonlinear model G at $(\bar{x}, \bar{u})$.

answered. (a) What is the exact form of the controller? (b) Under what conditions will this controller stabilize the original nonlinear system?

To answer the first question, observe that the linearized system (5.16) is developed in terms of deviations, $\tilde{x}$, $\tilde{u}$, and $\tilde{y}$ from (5.15), and that a linear feedback controller, K, designed based on the linearized model, $\tilde{G}$, as shown in the diagram in Fig. 5.15(a) correspond to

$$\tilde{u} = K(\bar{y} - \tilde{y}), \qquad \tilde{u} = u - \bar{u}, \qquad \tilde{y} = y - g(\bar{x}, \bar{u}).$$

For this reason the actual control that needs to be applied to the original nonlinear system, G, is therefore

$$u = \bar{u} + K(\bar{y} + g(\bar{x}, \bar{u}) - y). \tag{5.24}$$

The resulting closed-loop diagram is shown in Fig. 5.15(b). In practice, whenever possible, most control systems are linearized about a zero input $\bar{u}$ (can you explain why?), which further simplifies the diagram. Moreover, if integral control is used, one can often dispense with the input $\bar{u}$ if it is constant. Compare the block-diagram in Fig. 5.15(b) with the one in Fig. 4.11 and recall the discussion in Section 4.5 on how integral control can reject the constant "disturbance," $\bar{u}$. We will discuss integral control for nonlinear systems at the end of this section.

For a concrete example, consider the implementation of a linear proportional controller for the nonlinear simple pendulum from Section 5.5. We have obtained two linearized models around the equilibrium points $\theta = 0$ and $\theta = \pi$. The transfer-functions associated with these linearized models are

$$G_0(s) = \frac{1}{J_r s^2 + bs + mgr}, \qquad G_\pi(s) = \frac{1}{J_r s^2 + bs - mgr},$$

where J_r is given by (5.19). As discussed in Section 5.5, if all parameters are positive then G_0 is asymptotically stable and G_π is unstable. If a linear proportional controller is to be used to stabilize the pendulum around the unstable equilibrium $\theta = \pi$ then it must, at a minimum, stabilize the linearized model G_π. In other words, the poles of the sensitivity transfer-function

$$S_\pi(s) = \frac{1}{1 + KG_\pi(s)} = \frac{J_r s^2 + bs - mgr}{J_r s^2 + bs + K - mgr}$$

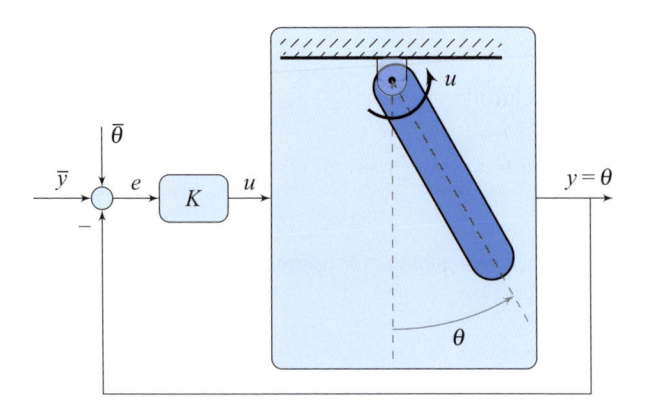

Figure 5.16 Linear control of the simple pendulum using the scheme of Fig. 5.15; $g(\bar{x}, \bar{u}) = \bar{\theta}$, $\bar{u} = 0$; $\bar{\theta} = 0$ around stable equilibrium, and $\bar{\theta} = \pi$ around unstable equilibrium.

must have negative real part. As we will discuss in detail in Sections 6.1 and 6.5, the linearized closed-loop system G_π is asymptotically stable if

$$K > mgr.$$

Following Fig. 5.15, this controller must be implemented as in Fig. 5.16 after setting $g(\bar{x}, \bar{u}) = \bar{\theta}$ and $\bar{u} = 0$.

One must be especially careful when a linear controller is used to move a nonlinear system away from its natural equilibrium point. If the reference $\bar{y}$ is constant, $K = K(s)$ is a dynamic linear controller, and the nonlinear system, G, is described in state-space by Equations (5.12), then the closed-loop equilibrium point must satisfy

$$f(\hat{x}, \hat{u}) = 0, \qquad \hat{u} = \bar{u} + K(0)(\bar{y} + g(\bar{x}, \bar{u}) - \hat{y}), \qquad \hat{y} = g(\hat{x}, \hat{u}), \quad (5.25)$$

where $K(0)$ is the steady-state gain of the controller.[15] The closed-loop equilibrium point, $\hat{x}$, is often not the same as the open-loop equilibrium point, $\bar{x}$.

When the reference input is zero, $\bar{y} = 0$, the controller is a regulator (see Section 4.5). From Lemma 5.1, we know that a regulator designed to stabilize a model linearized at $(\bar{x}, \bar{u})$ also stabilizes the original nonlinear system in the neighborhood of the equilibrium point $(\bar{x}, \bar{u})$. If the initial conditions of the closed-loop system are close enough to $\bar{x}$, then we should expect that x converges to $\bar{x}$, u converges to $\bar{u}$, and y converges to $y = g(\bar{x}, \bar{u})$, so that $\hat{x} = \bar{x}$ and $\hat{u} = \bar{u}$. This is a partial answer to question (b).

An interesting case is that of a controller which is designed to satisfy

$$K(0) = 0, \qquad (5.26)$$

which means that the controller has a zero at the origin. This effect is often obtained with the addition of a high-pass or *washout filter* of the form

$$\frac{s}{s+a}, \quad a > 0,$$

to the controller. Washout filters are used in applications where the controller should have no authority over the system's steady-state response but needs to act

---

[15] This is the case even if $K(s)$ is not asymptotically stable, as long as the closed-loop is internally stable as in Section 4.7. A formal analysis is given in Section 8.2.

during transients. A typical application is in the control of electric power systems, in which the steady-state response is dictated by the loads of the circuit and should not be affected by the controller [PSL96]. Another application is in flight control. See [FPE14, Section 10.3] for a complete design of a yaw damper for a Boeing 747 aircraft that uses a washout filter to preserve pilot authority.

When the controller is not a regulator, that is $\bar{y} \neq 0$, the closed-loop controller will typically modify the closed-loop equilibrium point. Tracking is all about modifying the natural (open-loop) equilibrium of systems. In light of Lemma 5.1, a necessary condition for stability of the closed-loop system when $\bar{y}$ is a constant is that the controller K stabilizes the closed-loop system linearized at the new equilibrium point $(\hat{x}, \hat{u})$, given in (5.25). If $\bar{y}$ is close to $g(\bar{x}, \bar{u})$ and the nonlinearities in f and g are *mild*, then one can *hope* that stabilization of the system linearized at $(\bar{x}, \bar{u})$ might also imply stabilization of the system linearized at $(\hat{x}, \hat{u})$, but there are no guarantees that can be offered in all cases. As with linear systems, there will likely be a nonzero steady-state error, i.e. $\bar{y} \neq \hat{y}$. In many cases, one can enforce a zero closed-loop steady-state error using integral control. With an integrator in the loop, one often dispenses with the constant inputs $g(\bar{x}, \bar{u})$ and $\bar{u}$ in the closed-loop diagram of Fig. 5.15(b), which reduces the controller to the standard block-diagram in Fig. 1.8:

$$u = K(\bar{y} - y).$$

An informal[16,17] argument to support integral control in nonlinear systems is as follows: when $\bar{y}$ is constant and K contains an integrator, $K(0) \to \infty$ and therefore one must have $\bar{y} \to y$ if the closed-loop system is to reach equilibrium. If the integrator is the first

---

[16] One problem with this argument is that there might not exist $(\hat{x}, \hat{u})$ such that

$$f(\hat{x}, \hat{u}) = 0, \qquad \bar{y} = g(\hat{x}, \hat{u}).$$

For example, consider the open-loop stable first-order linear system with an input saturation

$$\dot{x} = f(x, u) = -x + \tan^{-1}(u), \qquad y = g(x, u) = x.$$

In equilibrium $g(\bar{x}, \bar{u}) = \bar{x} = \tan^{-1}(\bar{u}) \in (-\pi/2, \pi/2)$, hence no controller can drive the system to produce a constant output that is equal to a reference $\bar{y}$ such that $|\bar{y}| > \pi/2$.

[17] A more subtle obstacle is that, even when there exists an equilibrium, it might not be reachable. Consider for example an integral controller in feedback with the nonlinear system

$$\dot{y} = \frac{1}{y-1} + u, \qquad \dot{u} = \bar{y} - y, \qquad \bar{y} = 2.$$

In state-space,

$$\dot{x} = f(x), \qquad x = \begin{pmatrix} y \\ u \end{pmatrix}, \qquad f(x) = \begin{pmatrix} 1/(x_1 - 1) + x_2 \\ 2 - x_1 \end{pmatrix}, \qquad g(x) = x_1,$$

which has an equilibrium point at $(\bar{x}_1, \bar{x}_2) = (2, -1)$. Because

$$A = \frac{\partial f}{\partial x}\bigg|_{\substack{x_1=2 \\ x_2=-1}} = \begin{bmatrix} -1/(x_1-1)^2 & 1 \\ -1 & 0 \end{bmatrix}_{\substack{x_1=2 \\ x_2=-1}} = \begin{bmatrix} -1 & 1 \\ -1 & 0 \end{bmatrix}$$

is Hurwitz this equilibrium point is asymptotically stable. For instance, with initial conditions $(x_1(0), x_2(0)) = (3, 0)$ the closed-loop system converges to $\bar{y} = \bar{x}_1 = 2$ as t grows. However, with initial condition $(x_1(0), x_2(0)) = (0, 0)$, the system never reaches this equilibrium. As $x_1(t)$ tries to grow continuously toward 2, the value of $x_2(t)$ grows linearly but the value of $1/(x_1 - 1)$ becomes more and more negative as $x_1(t)$ approaches 1. Indeed, $x_1(t)$ can never exceed one, so it never reaches equilibrium.

element in a controller realized in state-space, then

$$\dot{x}_{k1}(t) = \bar{y} - y(t),$$

where x_{k1} is the first state of the controller. Consequently

$$x_{k1}(t) = x_{k1}(0) + \int_0^t \bar{y} - y(\tau)d\tau.$$

If $\bar{y}$ and $\bar{x}_{k1}$ are constant, then $\hat{y} = g(\hat{x}, \hat{u}) = \bar{y}$. If the equilibrium point, $(\hat{x}, \hat{y})$, is asymptotically stable, then

$$\lim_{t \to \infty} x(t) = \hat{x}, \qquad \lim_{t \to \infty} u(t) = \hat{u}, \qquad \lim_{t \to \infty} y(t) = \hat{y} = \bar{y},$$

given that the initial conditions are close enough to $(\hat{x}, \hat{u})$. As with linear systems, the price to be paid is a more complex dynamic system to contend with. On the other hand, there is solace in the fact that neither $g(\bar{x}, \bar{u})$ nor $\bar{u}$ need be accurately estimated.

While a detailed analysis of nonlinear feedback systems is beyond the scope of this book, some methods to be introduced in Chapter 8 in the context of robustness can be used to rigorously analyze feedback systems with certain types of nonlinearities. For example, in Section 8.4, a rigorous analysis of the effectiveness of integral control will be provided in the case of the simple pendulum which takes into account the pendulum's full nonlinear model.

Whether we call it nonlinearity, uncertainty, or nature, we can never stress too much that predictions based on models, including closed-loop predictions, can be far off if not validated experimentally. One should be suspicious of predicted closed-loop performances that seem to be too good to be true. For example, in Chapters 2 and 4, aggressive controllers produced inputs that *saturated* the available throttle input in the car cruise controller. At a minimum, one should always look at the control signal produced by the controller to perform the desired task. This has been done in Chapters 2 and 4 first using linear models, then simulating the closed-loop system using a more refined model which was nonlinear. Ultimately, one would move on to experimentation to validate the predicted performance in a variety of operating conditions. That humans are prone to take models for reality more than we should is not news. Check out [Tal07] for some provocative discussion.

Problems

5.1 Write state-space equations for the block-diagrams in Fig. 5.17 and in each case compute the transfer-function from u to y.

5.2 Show that the connection of the state-space systems

$$\dot{x}_g = A_g x_g + B_g u_g,$$
$$y_g = C_g x_g + D_g u_g,$$

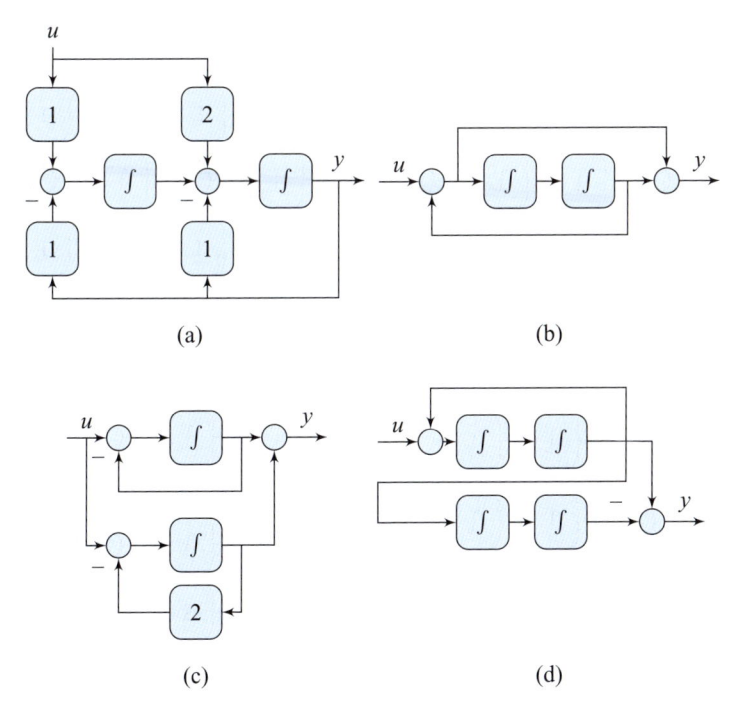

Figure 5.17 Block-diagrams for P5.1.

and

$$\dot{x}_k = A_k x_k + B_k u_k,$$
$$y_k = C_k x_k + D_k u_k$$

in feedback,

$$u_g = y_k, \qquad u_k = \bar{y} - y, \qquad y = y_g,$$

can be represented in state-space form (5.3) by the matrices

$$A = \begin{bmatrix} A_g - B_g(I + D_k D_g)^{-1} D_k C_g & B_g(I + D_k D_g)^{-1} C_k \\ -B_k(I + D_g D_k)^{-1} C_g & A_k - B_k(I + D_g D_k)^{-1} D_g C_k \end{bmatrix},$$

$$B = \begin{bmatrix} B_g(I + D_k D_g)^{-1} D_k \\ B_k(I + D_g D_k)^{-1} \end{bmatrix},$$

$$C = \begin{bmatrix} (I + D_g D_k)^{-1} C_g & D_g(I + D_k D_g)^{-1} C_k \end{bmatrix},$$

$$D = D_g(I + D_k D_g)^{-1} D_k.$$

Draw a block-diagram representing the connection.   *Hint: The letter I represents a square identity matrix, which for SISO systems is equal to the number* 1; *the formulas, as given, work also for MIMO systems.*

5.3 Verify that the state-space equations (5.3) with matrices as in (5.9) are controllable except when $\beta = 0$, i.e. there exist some nonzero vector z and some complex number λ such that

$$A^*z = \lambda z, \qquad B^*z = 0, \qquad z \neq 0, \qquad (5.27)$$

only when $\beta = 0$.

5.4 Verify that the state-space equations (5.3) with matrices as in (5.9) are observable except when $\alpha = 0$, i.e. there exist some nonzero vector x and scalar λ that solve Equations (5.10) only when $\alpha = 0$.

5.5 Show that the state-space equations (5.3) with matrices as in (5.11) are minimal. *Hint: Apply the conditions in P5.3 and (5.10) to show that they are controllable and observable.*

5.6 Compute the linearized equations for the pendulum in a cart model, Equations (5.20), developed in Section 5.6. Let $m_p = 2$ kg, $m_c = 10$ kg, $\ell = 1$ m, $b_p = 0.01$ kg m²/s, $b_c = 0.1$ km/s, $r = \ell/2$, and $J_p = m\ell^2/12$, and use MATLAB to compute the transfer-function from u to θ and from u to $\dot{x}_c$ around the equilibrium points calculated with $\bar{\theta} = 0$ and $\bar{\theta} = \pi$ and $\bar{u} = 0$. Are the equilibria asymptotically stable?

5.7 Verify that the linearized model of a steering car, Equations (5.23), developed in Section 5.7 is not controllable. Interpret this result on the basis of your knowledge of the physical behavior of this system. *Hint: Use P5.3.*

5.8 You have shown in P2.4 that the ordinary differential equation

$$m\dot{v} + bv = mg$$

is a simplified description of the motion of an object of mass m dropping vertically under constant gravitational acceleration, g, and linear air resistance, $-bv$. Let the gravitational force, mg, be the input and let the vertical velocity, v, be the output, and represent this equation in a block-diagram using only integrators. Rewrite the differential equation in state-space form.

5.9 Repeat P5.8 considering the vertical position, $x(t) = x(0) + \int_0^t v(\tau)d\tau$, as the output.

5.10 Repeat P5.8 considering the vertical acceleration, $\dot{v}$, as the output. *Hint: Use the original equation to obtain $\dot{v}$ as a function of v.*

5.11 The ordinary differential equation

$$m\dot{v} + bv|v| = mg$$

is a simplified description of the motion of an object of mass m dropping vertically under constant gravitational acceleration, g, and quadratic air resistance, $-bv|v|$. Let the gravitational force, mg, be the input and let the vertical velocity, v, be the output, and represent this equation in a block-diagram using only integrators. Rewrite the differential equation in state-space form.

5.12 Use the block-diagram obtained in P5.11 and MATLAB to simulate the velocity of an object with $m = 70$ kg, $b = 0.227$ kg/m, and $g = 10$ m/s^2 falling with zero initial velocity. Compare your solution with the one from P2.9.

5.13 Calculate the equilibrium points of the state-space representation obtained in P5.11. Linearize the state-space equations about the equilibrium points and compute the corresponding transfer-functions. Are the equilibrium points asymptotically stable?

5.14 Use P5.11 and P5.12 to redo P2.9.

5.15 You have shown in P2.10 and P2.12 that the ordinary differential equation

$$(J_1 r_2^2 + J_2 r_1^2)\dot{\omega}_1 + (b_1 r_2^2 + b_2 r_1^2)\omega_1 = r_2^2 \tau, \qquad \omega_2 = (r_1/r_2)\omega_1$$

is a simplified description of the motion of a rotating machine driven by a belt without slip as in Fig. 2.18(a), where ω_1 is the angular velocity of the driving shaft and ω_2 is the machine's angular velocity. Let the torque, τ, be the input and the machine's angular velocity, ω_2, be the output and represent this equation in a block-diagram using only integrators. Rewrite the differential equation in state-space form.

5.16 Repeat P5.15 considering the machine's angular acceleration, $\dot{\omega}_2$, as the output.

5.17 Repeat P5.15 considering the machine's angle, $\theta_2(t) = \theta_2(0) + \int_0^t \omega_2(\tau) d\tau$, as the output.

5.18 Use the block-diagram obtained in P5.15 and MATLAB to simulate the rotating machine's angular velocity, ω_2, with $\tau = 1$ N m, $r_1 = 25$ mm, $r_2 = 500$ mm, $b_1 = 0.01$ kg m^2/s, $b_2 = 0.1$ kg m^2/s, $J_1 = 0.0031$ kg m^2, $J_2 = 25$ kg m^2, and zero initial angular velocity. Compare your solution with the one from P2.13.

5.19 You have shown in P2.18 that the ordinary differential equation

$$(J_1 + J_2 + r^2(m_1 + m_2))\dot{\omega} + (b_1 + b_2)\omega = \tau + gr(m_1 - m_2), \qquad v_1 = r\omega$$

is a simplified description of the motion of the elevator in Fig. 2.18(b), where ω is the angular velocity of the driving shaft and v_1 is the elevator's load linear velocity. Let the torque, τ, and the gravitational torque, $gr(m_1 - m_2)$, be inputs and let the elevator's linear velocity, v_1, be the output, and represent this equation in a block-diagram using only integrators. Rewrite the differential equation in state-space form.

5.20 Compute the transfer-function from the two inputs, τ and $gr(m_1 - m_2)$, to the single output, v_1, in P5.19. The transfer-function will be a 1×2 matrix. *Hint: Use state-space and* (5.6).

5.21 Use the block-diagram obtained in P5.19 and MATLAB to simulate the elevator's linear velocity, v_1, with $g = 10$ m/s^2, $\tau = 0$ N m, $r = 1$ m, $m_1 = m_2 = 1000$ kg, $b_1 = b_2 = 120$ kg m^2/s, $J_1 = J_2 = 20$ kg m^2, and initial velocity $v_1(0) = 1$ m/s. Compare your solution with the one from P2.19.

5.22 Repeat P5.21 with $m_2 = 800\,\text{kg}$.

5.23 You have shown in P2.28 that the ordinary differential equation

$$m\ddot{x} + b\dot{x} + kx = mg\sin\theta$$

is a simplified description of the motion of the mass–spring–damper system in Fig. 2.19(b), where g is the gravitational acceleration and x_0 is equal to the spring rest length ℓ_0. Let the gravitational force, $mg\sin\theta$, be the input and the position, x, be the output and represent this equation in a block-diagram using only integrators. Rewrite the differential equation in state-space form.

5.24 Calculate the equilibrium points of the state-space representation obtained in P5.23 when $u = \bar{u} = mg\sin\theta$. Rewrite the state-space equations about the equilibrium points. Compare with P2.29.

5.25 Use the block-diagram obtained in P5.23 and MATLAB to simulate the mass–spring–damper position, x, with $g = 10\,\text{m/s}^2$, $m = 1\,\text{kg}$, $k = 1\,\text{N/m}$, $b = 0.1\,\text{kg/s}$, and initial position $x(0) = 10\,\text{cm}$.

5.26 You have shown in P2.32 that the ordinary differential equations

$$m_1\ddot{x}_1 + (b_1 + b_2)\dot{x}_1 + (k_1 + k_2)x_1 - b_2\dot{x}_2 - k_2 x_2 = 0,$$
$$m_2\ddot{x}_2 + b_2(\dot{x}_2 - \dot{x}_1) + k_2(x_2 - x_1) = f_2$$

constitute a simplified description of the motion of the mass–spring–damper system in Fig. 2.20(b), where x_1 and x_2 are displacements and f_2 is a force applied on the mass m_2. Let the force, f_2, be the input and let the displacement, x_2, be the output, and represent this equation in a block-diagram using only integrators. Rewrite the differential equations in state-space form.

5.27 Let $m_1 = m_2 = 1\,\text{kg}$, $b_1 = b_2 = 0.1\,\text{kg/s}$, $k_1 = 1\,\text{N/m}$, and $k_2 = 2\,\text{N/m}$. Use MATLAB to compute the transfer-function from the force f_2 to the displacement x_2 in P5.26. Is this system asymptotically stable? Use MATLAB to simulate the system assuming zero initial conditions and a constant force $f_2 = 1\,\text{N}$.

5.28 You have shown in P2.38 that the ordinary differential equation

$$R_1 C_2 \dot{v}_o + R_1 C_1 \dot{v} + v = 0$$

is an approximate model for the OpAmp-circuit in Fig. 2.23. Let the voltage v be the input and let the voltage v_o be the output, and represent this equation in a block-diagram using only integrators. Rewrite the differential equation in state-space form.

5.29 Let $R_1 = 1\,\text{M}\Omega$ and $C_1 = C_2 = 10\,\mu\text{F}$. Use MATLAB to simulate the circuit in P5.28 assuming zero initial conditions and a constant input voltage $v(t) = 10\,\text{V}, t \geq 0$, applied to the circuit.

5.30 You have shown in P2.41 that the ordinary differential equation

$$J\dot{\omega} + \left(b + \frac{K_t K_e}{R_a}\right)\omega = \frac{K_t}{R_a}v_a$$

is a simplified description of the motion of the rotor of the DC motor in Fig. 2.24, where ω is the rotor angular velocity. Let the armature voltage, v_a, be the input and let the angular velocity, ω, be the output, and represent this equation in a block-diagram using only integrators. Rewrite the differential equation in state-space form.

5.31 Repeat P5.30 considering the rotor's angle, $\theta(t) = \theta(0) + \int_0^t \omega(\tau)d\tau$, as the output.

5.32 Use the block-diagram obtained in P5.30 and MATLAB to simulate the DC motor angular velocity, ω, with $J = 227 \times 10^{-6}$ kg m², $K_t = 0.02$ N m/A, $K_e = 0.02$ V s/rad, $b = 289.4 \times 10^{-6}$ kg m²/s, $R_a = 7\,\Omega$, $v_a = 12$ V, and zero initial angular velocity.

5.33 From P2.41, the rotor torque is

$$\tau = K_t i_a,$$

where the armature current, i_a, is related to the armature voltage, v_a, and the rotor angular velocity, ω, through

$$v_a = R_a i_a + K_e \omega.$$

As in P5.30, let the armature voltage, v_a, be the input and let the torque, τ, be the output, and represent this equation in a block-diagram using only integrators. Rewrite the differential equation in state-space form and calculate the associated transfer-function. Compare your answer with that for P4.34.

5.34 As in in Section 2.8, the water level, h, in a rectangular water tank of cross-sectional area A can be modeled as the integrator

$$\dot{h} = \frac{1}{A}w_{in},$$

where w_{in} is the inflow rate. If water is allowed to flow out from the bottom of the tank through an orifice then

$$\dot{h} = \frac{1}{A}(w_{in} - w_{out}).$$

The outflow rate can be approximated by

$$w_{out} = \frac{1}{R}(p_t - p_a)^{1/\alpha},$$

where the *resistance* $R > 0$ and $\alpha > 0$ depend on the shape of the outflow orifice, p_a is the ambient pressure outside the tank, and

$$p_t = p_a + \rho g h$$

is the pressure at the water level, where ρ is the water density and g is the gravitational acceleration. Combine these equations to write a nonlinear differential equation in state-space relating the water inflow rate, w_{in}, to the water tank level, h, and represent this equation in a block-diagram using only integrators.

5.35 Determine a water inflow rate, w_{in}, such that the tank system in P5.34 is in equilibrium with a water level $h = \bar{h} > 0$. Linearize the state-space equations about this equilibrium point for $\alpha = 2$ and compute the corresponding transfer-function. Is the equilibrium point asymptotically stable?

5.36 You have shown in P2.49 that the temperature of a substance, T (in K or in °C), flowing in and out of a container kept at the ambient temperature, T_o, with an inflow temperature, T_i, and a heat source, q (in W), can be approximated by the differential equation

$$mc\dot{T} = q + wc(T_i - T) + \frac{1}{R}(T_o - T),$$

where m and c are the substance's mass and specific heat, and R is the overall system's thermal resistance. When the flow rate, w, is not constant, this model is nonlinear. Let the heat source, q, the flow rate, w, and the ambient and inflow temperatures, T_o and T_i, be the inputs and let the temperature, T, be the output, and represent this equation in a block-diagram using only integrators. Rewrite the differential equation in state-space form.

5.37 Determine a temperature, T, so that the substance in P5.36 is in equilibrium with a constant heat source, q, flow rate, w, and ambient and inflow temperatures, T_o and T_i. Linearize the state-space equations about this equilibrium point and compute the corresponding transfer-function. Is this equilibrium point asymptotically stable?

5.38 Assume that water's density and specific heat are $\rho = 997.1 \text{ kg/m}^3$ and $c = 4186 \text{ J/kg K}$. Design a feedback controller

$$q = K(\bar{T} - T)$$

for a 50 gal ($\approx 0.19 \text{ m}^3$) water heater rated at $\bar{q} = 40{,}000 \text{ BTU/h}$ ($\approx 12 \text{ kW}$) and thermal resistance $R = 0.27 \text{ K/W}$ at ambient temperature, $T_o = 77 °\text{F}$ ($\approx 25 °\text{C}$). Select K such that the closed-loop system calculated with the linearized model from P5.37 is asymptotically stable and so that a heater initially at ambient temperature never exceeds its maximum power with a flow

$$w(t) = \frac{\bar{w}}{2}(1 + \cos(\omega t))$$

and $\bar{T} = 140 °\text{F}$ ($\approx 60 °\text{C}$). Calculate the resulting closed-loop time-constant in hours and compare your answer with the open-loop time-constant. Is the closed-loop system capable of asymptotically tracking a constant reference temperature $\bar{T}(t) = \bar{T}, t \geq 0$? Use MATLAB to plot the temperature of the water during 3 days. Compute the average water temperature over the entire period and over the last 2 days. Compare your answer with that for P4.42.

5.39 The equations of motion of a rigid body with principal moments of inertia J_1, J_2, and J_3 are given by Euler's equations:

$$J_1\dot{\omega}_1 + \omega_2\omega_3(J_3 - J_2) = \tau_1,$$
$$J_2\dot{\omega}_2 + \omega_1\omega_3(J_1 - J_3) = \tau_2,$$
$$J_3\dot{\omega}_3 + \omega_1\omega_2(J_2 - J_1) = \tau_3.$$

Represent this set of equations in a block-diagram using only integrators and rewrite the differential equations in state-space, where

$$\omega = \begin{pmatrix} \omega_1 \\ \omega_2 \\ \omega_3 \end{pmatrix}, \quad \tau = \begin{pmatrix} \tau_1 \\ \tau_2 \\ \tau_3 \end{pmatrix}$$

are the angular velocity and torque vectors, ω is at the same time the state vector and the output, and τ is the input.

5.40 Verify that

$$\bar{\theta} = 0, \quad \bar{\omega} = (\bar{\omega}_1, \bar{\omega}_2, \bar{\omega}_3) = (\Omega, 0, 0)$$

is an equilibrium point for the rigid body in P5.39. Linearize the equations about this equilibrium point and show that if $J_2 < J_1 < J_3$ or $J_3 < J_1 < J_2$ then this is an unstable equilibrium point. Interpret this result.

5.41 The differential equations

$$m(\ddot{r} - r\omega^2) = u_\mathrm{r} - \frac{GMm}{r^2},$$
$$m(2\dot{r}\omega + r\dot{\omega}) = u_\mathrm{t}$$

are a simplified description of the motion of a satellite orbiting earth as in Fig. 5.18, where r is the satellite's radial distance from the center of the earth, ω is the satellite's angular velocity, m is the mass of the satellite, M is the mass of the earth, G is the universal gravitational constant, u_t is a force applied by a thruster in the tangential direction,

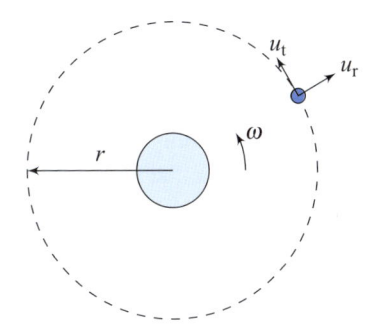

Figure 5.18 Satellite in orbit, P5.41.

and u_r is a force applied by a thruster in the radial direction. Represent these equations in a block-diagram using only integrators and rewrite the differential equations in state-space.

5.42 Show that if

$$\Omega^2 R^3 = GM$$

then $u_t(t) = u_r(t) = \dot{r}(t) = 0$, $r(t) = R$, and $\omega(t) = \Omega$ is an equilibrium point of the equations in P5.41. Linearize the equations about the equilibrium point and perform a change of coordinates to calculate the linearized system in state space:

$$\dot{\tilde{x}} = \begin{bmatrix} 0 & 1 & 0 \\ 3\Omega^2 & 0 & 2\Omega \\ 0 & -2\Omega & 0 \end{bmatrix} \tilde{x} + \begin{bmatrix} 0 & 0 \\ 0 & 1/m \\ 1/m & 0 \end{bmatrix} \tilde{u},$$

$$\tilde{y} = \begin{bmatrix} 1 & 0 & 0 \end{bmatrix} \tilde{x},$$

where

$$\tilde{x} = \begin{pmatrix} r - R \\ \dot{r} \\ R(\omega - \Omega) \end{pmatrix}, \quad \tilde{u} = \begin{pmatrix} u_t \\ u_r \end{pmatrix}, \quad \tilde{y} = r - R.$$

5.43 Consider the satellite model from P5.41 and P5.42. Letting $M \approx 6 \times 10^{24}$ kg be the mass of the earth, and $G \approx 6.7 \times 10^{-11}$ N m^2/kg^2, calculate the altitude, R, of a 1600 kg GPS satellite in medium earth orbit (MEO) with a period of 11 h. If earth's radius is approximately 6×10^3 km, calculate the satellite's altitude measured from the surface of the earth. Is the GPS satellite equilibrium point asymptotically stable? Is it unstable? Does it depend on the mass of the satellite?

5.44 Verify that the linearized model of a satellite in P5.42 is not controllable if only radial thrust is used. Interpret this result on the basis of your knowledge of the system. Hint: Use P5.3.

5.45 Verify that the linearized model of a satellite in P5.42 is controllable if tangential thrust alone is used. Interpret this result on the basis of your knowledge of the system. Hint: Use P5.3.

5.46 The exponential model

$$\dot{x} = rx$$

has been used by Malthus to study population growth. In this context, x is the current population and $r = b - m$ is the growth rate, where b is the birth rate and m is the mortality rate. Explain the behavior of this model when $b > m$, $b < m$, and $b = m$.

5.47 Verhulst suggested that the growth rate in the model of P5.46 often depends on the size of the current population:

$$r = r_0(1 - x/k), \quad r_0 > 0,$$

which leads to the *logistic model*:
$$\dot{x} = r_0(1 - x/k)x.$$

Calculate the equilibrium points for this model. Assume that all constants are positive, linearize about the equilibrium points, and classify the equilibria as asymptotically stable or unstable. Represent the equation in a block-diagram using integrators and use MATLAB to simulate a population with $r_0 = k = 1$ starting at $x(0) = x_0$ for x_0 equal to 0, 1/2, 1, and 2.

5.48 The Lotka–Volterra model,
$$\dot{x}_1 = rx_1 - ax_1x_2,$$
$$\dot{x}_2 = eax_1x_2 - mx_2,$$

where r, a, e, and m are positive constants, is used to model two populations where one of the species is the prey and the other is the predator, e.g. foxes and rabbits. The variable x_1 is the prey population, x_2 is the predator population, r is the intrinsic rate of prey population increase, a is the death rate of prey per predator encounter, e is the efficiency rate of turning prey into predators, and m is the intrinsic predator mortality rate. Calculate the equilibrium points for this model. Assume that all constants are positive, linearize about the equilibrium points, and classify the equilibria as asymptotically stable or unstable. Represent the equations in a block-diagram using integrators and use MATLAB to simulate the predator and prey populations for $r = 0.05$, $a = 0.05$, $m = 0.2$, and $e = 0.2$, with $x(0) = 9$ and $y(0) = 1$. Try other initial conditions and comment on your findings.

5.49 A simplified model for the level of *glucose*, y, in humans as a function of the *insulin concentration*, γ, is the following set of nonlinear ordinary differential equations (see [Ste+03]):
$$\dot{x}_1 = -ax_1 + b\gamma,$$
$$\dot{x}_2 = -(c + x_1)x_2 + d,$$

where $y = x_2$ and all constants are positive. Calculate the unique equilibrium point, $(\bar{x}_1, \bar{x}_2)$, when $\gamma = \bar{\gamma} > 0$ is constant. Represent the equations in a block-diagram using integrators.

5.50 Show that the insulin model from P5.49 linearized at its equilibrium point is
$$\dot{\tilde{x}} = \begin{bmatrix} -a & 0 \\ -\bar{x}_2 & -(c + \bar{x}_1) \end{bmatrix} \tilde{x} + \begin{bmatrix} b \\ 0 \end{bmatrix} \tilde{\gamma},$$
$$\tilde{y} = \begin{bmatrix} 0 & 1 \end{bmatrix} \tilde{x},$$

with transfer-function
$$\frac{\tilde{Y}(s)}{\tilde{\Gamma}(s)} = \frac{-b\bar{x}_2}{(s + a)(s + c + \bar{x}_1)}.$$

Explain the meaning of the negative sign in the numerator.

5.51 The authors of [Ste+03] have verified that when $\bar{x}_2 = 100$ mg/dL then $a = c + \bar{x}_1 = 1/33$ min^{-1} and $b\bar{x}_2/a^2 = 3.3$ mg/dL per µU/mL provide a good experimental fit. Calculate the poles and zeros of the linearized transfer-function model from P5.50 and classify the corresponding equilibrium point as asymptotically stable or unstable.

5.52 After insulin has been released in the plasma at a rate u, its concentration, γ, does not reach steady-state values instantaneously. Instead,

$$\dot{\gamma} = -f\gamma + gu.$$

Combine the results from P5.49–P5.51 to show that the transfer function from $\tilde{u}$ to $\tilde{y}$ is

$$\frac{\tilde{Y}(s)}{\tilde{U}(s)} = \frac{-gb\bar{x}_2}{(s+f)(s+a)(s+c+\bar{x}_1)}.$$

Use the value $g = 1/5$ min^{-1} and $f = 1/5$ from [Ste+03] and substitute numerical values from P5.51 to calculate the corresponding poles and zeros.

6 Controller Design

In this chapter we introduce a number of techniques that can be used to design controllers. Our main goal is to understand how the locations of the poles and zeros of the open-loop system influence the locations of the poles of the closed-loop system. We start with a thorough study of second-order systems and culminate with a graphic tool known as *root-locus*, in which the poles of the closed-loop system can be plotted in relation to the open-loop poles and zeros and the loop gain. Along the way we introduce derivative control and the ubiquitous proportional–integral–derivative controller.

6.1 Second-Order Systems

Stable first- and second-order systems epitomize the basic behaviors of stable dynamic linear systems: exponentially decaying potentially oscillatory responses. Consider a second-order system with characteristic equation

$$s^2 + 2\zeta\omega_n s + \omega_n^2 = 0, \qquad \omega_n > 0, \tag{6.1}$$

where the parameter ζ is the *damping ratio* and the parameter ω_n is the *natural frequency*. Most[1] second-order polynomials can be put in this form by an adequate choice of ω_n and ζ. We will provide concrete examples later. The nature of the response of a second-order system is controlled by the location of the roots of the characteristic equation (6.1):

$$s^2 + 2\zeta\omega_n s + \omega_n^2 = (s - s_1)(s - s_2), \tag{6.2}$$

where

$$s_1 = -\omega_n(\zeta + \sqrt{\zeta^2 - 1}),$$
$$s_2 = -\omega_n(\zeta - \sqrt{\zeta^2 - 1}).$$

Note that the parameter ω_n only *scales* the roots, and that the parameter ζ controls whether the roots are real or complex-conjugate: if $|\zeta| \geq 1$ the roots are real; if $|\zeta| < 1$ they are complex conjugates.

---

[1] The only second-order polynomials that are not encoded in (6.1) have the form

$$s^2 + 2\zeta\omega_n s - \omega_n^2 = 0, \qquad \omega_n > 0.$$

You will verify in P6.1 that in this case the roots are always real, with one of the roots always positive.

Controller Design

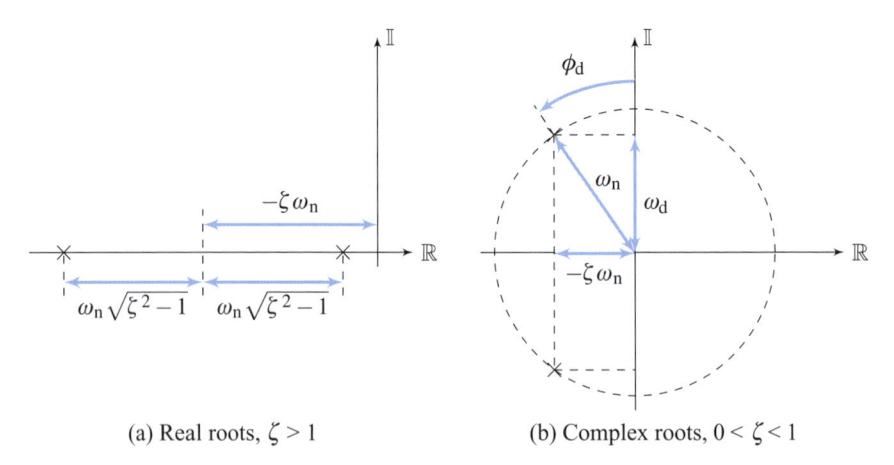

(a) Real roots, $\zeta > 1$ *(b) Complex roots, $0 < \zeta < 1$*

Figure 6.1 Locations of the roots of the second-order equation (6.1); roots are marked with "×."

When $|\zeta| \geq 1$ a second-order system has real roots and its response is the superposition of the response of two first-order systems. Indeed, the inverse Laplace transform of a second-order transfer-function with real poles has terms

$$\mathscr{L}^{-1}\left\{\frac{k_1}{s-s_1} + \frac{k_2}{s-s_2}\right\} = k_1 e^{s_1 t} + k_2 e^{s_2 t}, \quad t \geq 0,$$

where the exact values of k_1 and k_2 depend on the zeros of the transfer-function and can be computed from residues as in Chapter 3. In the complex plane, the roots are located on the real axis symmetrically about the point $s = -\zeta \omega_n$, as shown in Fig. 6.1(a). If $\zeta > 1$ the roots are on the left-hand side of the complex plane, hence the associated transfer-function is asymptotically stable. If $\zeta < -1$ the roots are on the right-hand side of the complex plane, and the associated transfer-function is unstable.

When $|\zeta| < 1$ a second-order system has complex-conjugate roots and its response is oscillatory. The inverse Laplace transform of such a second-order transfer-function (see Table 3.1) contains the terms

$$\mathscr{L}^{-1}\left\{\frac{k}{s+\zeta\omega_n - j\omega_d} + \frac{k^*}{s+\zeta\omega_n + j\omega_d}\right\} = 2|k|e^{-\zeta\omega_n t}\cos(\omega_d t + \angle k),$$

where

$$\omega_d = \omega_n\sqrt{1-\zeta^2}. \tag{6.3}$$

The following facts can be used to locate the roots in the complex plane:

(a) the real part of the roots is equal to $-\zeta\omega_n$;
(b) the absolute value of both roots is equal to

$$\omega_n\left|-\zeta \pm j\sqrt{1-\zeta^2}\right| = \omega_n\sqrt{\zeta^2 + (1-\zeta^2)} = \omega_n;$$

(c) the angle measured between the root and the imaginary axis is

$$\phi_d = \sin^{-1}\left(\frac{\zeta\omega_n}{\omega_n}\right) = \sin^{-1}\zeta = \tan^{-1}\left(\frac{\zeta}{\sqrt{1-\zeta^2}}\right) = \tan^{-1}\left(\frac{\zeta\omega_n}{\omega_d}\right). \quad (6.4)$$

These relationships can be visualized in the diagram of Fig. 6.1(b).

The damping ratio, ζ, controls how *damped* the system response is. The name and the following terminology are borrowed from the standard analysis of the *harmonic oscillator*: when $\zeta = 0$ the roots are purely imaginary and the response contains an oscillatory component at the natural frequency, $\omega_n = \omega_d$; when $0 < \zeta < 1$ the system is said to be *underdamped* as any response will be oscillatory with *damped natural frequency* $\omega_d < \omega_n$; when $\zeta = 1$ the system is said to be *critically damped*, with two repeated real roots; when $\zeta > 1$ the system has two real roots with negative real part and is said to be *overdamped*.

For example, consider the step response of the second-order model:

$$G(s) = \frac{\omega_n^2}{s^2 + 2\zeta\omega_n s + \omega_n^2}, \qquad \omega_n > 0. \quad (6.5)$$

The choice of ω_n^2 as the numerator is for normalization, so that $G(0) = 1$. The most interesting case is when G is asymptotically stable but has complex-conjugate poles, that is, $0 < \zeta < 1$. Using the methods of Section 3.5, you will show in P6.2 after a bit of algebra that the step response of a system with transfer-function (6.5) is

$$y(t) = \mathcal{L}^{-1}\left\{\frac{G(s)}{s}\right\} = 1 - \frac{1}{\sqrt{1-\zeta^2}} e^{-\zeta\omega_n t} \sin(\omega_d t + \pi/2 - \phi_d), \quad t \geq 0, \quad (6.6)$$

where the damped natural frequency, ω_d, is given by (6.3), and ϕ_d is given by (6.4). We plot the step response for various values of the damping ratio, ζ, in Fig. 6.2 [2] The plots are normalized by the parameter

$$t_n = \frac{2\pi}{\omega_n}, \quad (6.7)$$

so that the resulting function, $y(t/t_n)$, is a function of ζ alone. The role of ω_n is therefore controlling the *speed* of the response: the larger ω_n the faster the response. Insofar as ζ is concerned, the larger ζ the faster the rate of decay, that is, the *damping* of the response.

It is useful to define and quantify some terminology with respect to the step response of the standard second-order system (6.5) when $0 < \zeta < 1$. As can be seen in Fig. 6.2, there is always some *overshoot*, that is, the response exceeds its steady-state value. This is a new dynamic characteristic, since strictly proper first-order systems never overshoot. By differentiating the response with respect to t and looking for peaks, you will determine in P6.3 that

$$t_p = \frac{\pi}{\omega_d}, \qquad y_p = 1 + e^{-\zeta\pi/\sqrt{1-\zeta^2}} \quad (6.8)$$

---

[2] These curves were also used to generate the three-dimensional figure on the cover of this book.

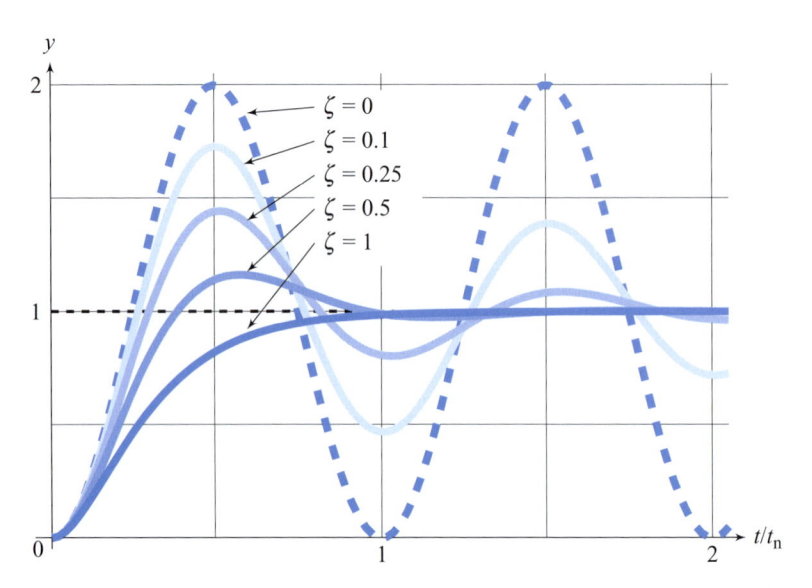

Figure 6.2 Normalized step responses $y(t/t_n)$ for second-order transfer-function $G(s) = \omega_n^2/(s^2 + 2\zeta\omega_n s + \omega_n^2)$ with $0 < \zeta < 1$, $t_n = 2\pi/\omega_n$.

are the time and value of the first peak. It is common to quantify overshoot in terms of the *percentage overshoot*:

$$\text{PO} = 100 e^{-\zeta\pi/\sqrt{1-\zeta^2}}. \tag{6.9}$$

Another useful metric is the *settling-time*, t_s, which is the time it takes the step response to become confined within 2% of its steady-state value. You will show in P6.4 that the settling-time can be approximated by

$$t_s = \frac{\log(50)}{\zeta\omega_n} \approx \frac{3.9}{\zeta\omega_n}. \tag{6.10}$$

These figures of merit are illustrated in Fig. 6.3. The *rise-time*, t_r, and the *time-constant*, τ, are also shown in this figure. They were computed in the exact same way as for first-order systems (see Section 2.3): the rise-time is the time it takes for the response to go from 10% to 90% of its steady-state value, and the time-constant is the time it takes the response to reach $100e^{-1} \approx 63\%$ of its steady-state value. Unfortunately it is not possible to have simple formulas for the rise-time and the time-constant for second-order systems. The approximate formulas

$$\frac{t_r}{t_n} \approx 0.16 + 0.14\zeta + 0.24\zeta^3, \quad \frac{\tau}{t_n} \approx 0.19 + 0.1\zeta + 0.054\zeta^3 \tag{6.11}$$

have a maximum relative error of about 1% in the range $0 < \zeta < 1$ (see P6.5).

6.1 Second-Order Systems

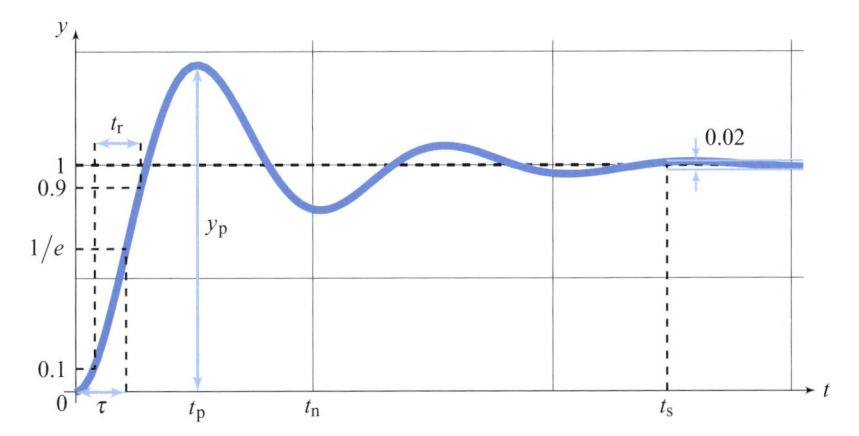

Figure 6.3 Step response $y(t)$ showing the peak-time, t_p, settling-time, t_s, rise-time, t_r, peak-response, y_p, and time-constant, τ.

With the above facts in mind, we revisit the cruise control solution with integral control discussed in Section 4.2. The closed-loop characteristic equation obtained in (4.12) can be put in the form (6.1) after setting

$$\omega_n^2 = \frac{p}{m}K_i, \qquad 2\zeta\omega_n = \frac{b}{m}.$$

In other words,

$$\omega_n = \sqrt{\frac{p}{m}K_i}, \qquad \zeta = \frac{b}{2\sqrt{pmK_i}} > 0.$$

From these relations it is clear that one cannot raise K_i in order to increase ω_n, i.e. the speed of the response, without compromising the damping ratio, ζ. By contrast, if proportional–integral control is used, the closed-loop poles are governed by the characteristic equation (4.17), meaning that they are in the form (6.1) with

$$\omega_n = \sqrt{\frac{p}{m}K_i}, \qquad \zeta = \frac{1}{2\omega_n}\left(\frac{b}{m} + \frac{p}{m}K_p\right) > 0.$$

One can now choose K_i to set the natural frequency, ω_n, to any desired value and then choose K_p to set the damping ratio, ζ, independently.

Consider as a second example the simple pendulum introduced in Section 5.5. We will design and analyze more sophisticated controllers for the simple pendulum in Sections 6.5, 7.8, and 8.4. But we can already take first steps based on our fresh analysis of second-order systems. The model derived for the simple pendulum in Section 5.5 is nonlinear and we have calculated two linearized models around the equilibrium points $\theta = 0$ and $\theta = \pi$. The transfer-functions associated with these linearized models are

$$G_0(s) = \frac{1}{J_r s^2 + bs + mgr}, \qquad G_\pi(s) = \frac{1}{J_r s^2 + bs - mgr}.$$

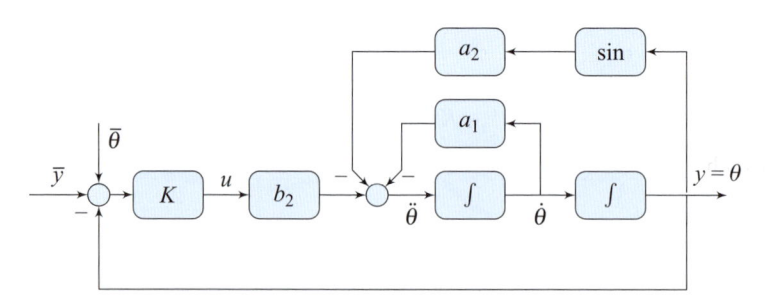

Figure 6.4 Linear proportional control of the simple pendulum; $\bar{\theta} = 0$ around stable equilibrium, and $\bar{\theta} = \pi$ around unstable equilibrium.

As in Section 5.5, if all parameters are positive then the model linearized around $\theta = u = 0$ is asymptotically stable and the model linearized around $u = 0$, $\theta = \pi$, is unstable. Our first task is to develop a controller that can stabilize the unstable linearized system G_π. Recalling the discussion in Section 5.8, a linear proportional controller that stabilizes a nonlinear system around its equilibrium point will also stabilize the nonlinear system, at least in the neighborhood of the equilibrium.

The connection of the nonlinear model of the pendulum derived in Section 5.5 with the proportional controller $K(s) = K$ is depicted in the block-diagram in Fig. 6.4. The feedback connection of the pendulum with this controller linearized in the neighborhood of $\theta = \pi$ has as closed-loop transfer-function[3]

$$H_\pi(s) = \frac{KG_\pi(s)}{1 + KG_\pi(s)} = \frac{K}{J_r s^2 + bs + K - mgr}.$$

This is a second-order system with characteristic equation (6.1), where

$$\omega_{n_\pi} = \sqrt{\frac{K - mgr}{J_r}}, \quad \zeta_\pi = \frac{b}{2\sqrt{J_r(K - mgr)}}.$$

These expressions are similar to the ones found in the case of the integral cruise control of the car, where increasing K improves the speed of response, ω_n, at the expense of decreasing the damping ratio, ζ. A complication is the fact that K needs to be chosen high enough to stabilize the pendulum:

$$K > mgr.$$

It is desirable to have a controller that can work around both equilibrium points, that is, near both $\theta = 0$ and $\theta = \pi$. A controller that is capable of operating a system in a variety of conditions is a *robust* controller. We will discuss *robustness* in more detail later in Sections 7.7 and 8.2. For now, we are happy to verify that the same proportional control does not de-stabilize the system when operated near the stable equilibrium $\theta = 0$. Repeating the same steps as above, we obtain the closed-loop

---

[3] You will work this out in detail in P6.9.

transfer-function

$$H_0(s) = \frac{KG_0(s)}{1 + KG_0(s)} = \frac{K}{J_r s^2 + bs + K + mgr},$$

for which

$$\omega_{n_0} = \sqrt{\frac{K + mgr}{J_r}}, \qquad \zeta_0 = \frac{b}{2\sqrt{J_r(K + mgr)}}.$$

We can see that any value of the gain K that stabilizes the unstable transfer-function G_π does not de-stabilize the asymptotically stable transfer-function G_0. This is because $K + mgr > 2mgr > 0$. For example, the selection of

$$K = 3mgr \tag{6.12}$$

implies

$$\omega_{n_\pi} = \sqrt{2}\omega_{n_{ol}}, \qquad \omega_{n_0} = 2\omega_{n_{ol}}, \qquad \omega_{n_{ol}} = \sqrt{\frac{mgr}{J_r}},$$

where $\omega_{n_{ol}}$ is the natural frequency of the stable transfer-function G_0 in open-loop, i.e. without control. The choice of gain (6.12) doubles the natural frequency near the stable equilibrium $\bar{\theta} = 0$ while stabilizing the unstable equilibrium $\bar{\theta} = \pi$ by making its closed-loop natural frequency be approximately 40% higher than $\omega_{n_{ol}}$. Insofar as the damping ratio is concerned, a simple calculation reveals

$$\zeta_0 = \frac{1}{2}\zeta_{ol}, \qquad \zeta_\pi = \frac{1}{\sqrt{2}}\zeta_{ol}, \qquad \zeta_{ol} = \frac{b}{2\sqrt{J_r mgr}}.$$

The choice (6.12) halves the damping ratio near the stable equilibrium $\bar{\theta} = 0$ while providing a damping ratio of about 70% of the open-loop damping ratio, ζ_{ol}, around the unstable equilibrium $\bar{\theta} = \pi$. In both cases the controlled pendulum has a damping ratio that is smaller than that of the open-loop pendulum. The only way to improve the damping ratio is to reduce K, which makes the closed-loop system response slower. In the car cruise control problem, a better solution was proposed in Section 4.2 in the form of a more complex controller: a proportional–integral controller enabled us to set the damping ratio *and* the natural frequency by a choice of two independent gains, K_p and K_i. As we will discuss in detail in the next section, the key in the case of the pendulum is the addition of a zero in the controller transfer-function: instead of an integrator we need a differentiator.

We close this section with a note on the behavior of some systems that have order higher than second. Even though it is not possible to provide simple formulas for the response of such systems, asymptotically stable higher-order systems that have a real pole or a pair of complex-conjugate poles with real part much smaller than the system's other poles tend to behave like a first- or second-order system. Indeed, the response quickly becomes *dominated* by the slower response of such poles: the contribution of faster and highly damped poles approaches zero quicker and leaves the response appearing to be that of a first- or second-order system, the slow poles being called *dominant*

Controller Design

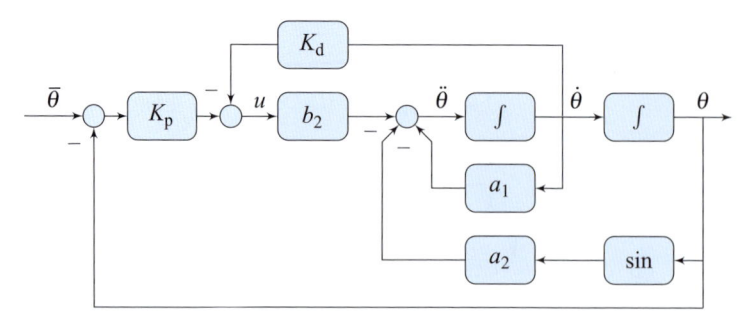

Figure 6.5 Proportional–derivative control of the simple pendulum.

poles. For this reason, it is common to see performance specifications of open- and closed-loop systems in terms of first- or second-order figures of merit.

6.2 Derivative Action

We have seen in previous sections that increasing the control gain may have a detrimental effect on the damping properties of a feedback system. This has been the case in the examples of integral cruise control as well as of proportional control of the pendulum. In order to improve damping we generally need to increase the complexity of the controller. In a mechanical system, the idea of damping is naturally associated with a force that opposes and increases with the system velocity, for instance, viscous friction in the simple pendulum model, $-b\dot{\theta}$. Using velocity or, more generally, the derivative of a signal, is an effective way to improve damping. However, as will become clear later, velocity feedback alone may not be enough to asymptotically stabilize an unstable system, and hence derivative action is often combined with proportional control.

We introduce the idea of proportional–derivative control using the simple pendulum as an example. Let the proportional–derivative controller

$$u(t) = K_\text{p}\left(\bar{\theta}(t) - \theta(t)\right) - K_\text{d}\dot{\theta}(t) \tag{6.13}$$

be applied to the simple pendulum model derived in Section 5.5. This controller can be implemented as shown in the block-diagram of Fig. 6.5, in which the controller makes use of two measurements: the angular position, θ, and the angular velocity, $\dot{\theta}$. If we are not able to directly measure the velocity, $\dot{\theta}$, then we have to be careful when building the controller, given the difficulties involved in physically implementing a derivative block, as discussed in Chapter 5. Either way, in order to avoid working with two outputs, it is convenient to introduce a derivative block component during controller analysis and design. With this caveat in mind, we can study the single-output block-diagrams in Fig. 6.6, where we use the Laplace transform variable "s" to represent an idealized derivative operation.[4]

---

[4] Recall from Table 3.2 that $\mathscr{L}(\dot{f}(t)) = sF(s) - f(0)$.

6.2 Derivative Action

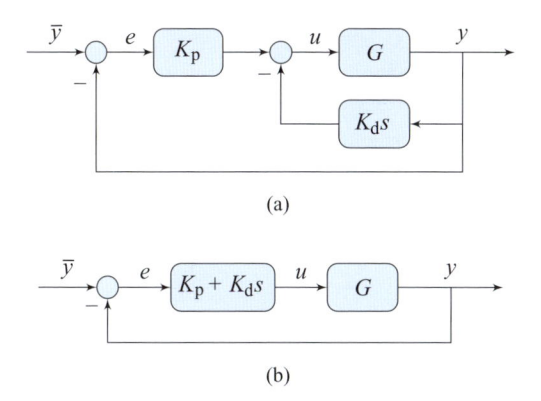

Figure 6.6 Block-diagrams for proportional–derivative control.

The closed-loop systems in Figs. 6.6(a) and (b) are not equivalent and behave slightly differently if $\bar{y}$ is not constant. The diagram in Fig. 6.6(a) corresponds to Fig. 6.5, which implements the control law (6.13). The diagram in Fig. 6.6(b) is the standard *proportional–derivative* controller, or PD controller:

$$u(t) = K_p e(t) + K_d \dot{e}(t), \qquad e(t) = \bar{y}(t) - y(t). \qquad (6.14)$$

From an implementation perspective, the diagram in Fig. 6.6(a) offers some advantages, specially when $\bar{y}$ has discontinuities or is rich in high-frequency components. Note that y is continuous even if $\bar{y}$ is not. For this reason, in the presence of discontinuities, one should expect to see short-lived large signal spikes[5] in the control input u in Fig. 6.6(b) in comparison with Fig. 6.6(a). This is due to the differentiation of $\bar{y}$ in Fig. 6.6(b). In terms of transfer-functions, this difference amounts to an extra zero, with both diagrams having the same closed-loop poles. To see this, let $G = N_L/D_L$, and verify that (see P6.6)

$$H(s) = \frac{K_p N_L(s)}{D_L(s) + (K_p + K_d s) N_L(s)} \qquad (6.15)$$

for the loop in Fig. 6.6(a), whereas

$$H(s) = \frac{(K_p + K_d s) N_L(s)}{D_L(s) + (K_p + K_d s) N_L(s)} \qquad (6.16)$$

in Fig. 6.6(b). Since Fig. 6.6(b) is a better fit for the standard feedback diagram of Fig. 4.11 that we have analyzed earlier, we shall use Fig. 6.6(b) in the rest of this section.

Consider the control of the simple pendulum by the proportional–derivative controller in Fig. 6.6(b). Linearizing about $u = 0$ and $\theta = \pi$ we obtain the linearized closed-loop transfer-function (see P6.10)

$$H_\pi(s) = \frac{(K_p + K_d s) G_\pi(s)}{1 + (K_p + s K_d) G_\pi(s)} = \frac{K_p + K_d s}{J_r s^2 + (K_d + b) s + (K_p - mgr)}.$$

---
[5] See the discussion in Section 5.1 about impulses appearing in the response due to differentiation.

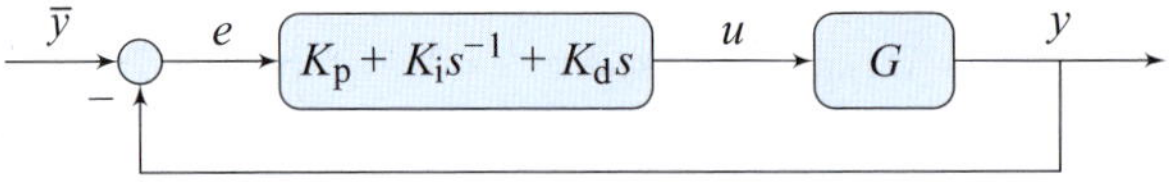

Figure 6.7 Block-diagram for proportional–integral–derivative (PID) control.

Because the controller does not have any poles, this is also a second-order system with characteristic equation (6.1) for which

$$\omega_{n_\pi} = \sqrt{\frac{K_p - mgr}{J_r}}, \qquad \zeta_\pi = \frac{K_d + b}{2\sqrt{J_r(K_p - mgr)}}.$$

The same calculations repeated for the model linearized about $u = 0$ and $\theta = 0$ produce

$$\omega_{n_0} = \sqrt{\frac{K_p + mgr}{J_r}}, \qquad \zeta_0 = \frac{K_d + b}{2\sqrt{J_r(K_p + mgr)}}.$$

Note how the values of ω_n are not affected by the introduction of the derivative gain, K_d, which leads to the important conclusion that damping alone (derivative action), would not have been able to stabilize the pendulum. The effect of K_d is confined to the damping ratio, ζ.

For example, after stabilizing the closed-loop pendulum with the choice of $K_p = 3mgr$ from (6.12) we can choose K_d to provide any desired level of damping, say

$$\zeta_\pi = \frac{\sqrt{2}}{2} \approx 0.7$$

around the equilibrium $\theta = \pi$. This corresponds to setting

$$K_d = 2\zeta_\pi \sqrt{J_r(K_p - mgr)} - b = 2\sqrt{J_r mgr} - b.$$

For such K_p and K_d, the damping ratio around the equilibrium $\theta = 0$ is

$$\zeta_0 = \frac{2\sqrt{J_r mgr}}{2\sqrt{4 J_r mgr}} = \frac{1}{2},$$

which is a bit smaller than ζ_π, but still acceptable. We will continue to improve this controller in Sections 6.5 and 7.8.

6.3 Proportional–Integral–Derivative Control

Derivative action can help improve the damping properties of a feedback system, as seen in the last section. If integral action is required to track constant or low-frequency references, or reject low-frequency input disturbances, a controller can be constructed by combining a proportional term with integral *and* derivative terms to form a *proportional–integral–derivative* controller, or PID controller. A generic block-diagram of a feedback system with a PID controller is shown in Fig. 6.7. PID control is arguably the most popular form of control and is widely used in industry. Manufacturers of instrumentation and control hardware offer implementations of PID controllers where the gains K_p, K_i, and K_d can be tuned by the user to fit the controlled process. Most modern hardware comes also with algorithms for identifying the process model and automatically tuning the gains, and other bells and whistles.

The PID controller in Fig. 6.7 implements the control law

$$u(t) = K_p e(t) + K_d \dot{e}(t) + K_i \int_0^t e(\tau)d\tau, \qquad e(t) = \bar{y}(t) - y(t), \qquad (6.17)$$

which is associated with the transfer-function

$$K(s) = K_p + K_d s + \frac{K_i}{s} = \frac{K_d s^2 + K_p s + K_i}{s}. \qquad (6.18)$$

This transfer-function has two zeros and one pole at the origin. When the derivative gain K_d is not zero, this transfer-function is not proper and cannot be physically implemented, an issue that we will address later, in Section 6.5, by introducing extra poles in the controller. Given the importance and widespread use of PID control, there is an extensive literature on how to select the control gains depending on assumptions on the underlying model of the system being controlled. We do not study such rules here, as they can be found in standard references, e.g. [FPE14; DB10]. These *tuning rules* might often provide a useful and quick way to have a feedback system up and running by relying on a few parameters that can be obtained experimentally, but do little to elucidate the role of the controller and its relation with the controlled system in a feedback loop. For this reason we shall spend our time studying methods that can help guide the selection of the PID control gains but require a more complete model of the system.

One difficulty is that the controller (6.18) is of order one, and application to a second-order system generally leads to a third-order closed-loop system. The situation gets much more complicated if additional poles are introduced to make (6.18) proper and therefore implementable, in which case the order of the PID controller is at least two. Calculating the roots of a closed-loop characteristic polynomial of order three or four in terms of the parameters K_p, K_i, and K_d is possible but requires working with highly convoluted formulas. In addition to that, it is impossible to come up with any type of analytic formula for computing the roots of a polynomial of order five or higher except in very special cases.[6] Instead, we will rely on numerical computations and auxiliary graphic tools that we will introduce in the next sections and following chapters.

6.4 Root-Locus

In this section we introduce a technique for controller design that is based on a graphic known as *root-locus*. The root-locus is a plot showing the poles of a system with transfer-function L in feedback with a static gain, α, as shown in Fig. 6.8. That is a plot of the roots of the characteristic equation

$$1 + \alpha L(s) = 0, \qquad \alpha \geq 0,$$

---

[6] This is the Abel–Ruffini theorem [Wae91, Section 8.7].

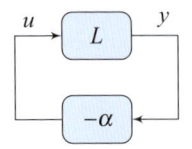

Figure 6.8 Closed-loop feedback configuration for root-locus; $\alpha \geq 0$.

obtained as α varies from 0 to infinity. The root-locus plot is used to study general SISO feedback systems after properly grouping all dynamic elements into the *loop transfer-function*, L.

Take for instance the diagram of Fig. 4.18. As shown in Chapter 4, the closed-loop poles are the zeros of the characteristic equation

$$1 + G(s)K(s)F(s) = 0.$$

The transfer-function $G(s)K(s)F(s)$ is the product of all transfer-functions appearing in the feedback loop. Now let the controller $K(s)$ be divided into two parts: a *positive* static gain α and a dynamic part $C(s)$, that is,

$$K(s) = \alpha\, C(s), \qquad \alpha > 0.$$

In this way, the characteristic equation is rewritten as

$$1 + \alpha\, L(s) = 0, \qquad L(s) = G(s)C(s)F(s),$$

and the root-locus plot is used for the determination of a suitable gain $\alpha > 0$.

As mentioned earlier, one reason for working with a graphic rather than an algebraic tool is that it is not possible to find a closed-form expression for the roots of a polynomial of arbitrary degree as a function of its coefficients, even in this simple case where the roots depend on only one parameter, α. It is, however, relatively simple to *sketch* the location of the roots with respect to α without computing their exact location. This can even be done *by hand* using a set of *rules* that depend only on the computation of the zeros and poles of $L(s)$. With the help of a computer program, such as MATLAB, very accurate root-locus plots can be traced without effort. Knowledge of some but not necessarily all of the root-locus rules can help one predict the effect of moving or adding extra poles or zeros to L, which is a very useful skill to have when designing controllers and analyzing feedback systems. It is with this intent that we introduce some of the root-locus rules. Readers are referred to standard references, e.g. [FPE14; DB10], for a complete set of rules including a detailed discussion of the case $\alpha < 0$.

A key observation behind the root-locus is the following property: suppose the complex number s is a root of $1 + \alpha L(s)$ and $\alpha > 0$.[7] In this case,

$$L(s) = -\frac{1}{\alpha} < 0 \quad \Longrightarrow \quad \angle L(s) = \pi. \qquad (6.19)$$

---

[7] If $\alpha < 0$ then $\angle L(s) = 0$.

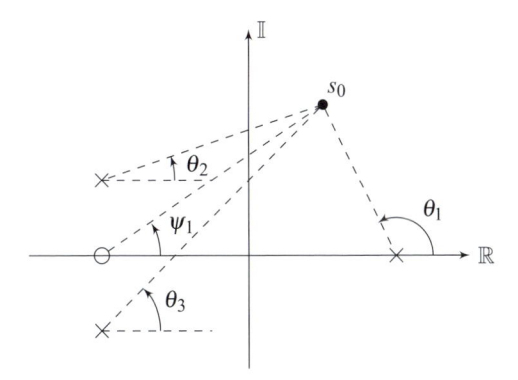

Figure 6.9 Measuring the phase of $L(s)$ at $s = s_0$; "×" denotes a pole of $L(s)$ and "∘" denotes a zero of $L(s)$; $\angle L(s_0) = \psi_1 - \theta_1 - \theta_2 - \theta_3$; the point s_0 shown is not a root of $1 + \alpha L(s)$ because $\angle L(s_0) \neq \pi$.

In other words, $L(s)$ is a negative real number and hence with phase equal to π. Conversely, suppose that $\angle L(s) = \pi$ for some s, then $L(s)$ is a negative real number; that is, $L(s) = -\alpha^{-1}$ for some $\alpha > 0$ and $1 + \alpha L(s) = 0$. It is much easier to locate points in the complex plane for which $L(s)$ has phase equal to π than to compute the roots of $1 + \alpha L(s)$ as a function of α. This is especially simple when $L(s)$ is rational, as discussed below.

Let $L = \beta N_L / D_L$ be a rational transfer-function where $\beta \neq 0$ is a scalar, n is the degree of the denominator, D_L, and m is the degree of the numerator, N_L. Assume that N_L and D_L are monic[8] and factor N_L and D_L into the product of their possibly complex roots:

$$N_L = (s - z_1)(s - z_2) \ldots (s - z_m),$$
$$D_L = (s - p_1)(s - p_2) \ldots (s - p_n).$$

Using the fact that

$$\angle L = \angle \beta + \angle N_L - \angle D_L = \angle \beta + \sum_{i=1}^{m} \angle(s - z_i) - \sum_{k=1}^{n} \angle(s - p_k),$$

the phase of L, i.e. $\angle L$, can be quickly computed by evaluating angles measured from the zeros $z_1, z_2, \ldots, z_m$ and poles $p_1, p_2, \ldots, p_n$. This is the key idea behind sketching the root-locus plot and is illustrated in Fig. 6.9.

Because L and $\angle L$ are continuous functions of s at any point that is not a pole of L, that is a root of D_L, one should expect the root-locus to be a set of *continuous curves* in the complex plane. Indeed, this fact and more is part of our first *root-locus rule*:

---

[8] The coefficient of the monomial with highest power is one.

Controller Design

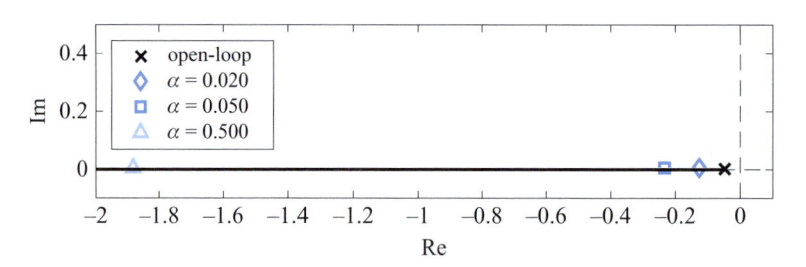

Figure 6.10 Root-locus for the proportional control of the car speed model (2.3) with parameters $b/m = 0.05$ and $p/b = 73.3$; symbols show roots for the gains indicated in the legend. Compare this with the step responses in Fig. 2.10.

If the rational function $L = \beta N_L/D_L$, $\beta \neq 0$, is strictly proper,[9] $n > m$, and there is no pole–zero cancellation between N_L and D_L, the characteristic equation $1 + \alpha L(s) = 0$ has exactly n roots for any $\alpha > 0$. These roots belong to n continuous curves in the complex plane (the root-locus). When $\alpha \to 0$ these roots are the roots of the denominator of D_L, and when $\alpha \to \infty$ these roots are the roots of the numerator of N_L plus $n - m$ roots at infinity.

The roots *depart* from the poles and *arrive* at the zeros of L because

$$\lim_{\alpha \to 0} D_L(s)(1 + \alpha L(s)) = \lim_{\alpha \to 0} D_L(s) + \alpha \beta N_L(s) = D_L(s),$$

$$\lim_{\alpha \to \infty} \alpha^{-1} D_L(s)(1 + \alpha L(s)) = \lim_{\alpha \to \infty} \alpha^{-1} D_L(s) + \beta N_L(s) = \beta N_L(s),$$

for any s such that $D_L(s) \neq 0$, that is, any s that is not a pole of L.

Take for example the linear model we have obtained for the speed of the car in (2.3) and the parameters $b/m = 0.05$ and $p/b = 73.3$ estimated in Chapter 2. This model has an open-loop transfer-function (3.16):

$$G(s) = \frac{p/m}{s + b/m} = \frac{3.7}{s + 0.05}.$$

The stability of the feedback loop with proportional control in Fig. 2.7 can be studied with the root-locus plot after setting

$$\alpha = K, \qquad L = G = \frac{\beta}{D_L}, \qquad \beta = 3.7, \qquad D_L = s + 0.05.$$

The root-locus plot in this case is very simple, namely the black solid segment of line in Fig. 6.10. Because $n = 1$, $m = 0$, it consists of a single curve starting at the open-loop root of D_L, marked with a cross, and ending at the $n - m = 1$ zero of N_L, which is at minus infinity. The markers in Fig. 6.10 locate selected values of α that correspond to the control gains used to generate the responses in Fig. 2.8. There are no other curves in this root locus. It is not possible for the closed-loop system to have complex roots because L is of order one, and the root-locus remains entirely on the real axis.

---

[9] If L is proper but not strictly proper then the characteristic equation $1 + \alpha L(s) = 0$ may have fewer than n roots for some values of α. For example, if $L(s) = (1 - s)/(s + 1)$ then $1 + L(s) = 2/(s + 1)$, which has no roots when $\alpha = 1$.

6.4 Root-Locus

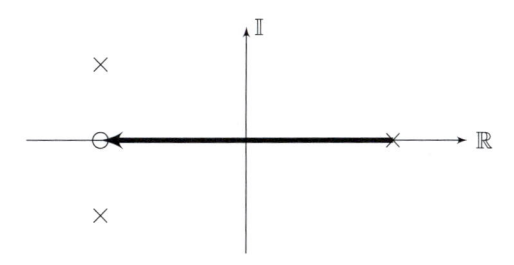

Figure 6.11 Locus of real roots; complex locus is not shown; see Fig. 6.13 for complete root-locus.

In general, when L is a rational function with real coefficients, it is possible to quickly determine which parts of the real axis belong to the root-locus. Consider first the case where L has only real poles and real zeros and $\beta > 0$. If s is a point on the real axis, then zeros and poles to the left of s contribute nothing to the phase of L, whereas zeros and poles to the right of s contribute either π or $-\pi$. Therefore, whenever there is an odd number of real poles or real zeros to the right of the real point s, this point should be part of the root-locus. If there is an even number of real poles and real zeros to the right of the real point s, this point should not be part of the root-locus. After observing that complex zeros or poles always come in complex-conjugate pairs and that the net contribution of a complex-conjugate pair of zeros or poles to the phase of L for a point s on the real axis is always 0 if the zero or pole is to the left of s or $\pm 2\pi$ if the zero or pole is to the right[10] of s, we can come up with the following rule:

The real point s is part of the root-locus of the rational function $L = \beta\, N_L/D_L$, $\beta > 0$, where N_L and D_L are monic polynomials with real coefficients if and only if L has an odd number of poles and zeros with real part greater than or equal to s.

This rule is all that is needed to completely explain the root-locus of Fig. 6.10. For another example, take the poles and zeros in Fig. 6.9. Application of the rule leads to the real segment shown in Fig. 6.11 being the only part of the root-locus on the real axis. The arrow indicates the fact that the locus departs from the pole and arrives at the zero. We will add the remaining curves to this plot later. Note that the above rule applies only[11] to the case when $\beta > 0$.

When L is of order 2 or higher the root-locus will generally have complex roots. Consider for example the car model

$$G(s) = \frac{p/m}{s + b/m} = \frac{3.7}{s + 0.05},$$

this time in closed-loop with the integral controller shown in Fig. 4.3. In this case, $K(s) = KC(s)$, $C(s) = s^{-1}$, and the root-locus computed with

$$\alpha = K, \qquad L = GC = \frac{\beta}{D_L}, \qquad \beta = 3.7, \qquad D_L = s(s + 0.05),$$

---

[10] Imagine what happens if the point s_0 is placed on the real axis in Fig. 6.9.
[11] Can you figure out the correct rule when $\beta < 0$?

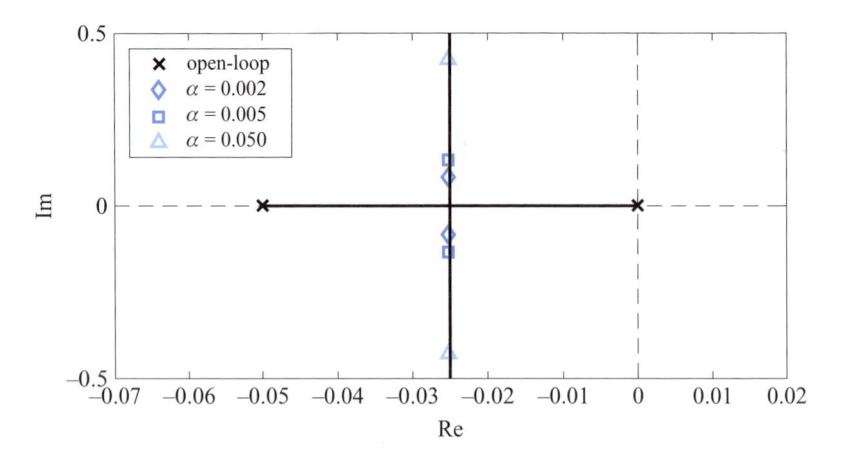

Figure 6.12 Root-locus for the integral control of the car speed model (2.3) with parameters $b/m = 0.05$ and $p/b = 73.3$; symbols show roots for the gains indicated in the legend. Compare this with the step responses in Fig. 4.4.

is shown in Fig. 6.12. In this example $n = 2$, $m = 0$, and we have two curves (solid segments of line) starting at the roots of D_L, $s = 0$ and $s = -0.05$, and ending at the $n - m = 2$ zeros at infinity. The two curves meet at the *break-away point* $s = -0.025$, after which the two roots become complex-conjugate pairs. The markers locate values of α that correspond to the control gains used to generate the responses in Fig. 4.4. The root-locus helps us understand why this example's transient performance with integral-only control is much worse than with proportional-only control: in the I case, the closed-loop roots can never have real part more negative than the open-loop pole, $s = -0.05$, whereas in the P case, the closed-loop roots always have real part more negative than the open-loop pole. Compare Fig. 6.12 with Fig. 6.10. The behavior after the roots have broken away from the real line is explained by the next rule:

If the rational function $L = \beta N_L/D_L$, $\beta \neq 0$, where N_L and D_L are monic polynomials with real coefficients, is strictly proper and $n > m + 1$, then there are $n - m$ roots that converge to a set of $n - m$ straight-line asymptotes which intersect at the real point

$$c = \frac{\sum_{k=1}^{n} p_k - \sum_{i=1}^{m} z_i}{n - m}.$$

If $\beta > 0$, the asymptotes intersect the real axis at angles[12]:

$$\phi_k = \frac{\pi + k 2\pi}{n - m}, \quad k = 0, \ldots, n - m - 1.$$

In Fig. 6.12 we have $n - m = 2$ and two straight-line asymptotes intersecting at

$$c = \frac{-0.05 + 0}{2} = -0.025$$

---

[12] If $\beta < 0$, drop π from the formula.

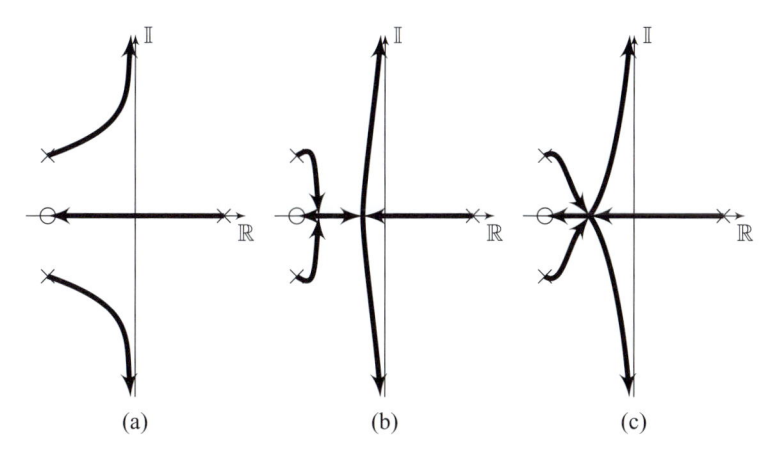

Figure 6.13 Possible root-locus for transfer-function with three poles and one zero.

at angles

$$\phi_0 = \frac{\pi}{2}, \qquad \phi_1 = \frac{3\pi}{2}.$$

The behavior away from the real axis in Fig. 6.12 is completely determined by the asymptotes. In a more complicated example, the finite portion of the locus depends on the relative location of the poles and zeros. For the pole–zero configuration in Fig. 6.9 we have $n = 3$, $m = 1$, and hence $n - m = 2$ asymptotes. The exact values of the poles and zeros are not given in the figure, but we can assume by symmetry that

$$p_1 = a, \qquad p_2 = -a + jb, \qquad p_2 = -a - jb, \qquad z_1 = -a,$$

for some real $a > 0$ and $b > 0$. Hence, the asymptotes should intersect at

$$c = \frac{(-a - jb) + (-a + jb) + (a) - (-a)}{3 - 1} = 0,$$

which is the origin. That is, the two asymptotes coincide with the imaginary axis. The path followed by the complex root-locus will, however, depend on the particular values of a and b. Figure 6.13 illustrates three possibilities for $\beta > 0$: in Fig. 6.13(a) the complex locus never intersects the real locus; in Fig. 6.13(b) the complex locus and the real locus merge at break-away points where the roots have multiplicity two; finally, in Fig. 6.13(c) the complex locus and the real locus touch at a single break-away point where the roots have multiplicity three. Which one will take place depends on the particular values of the open-loop poles and zeros. There exist rules to determine the existence of multiple roots on the real axis which allow one to distinguish among these three cases. These days, however, the best practical way to determine which case occurs is with the help of computer software, such as MATLAB. Note that, intuitively, the root paths tend to *attract* each other. So, if the imaginary component of the complex-conjugate poles is large, Fig. 6.13(a) is likely to apply, e.g. $b/a = 1$; whereas if the imaginary component is small Fig. 6.13(b) is likely to apply, e.g. $b/a = 1/4$; finally, Fig. 6.13(c) represents

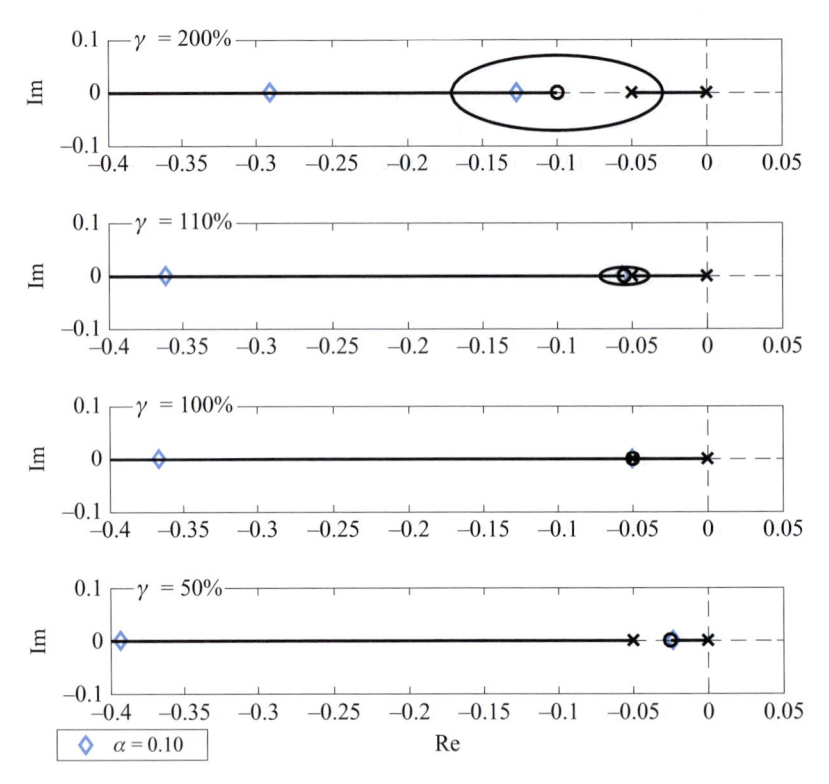

Figure 6.14 Root-locus for proportional–integral control of the car speed model (2.3) with parameters $b/m = 0.05$ and $p/b = 73.3$; diamonds show roots for a proportional gain of $K_p = \alpha = 0.1$ and integral gain $K_i/K_p = \gamma b/m$; when $\gamma = 1$ (100%) the controller performs an exact pole–zero cancellation. Compare this with the step responses in Fig. 4.19.

the transition from Fig. 6.13(a) to Fig. 6.13(b), and applies only for very special values of a and b, in this example only if $b/a = 2\sqrt{3}/9 \approx 0.38$.

Finally, the car cruise control model with the proportional and integral (PI) controller in Fig. 4.5 has

$$G = \frac{p/m}{s + b/m} = \frac{3.7}{s + 0.05}, \qquad K(s) = K_p + \frac{K_i}{s}.$$

We let $K(s) = K_p C(s)$, $C(s) = s^{-1}(s + K_i/K_p)$, and

$$\alpha = K_p, \qquad L = GC = \frac{\beta N_L}{D_L}, \qquad N_L = s + \frac{K_i}{K_p}, \qquad D_L = s(s + 0.05), \qquad \beta = 3.7,$$

and compute the root-locus after setting

$$\frac{K_i}{K_p} = \gamma \frac{b}{m} = \gamma \times 0.05$$

for various values of γ. The pole–zero cancellation choice of (4.18) corresponds to $\gamma = 1$ (100%). With cancellation, the order of the closed-loop system is reduced to first order, and the root-locus is a straight line, as can be seen in the third plot in Fig. 6.14

(the canceled pole–zero pair is still displayed). Other plots in Fig. 6.14 show the root-locus when the cancellation is not exact, that is, $\gamma \neq 1$. When we overestimate the value of the pole the controller zero is placed on the left of the pole and the root-locus must branch out of the real axis to reach the zeros, one of which is at infinity. When we underestimate the value of the pole the controller zero is placed on the right of the pole and the root-locus has a root which is necessarily slower than the value of this zero. The chosen roots marked with diamonds correspond to the controllers used to plot the responses in Fig. 4.19.

6.5 Control of the Simple Pendulum – Part I

We will now work a complete design example: the control of the simple pendulum. The goal is to design a controller that can drive the pendulum to any desired angular position. That is, we want to track a given reference angle around both stable and unstable equilibrium points. We start as we left off in Section 6.2, where we designed a proportional–derivative (PD) controller with transfer-function

$$C_{\text{PD}}(s) = K_{\text{d}}(s+z), \qquad z = \frac{K_{\text{p}}}{K_{\text{d}}},$$

that was able to stabilize the simple pendulum around both equilibrium points. The choice of gains

$$K_{\text{p}} = 3mgr, \qquad K_{\text{d}} = 2\sqrt{J_{\text{r}}mgr} - b$$

was shown to produce a stable second-order closed-loop system around the unstable equilibrium $\theta = \pi$ with natural frequency $\omega_{n_\pi} = \sqrt{2}\,\omega_{n_{\text{ol}}}$ and a damping ratio $\zeta_\pi = \sqrt{2}/2$. In order to better quantify the performance of this controller we consider the following numerical parameters:

$$m = 0.5\,\text{kg}, \quad \ell = 0.3\,\text{m}, \quad r = \ell/2 = 0.15\,\text{m}, \tag{6.20}$$

$$b = 0\,\text{kg/s}, \quad g = 9.8\,\text{m/s}^2, \quad J = \frac{m\ell^2}{12} = 3.75 \times 10^{-3}\,\text{kg m}^2, \quad J_{\text{r}} = J + mr^2. \tag{6.21}$$

Note that we are designing a controller for a frictionless pendulum model, $b = 0$, so all damping must be provided by the controller. We compute the open-loop natural frequency around the stable equilibrium $\theta = 0$:

$$\omega_{n_{\text{ol}}} = \sqrt{\frac{mgr}{J_{\text{r}}}} = 7$$

and the controller

$$K_{\text{d}} = 2\sqrt{J_{\text{r}}mgr} - b \approx 0.21, \quad K_{\text{p}} \approx 2.21, \quad z = \frac{K_{\text{p}}}{K_{\text{d}}} \approx 10.5, \quad C_{\text{PD}}(s) \approx 0.21(s+z).$$

A root-locus diagram corresponding to this controller is obtained with

$$\alpha = K_{\text{d}}, \qquad L = \frac{1}{K_{\text{d}}} G_\pi C_{\text{PD}},$$

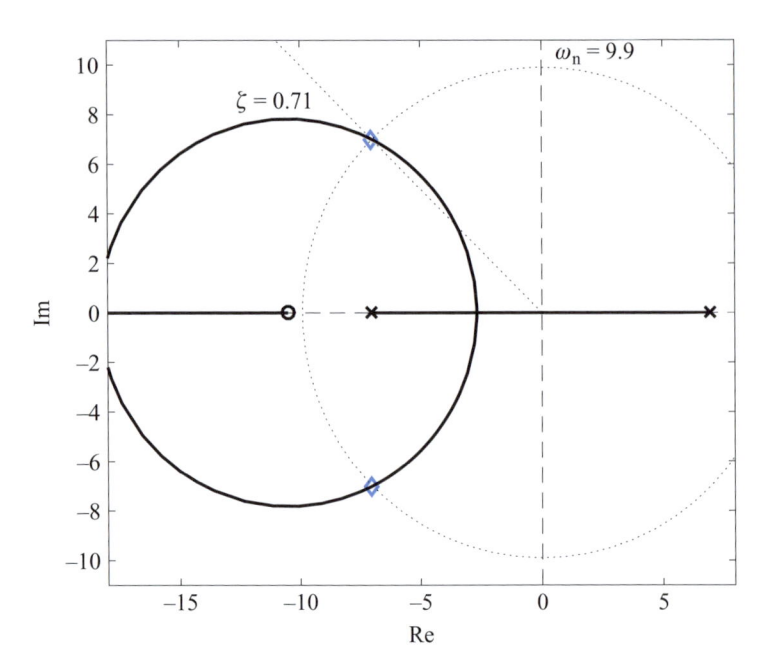

Figure 6.15 Root-locus for the proportional–derivative control of the simple pendulum; diamonds mark the choice of gains from Section 6.2.

in Fig. 6.15. Diamond markers corresponding to $K_d = \alpha \approx 0.21$ confirm the predicted closed-loop natural frequency and damping ratio:

$$\omega_{n_\pi} = \sqrt{2}\,\omega_{n_{ol}} \approx 9.90, \qquad \zeta_\pi = \frac{1}{\sqrt{2}} \approx 0.71.$$

As commented in Section 6.2, in order to physically implement this PD controller we need to have a direct measurement of the pendulum angular velocity, since the controller transfer-function is not proper. If this is not an option, then we need an alternative strategy. Recalling our ultimate goal of achieving arbitrary angular tracking, one might be tempted to simply add a pole at zero to the controller, obtaining the PI controller:

$$\alpha \frac{s+z}{s}, \qquad z > 0.$$

The problem with the introduction of the extra pole at the origin is that this controller no longer stabilizes the closed-loop system for any value of $\alpha > 0$. The introduction of the pole means that the locus has $n - m = 3 - 1 = 2$ asymptotes centered at

$$c = \frac{\sqrt{mgr} - \sqrt{mgr} + 0 - (-z)}{n - m} = \frac{z}{2} > 0$$

and most likely the roots on the right-hand side of the complex plane will simply converge toward the two vertical asymptotes which have positive real part when $\alpha > 0$. This is illustrated by the root-locus in Fig. 6.16.

A conclusion from the analysis of the asymptotes is that if a pole, $-p$, is to be added to make $C_{PD}(s)$ a proper controller it should be added to the left of the zero, $-z$. In this

6.5 Control of the Simple Pendulum – Part I

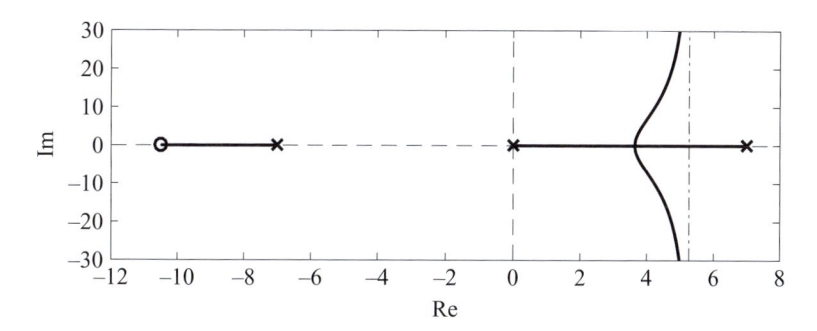

Figure 6.16 Root-locus for proportional–integral control of the simple pendulum; no positive choice of gain makes the closed-loop stable.

way $p > z$ and the pair of asymptotes is now centered at

$$c = \frac{\sqrt{mgr} - \sqrt{mgr} + (-p) - (-z)}{n - m} = \frac{z - p}{2} < 0,$$

on the left-hand side of the complex plane. This is the controller of the form

$$K \frac{s + z}{s + p}, \qquad p > z > 0, \qquad K > 0.$$

For reasons that will be clear in Chapter 7, the above controller is known as a *lead compensator*. We shall use the root-locus diagram to help us select the position of the pole, that is, the value of p. This is accomplished by defining the loop transfer-function:

$$\alpha = \frac{K}{pK_d}, \qquad L = G_\pi C_{\text{lead}}, \qquad C_{\text{lead}}(s) = pK_d \frac{s + z}{s + p}, \qquad p > z = \frac{K_p}{K_d} > 0.$$

Note that

$$C_{\text{lead}}(s) \approx K_p + sK_d = C_{\text{PD}}(s) \qquad \text{when } s \approx 0,$$

which helps to preserve the low-frequency behavior of the PD controller designed before. For this reason, we expect to be able to select a root-locus gain α close to one. Moreover,

$$C_{\text{lead}}(s) \approx K_p + sK_d = C_{\text{PD}}(s) \qquad \text{when } p \text{ is large.}$$

Hence, if we locate the controller pole, $s = -p$, far enough to the left of the controller zero, $s = -z$, we should expect that the new pole will not disturb the behavior of the closed-loop system. Indeed, this is what we see when placing the pole eight times as far to the left as z, i.e. $p = 8z$, on the root-locus in Fig. 6.17(a). Note that the region close to the origin is essentially the same as in Fig. 6.15, with the new asymptotes staying on the far left side of the plot. It is still possible to select a gain, α, that places the dominant closed-loop poles with a natural frequency, ω_n, and damping ratio, ζ, which are very close to the ones obtained with the PD controller. This choice of gain is marked with diamonds.

Placing the pole, $-p$, too far to the left means that the controller still has to deliver high gains in a wide region of the frequency spectrum. We compare the magnitude of the frequency response of the above controllers, that is $20 \log_{10} |C(j\omega)|$

Controller Design

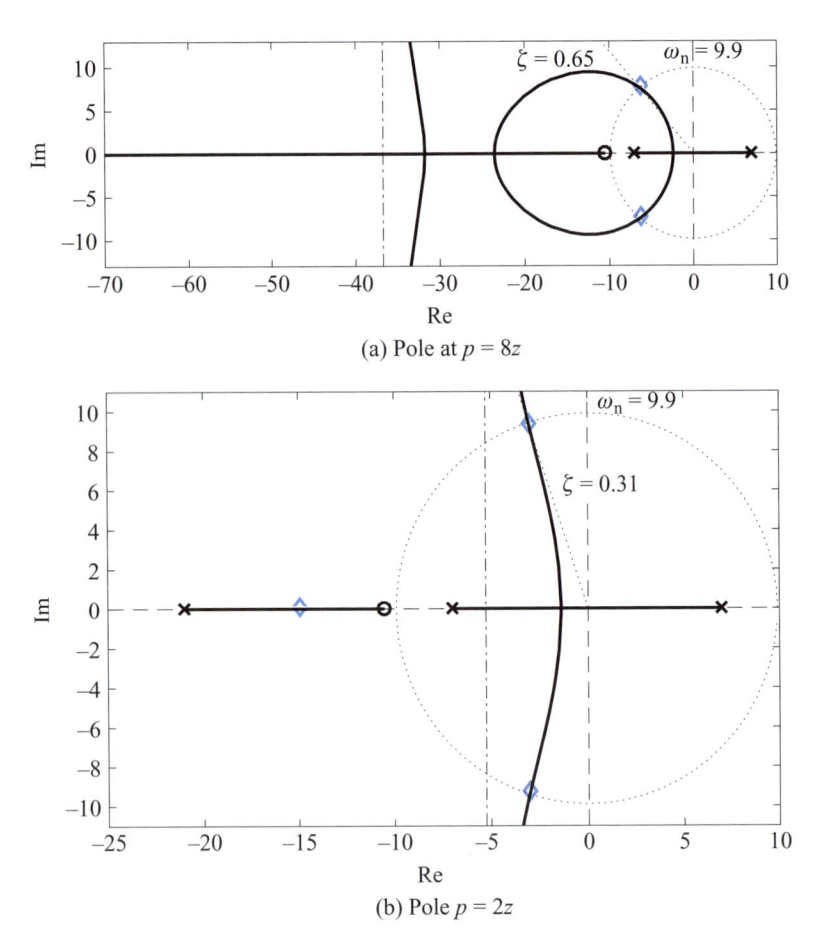

Figure 6.17 Root-locus for the lead control of the simple pendulum; (a) when the pole is placed far enough on the left of the zero, $p = 8z$, the root-locus near the origin is similar to Fig. 6.15; (b) as the pole is brought closer to the zero, $p = 2z$, the root-locus behavior changes; for the same closed-loop natural frequency, $\omega_n \approx 9.9$ rad/s, the closed-loop damping ratio, $\zeta \approx 0.31$, achievable for $p = 2z$ is much smaller than the damping ratio, $\zeta \approx 0.65$, achievable for $p = 8z$.

(dB) versus $\log_{10} \omega$, in Fig. 6.18. We will be much better equipped to understand such plots, called Bode plots, in Chapter 7. For now we just want to compare the magnitude of the frequency response of the PD controller, $(s + z)$ in Fig. 6.18, with the lead controller with the pole at $p = 8z$, $(s + z)/(s + 8z)$ in Fig. 6.18. The lead controller still requires large amplification at high frequencies: the magnitude of the frequency response of the controller at $\omega = 10^3$ rad/s is about 10 times higher than at $\omega = 1$ rad/s (≈ 20 dB, see Section 7.1). Because of the high gain at high frequencies, we will attempt to bring the controller pole *closer* to the controller zero.

On placing the pole twice as far to the left as z, that is, at $p = 2z$, the resulting controller has to deliver much less gain at high frequencies, about 5 dB or 1.8 times higher than at low-frequencies, as seen for $(s + z)/(s + 2z)$ in Fig. 6.18. However, the closed-loop behavior is much impacted, with the root-locus taking a completely different shape,

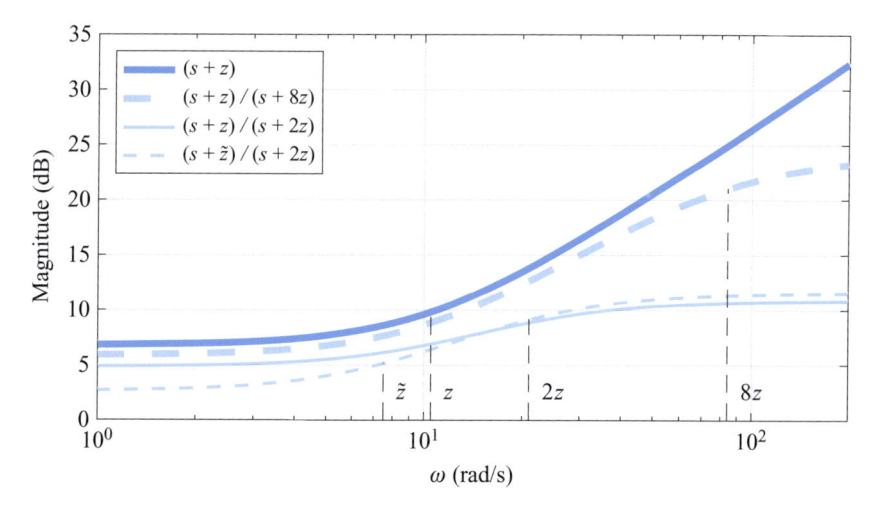

Figure 6.18 Magnitude of the frequency response of the proportional–derivative controller and lead controllers for different choices of the pole and zero.

Fig. 6.17(b). The asymptotes are now located between the two system poles, which results in much less damping, $\zeta \approx 0.31$, if ω_n is to stay close to $\omega_n \approx 9.9$. A comparable damping ratio is now possible only if the closed-loop system operates at a much slower natural frequency, which is not what one would want in this case.

In order to improve the closed-loop damping ratio we fix the pole at $p = 2z$ and shift the zero of the controller toward the right in an attempt to move the asymptotes more to the left. This forces the intersection of the locus with the circle of radius $\omega_n = 9.90$ to happen at higher damping ratios. After moving the zero from $z \approx 10.5$ to

$$\tilde{z} \approx 7.5$$

we obtain the root-locus shown in Fig. 6.19. It is now possible to select

$$K \approx 3.83$$

such that $\omega_n \approx 9.9$ and still obtain a damping ratio of about 0.60, which is a huge improvement when compared with the damping ratio of 0.31 obtained with the zero in its original location. As a side-effect, as seen for $(s + \tilde{z})/(s + 2z)$ in Fig. 6.18, the resulting controller operates with lower gains at lower frequencies. Interestingly, after moving the zero we almost performed a pole–zero cancellation of the stable pole of the pendulum! The result is the controller

$$C_{\text{lead}}(s) \approx 3.83 \frac{s + 7.5}{s + 21}. \tag{6.22}$$

The controller (6.22) was designed to achieve good performance around the unstable equilibrium $\theta = \pi$. Before proceeding we shall also evaluate its performance around the stable equilibrium $\theta = 0$. Figure 6.20 shows the root-locus obtained with controller (6.22) in closed-loop with the pendulum model linearized around $\theta = 0$. The root-locus with the pendulum model linearized around $\theta = \pi$ is shown as a dashed line

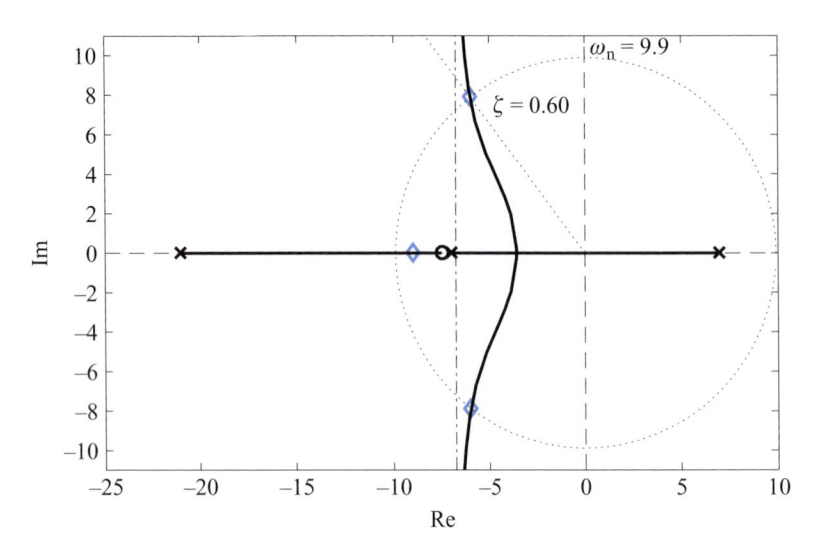

Figure 6.19 Root-locus for the lead control of the simple pendulum after fixing the pole and adjusting the zero in order to improve the closed-loop damping ratio, $\zeta \approx 0.60$.

for comparison. Note that the closed-loop natural frequency is close to the one predicted before with the proportional–derivative controller, i.e. $\omega_n \approx 14.4 \approx 2\omega_{n_{ol}}$, but this controller displays much less closed-loop damping, $\zeta \approx 0.24$. This means that one should be careful if the same controller is to be used around both equilibria, even though, in

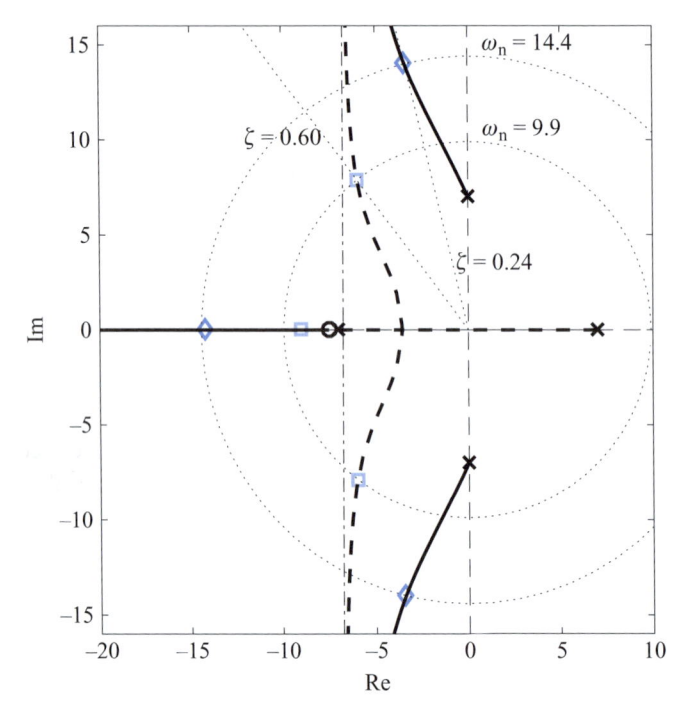

Figure 6.20 Root-locus for the lead control (6.22) of the simple pendulum in closed-loop with the model linearized around the stable equilibrium $\theta = 0$; the root-locus for the model linearized around the unstable equilibrium $\theta = \pi$ is shown as a dashed line for comparison.

6.5 Control of the Simple Pendulum – Part I

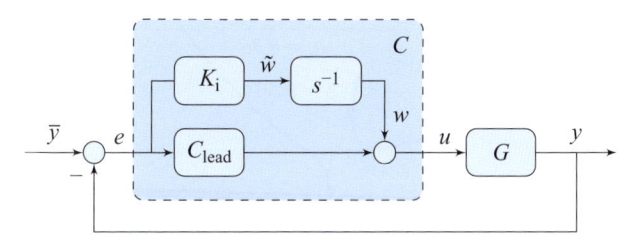

Figure 6.21 Block-diagram for controller with additional integral action.

practice, additional mechanical damping, which we chose to ignore during the design, might improve the overall damping. We will justify this later in Section 8.2 when we perform a more complete analysis of the closed-loop system.

Let us turn our attention once again to the issue of tracking. As discussed earlier in Section 5.8, one should be careful when using a model linearized around equilibrium as the basis for a tracking controller that attempts to move the system away from equilibrium: if the model near the new equilibrium is too different than the one for which the controller was originally designed then the design might no longer be valid. Having said that, we cautiously proceed using the model linearized around the unstable equilibrium, $\theta = \pi$, ready to perform a redesign if necessary.

Our goal is to add integral action to the loop. Our previous attempt to simply add a pole at the origin, Fig. 6.16, revealed serious problems with closed-loop stability. For this reason we must take a different route. Instead of adding the integrator in *series* with the controller, we add it in *parallel*, the resulting controller being very close to an implementable PID controller, as we will see. We assume that $C_{\text{lead}}(s)$ has already been designed to provide regulation performance, our goal being to design K_i to achieve adequate levels of integral action. The feedback connection is illustrated in Fig. 6.21. The controller resulting from this diagram has the transfer-function

$$C(s) = \frac{K_i}{s} + C_{\text{lead}}(s) = K\frac{s^2 + (z + K_i/K)s + pK_i/K}{s(s+p)}, \qquad (6.23)$$

which has not only an extra pole at the origin but also an extra zero, because both the denominator and the numerator of the controller are of degree 2.

At first it might not be obvious what loop transfer-function, L, to choose for a root-locus analysis of the diagram in Fig. 6.21. Our goal is to visualize the closed-loop poles as a function of K_i so we compare Fig. 6.21 with Fig. 6.8 to conclude[13] that "$-L$" must be the transfer-function between the input $\tilde{w}$ and the error signal e. From Figs. 6.21 and 4.11, the closed-loop transfer-function from w to e has already been computed in (4.25) and (4.26):

$$e = -Dw, \qquad D = \frac{G_\pi}{1 + G_\pi C_{\text{lead}}},$$

---

[13] Recall that α is a *negative* feedback gain in Fig. 6.21.

Controller Design

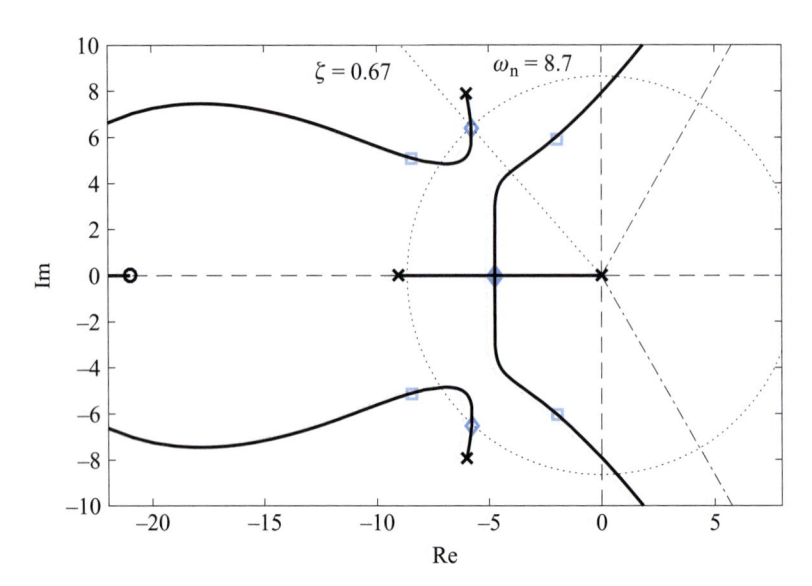

Figure 6.22 Root-locus for the lead control with additional integral action (6.23); integral gain is chosen for the closed-loop system to have double real poles and a slightly smaller natural frequency, $\omega_n \approx 8.7$ rad/s, marked with diamonds; alternative choice of gain that matches $\omega_n \approx 9.9$ rad/s leads to highly underdamped poles close to the imaginary axis, marked with squares; higher integral gains eventually lead to instability.

with which

$$e = -L\tilde{w}, \qquad L = s^{-1}D = \frac{s^{-1}G_\pi}{1 + G_\pi C_{\text{lead}}}.$$

It follows that L has as poles the closed-loop poles of the connection of G_π with C_{lead} plus the origin,[14] and has as zero the single pole of C_{lead}. The corresponding root-locus is plotted in Fig. 6.22.

Intuitively, if C_{lead} is able to internally stabilize the closed-loop then a small enough K_i should still preserve stability. Too large a K_i will generally lead to instability. Too small a K_i, however, will produce a closed-loop system that has a pole too close to the origin, making the closed-loop system too slow. The presence of three asymptotes in Fig. 6.22 explains why too much integral gain will lead to instability. The real path departing from the origin explains why too little integral gain will leave a pole too close to the origin.

Roots on the path that depart from the pair of complex poles have more and more damping as K_i grows. It is the pair of roots departing from the real poles that become dominant and need special attention. With this in mind we select as integral gain the root-locus gain

$$\alpha = K_i \approx 1.20, \tag{6.24}$$

---

[14] Verify the location of the poles and zeros in the root-locus diagram in Fig. 6.22!

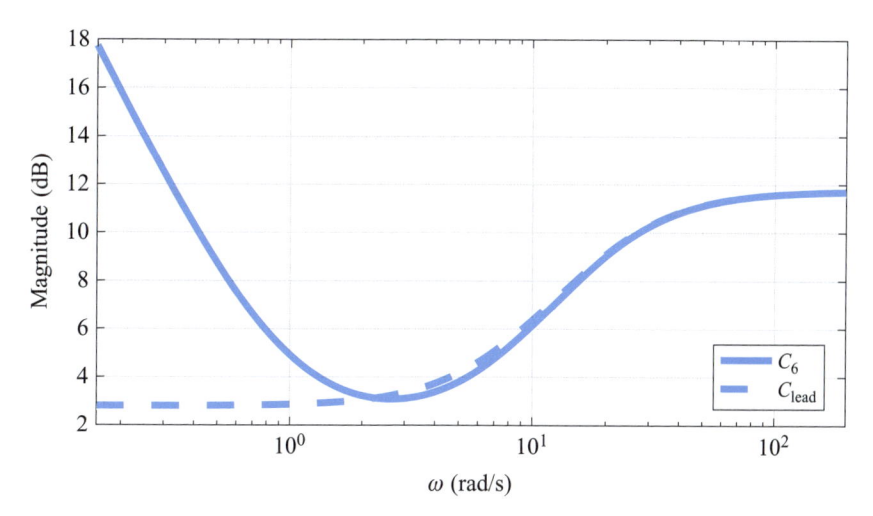

Figure 6.23 Magnitude of the frequency response of the lead controller, C_{lead} from (6.22), and the lead controller with integral action, C_6 from (6.25).

which corresponds to closed-loop poles at the break-away point from the real axis, marked in Fig. 6.22 by diamonds. This implies the fastest possible response from the dominant poles that is not oscillatory.[15] Note that the alternative choice of gain $K_i \approx 2.8$ leads to the only possible stable roots that intersect the circle $\omega_n \approx 9.9$, marked in Fig. 6.22 by squares, but we do not opt for this choice because it produces a pair of highly underdamped complex poles on the dominant branch.

The controller obtained after selecting K_i as in (6.24) results in the transfer-function

$$C_6(s) \approx 3.83 \frac{(s+6.86)(s+0.96)}{s\,(s+21)}. \tag{6.25}$$

Note how the magnitude of the frequency response of the final controller C_6 and that of the lead controller C_{lead}, plotted in Fig. 6.23, are very close at high-frequencies. At low-frequencies, C_6 displays the typical unbounded low-frequency gain provided by the integrator.

Of course one may take different paths to arrive at a suitable controller. For instance, one might have concluded after the root-locus plotted in Fig. 6.16 that a second zero was needed and designed a second-order controller by placing the additional zero and pole and studying the resulting root-locus diagram. The diagram in Fig. 6.24 shows the root-locus one would study if a second-order controller with the same poles and zeros as C_6 from (6.25) were used. The gain corresponding to the closed-loop poles marked by diamonds in Fig. 6.24 recovers those shown in Fig. 6.22. Note how arguments similar to the ones brought when discussing previous root-locus plots can be used to help locate the poles and zeros in this direct design. One might also enjoy the complementary discussion in Section 7.8, in which we will revisit the control of the simple pendulum using frequency domain techniques.

---

[15] The other two poles still contribute oscillatory components to the response.

Controller Design

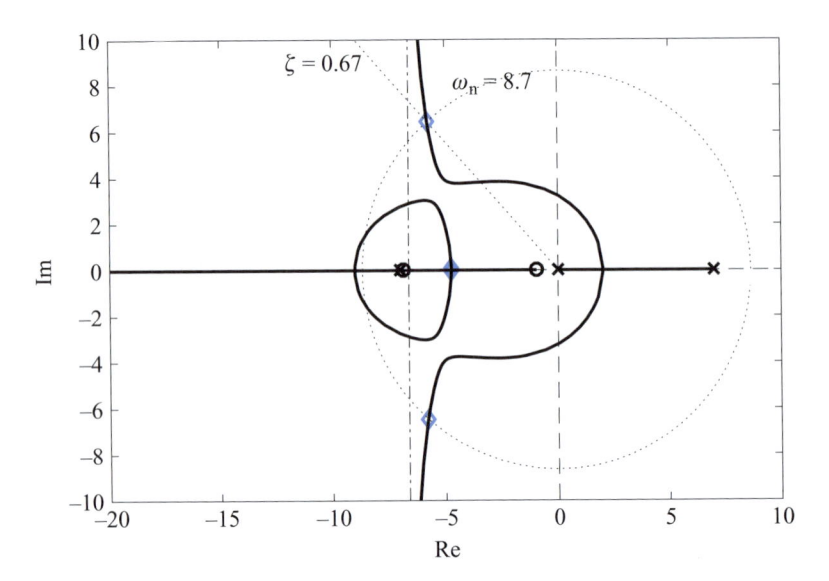

Figure 6.24 Alternative root-locus for the direct study of the control of the pendulum in closed-loop with a second-order controller with the same poles and zeros as (6.25); diamond marks indicate the choice of gain that leads to the same closed-loop poles as in the root-locus in Fig. 6.22.

Problems

6.1 Show that the roots of the second-order polynomial

$$s^2 + 2\zeta\omega_n s - \omega_n^2 = 0, \qquad \omega_n > 0,$$

are always real and at least one of the roots is positive.

6.2 Show that the step response of an underdamped second-order system with transfer-function

$$G(s) = \frac{\omega_n^2}{s^2 + 2\zeta\omega_n s + \omega_n^2}, \qquad \omega_n > 0,$$

is

$$y(t) = 1 - \frac{e^{-\zeta\omega_n t}}{\sqrt{1-\zeta^2}} \sin(\omega_d t + \pi/2 - \phi_d),$$

where

$$\omega_d = \omega_n\sqrt{1-\zeta^2}, \qquad \phi_d = \tan^{-1}\left(\frac{\zeta}{\sqrt{1-\zeta^2}}\right),$$

for $t \geq 0$, $0 < \zeta < 1$.

6.3 Maximize $y(t)$ to show that t_p and y_p given in (6.8) are the time and value of the first peak of the step response of an underdamped second-order system with transfer-function $G(s)$ as in P6.2 with $0 < \zeta < 1$. *Hint: Differentiate $y(t)$ and solve $\dot{y}(t) = 0$.*

6.4 Use the approximation $\omega_d t \approx 2k\pi$, $k \in \mathbb{Z}$, formula (6.6), and solve for $y(t_s) - 1 \approx 0.02$ to establish the settling time approximate formula (6.10).

6.5 Verify graphically that the formulas in (6.11) have a maximum relative error of less than 1% in the range $0 < \zeta < 1$.

6.6 Calculate (6.15) and (6.16).

6.7 Sketch the root-locus for the SISO systems with poles and zeros shown in Fig. 6.25 then use MATLAB to verify your answer.

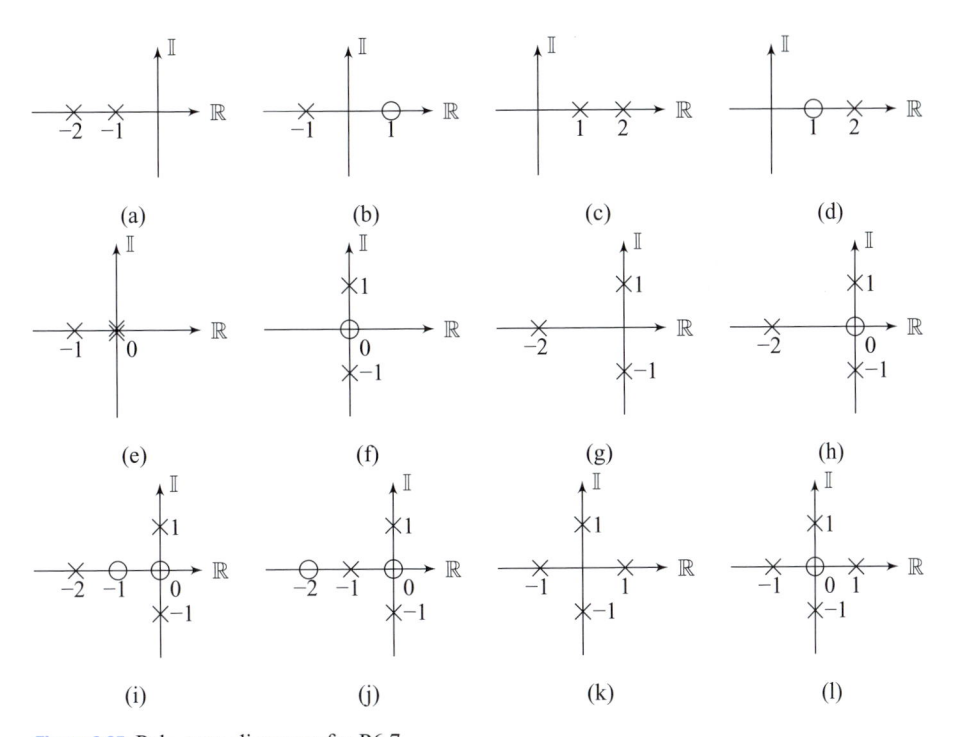

Figure 6.25 Pole–zero diagrams for P6.7.

6.8 Use the root-locus method to determine a proper feedback controller that can stabilize the SISO systems with poles and zeros shown in Fig. 6.25. Is the transfer-function of the controller asymptotically stable? Recall that you should never perform a pole–zero cancellation on the right-hand side of the complex plane. *Note: Some are not trivial!*

6.9 You were shown in Section 5.5 that the nonlinear differential equation

$$J_r \ddot{y} + b\dot{y} + mgr\sin\theta = u$$

is an approximate model for the motion of the simple pendulum in Fig. 5.11 and that $(\bar{\theta}, \bar{u}) = (0, 0)$ and $(\bar{\theta}, \bar{u}) = (\pi, 0)$ are the pendulum equilibrium points. Calculate the nonlinear differential equation obtained in closed-loop with the linear proportional

Controller Design

controller

$$u = Ke, \qquad e = \bar{\theta} - \theta.$$

Show that $(\bar{\theta}, \bar{u}) = (0, 0)$ and $(\bar{\theta}, \bar{u}) = (\pi, 0)$ are still equilibrium points. Linearize the closed-loop system linearized about $(\bar{\theta}, \bar{u}) = (0, 0)$ and $(\bar{\theta}, \bar{u}) = (\pi, 0)$ and calculate the associated transfer-function. Assuming all constants are positive, find the range of values of K that stabilize both equilibrium points.

6.10 Repeat P6.9 with the linear proportional–derivative controller:

$$u = K_p e - K_d \dot{e}, \qquad e = \bar{\theta} - \theta.$$

6.11 You have shown in P2.10 and P2.12 that the ordinary differential equation

$$\left(J_1 r_2^2 + J_2 r_1^2\right) \dot{\omega}_1 + \left(b_1 r_2^2 + b_2 r_1^2\right) \omega_1 = r_2^2 \tau, \qquad \omega_2 = (r_1/r_2)\omega_1$$

is a simplified description of the motion of a rotating machine driven by a belt without slip as in Fig. 2.18(a), where ω_1 is the angular velocity of the driving shaft and ω_2 is the machine's angular velocity. Let $r_1 = 25$ mm, $r_2 = 500$ mm, $b_1 = 0.01$ kg m²/s, $b_2 = 0.1$ kg m²/s, $J_1 = 0.0031$ kg m², $J_2 = 25$ kg m². Use the root-locus method to design an I controller:

$$\tau(t) = K_i \int_0^t e(\sigma) d\sigma, \qquad e = \bar{\omega}_2 - \omega_2,$$

and select K_i so that both closed-loop poles are real and as negative as possible. Is the closed-loop capable of asymptotically tracking a constant reference input $\bar{\omega}_2$? Is the closed-loop capable of asymptotically rejecting a constant input torque disturbance?

6.12 Repeat P6.11 using the PI controller

$$\tau(t) = K_p e(t) + K_i \int_0^t e(\sigma) d\sigma, \qquad e = \bar{\omega}_2 - \omega_2,$$

and select the gains K_p and K_i such that both closed-loop poles have negative real part more negative than the pole of the machine without performing a pole–zero cancellation. Is the closed-loop capable of asymptotically tracking a constant reference input $\bar{\omega}_2$?

6.13 The rotating machine in P6.11 is connected to a piston that applies a periodic torque that can be approximated by $\tau_2(t) = h \cos(\sigma t)$, where the angular frequency σ is equal to the angular velocity ω_2. Show that the modified equation including this additional torque is given by

$$\left(J_1 r_2^2 + J_2 r_1^2\right) \dot{\omega}_1 + \left(b_1 r_2^2 + b_2 r_1^2\right) \omega_1 = r_2(r_2 \tau + r_1 \tau_2), \qquad \omega_2 = (r_1/r_2)\omega_1.$$

Use the root-locus method to design a dynamic feedback controller that uses τ as control input and ω_2 as the measured output so that the closed-loop system is capable of asymptotically tracking a constant reference input $\bar{\omega}_2(t) = \bar{\omega}_2 = 4\pi, t \geq 0$, and asymptotically rejecting the torque perturbation $\tau_2(t) = h \cos(\sigma t)$ when $\sigma = \bar{\omega}_2$.

6.14 You have shown in P2.18 that the ordinary differential equation

$$\left(J_1 + J_2 + r^2(m_1 + m_2)\right)\dot{\omega} + (b_1 + b_2)\omega = \tau + gr(m_1 - m_2), \qquad v_1 = r\omega,$$

is a simplified description of the motion of the elevator in Fig. 2.18(b), where ω is the angular velocity of the driving shaft and v_1 is the elevator's load linear velocity. Let $r = 1$ m, $m_1 = m_2 = 1000$ kg, $b_1 = b_2 = 120$ kg m²/s, $J_1 = J_2 = 20$ kg m², and $g = 10$ m/s². Use the root-locus method to design a dynamic feedback controller that uses τ as control input and the elevator's load vertical position

$$x_1(t) = x_1(0) + \int_0^t v_1(\tau)d\tau$$

as the measured output so that the closed-loop is capable of asymptotically tracking a constant position reference $\bar{x}_1(t) = \bar{x}_1, t \geq 0$.

6.15 Repeat P6.14 with $m_2 = 800$ kg.

6.16 You have shown in P2.28 that the ordinary differential equation:

$$m\ddot{x} + b\dot{x} + kx = f + mg\sin\theta$$

is a simplified description of the motion of the mass–spring–damper system in Fig. 2.19(b), where g is the gravitational acceleration and x_0 is equal to the spring rest length ℓ_0. The additional force, f, will be used as a control input. Let $g = 10$ m/s², $m = 1$ kg, $k = 1$ N/m, and $b = 0.1$ kg/s, and use the root-locus method to design a dynamic feedback controller that uses f as control input and x as the measured output and that can regulate the position, x, at zero for any constant possible value of inclination $\theta \in (-\pi/2, \pi/2)$. *Hint: Treat the inclination as a disturbance.*

6.17 You have shown in P2.32 that the ordinary differential equations

$$m_1\ddot{x}_1 + (b_1 + b_2)\dot{x}_1 + (k_1 + k_2)x_1 - b_2\dot{x}_2 - k_2 x_2 = f_1,$$
$$m_2\ddot{x}_2 + b_2(\dot{x}_2 - \dot{x}_1) + k_2(x_2 - x_1) = f_2$$

constitute a simplified description of the motion of the mass–spring–damper system in Fig. 2.20(b), where x_1 and x_2 are displacements, and f_1 and f_2 are forces applied on the masses m_1 and m_2. Let the force, f_2, be the control input and the displacement, x_2, be the measured output. Let $m_1 = m_2 = 1$ kg, $b_1 = b_2 = 0.1$ kg/s, $k_1 = 1$ N/m, and $k_2 = 2$ N/m. Use the root-locus method to design a dynamic feedback controller that uses f_2 as control input and x_2 as the measured output and that can regulate the position, x_2, at zero for any constant possible value of force f_1. *Hint: Treat the force f_1 as a disturbance.*

6.18 Repeat problem P6.17 for a force $f_1 = \cos(2\pi t)$.

6.19 The mass–spring–damper diagram in Figure 6.26(a) is used to model the suspension dynamics of a car [Jaz08, Chapter 14], and is known as the *one-eighth-car model*. The mass m represents $1/4$ of the total mass of the car. The constants k and b

are the stiffness and damping coefficient of the spring and shock absorber. The ordinary differential equation

$$m\ddot{x} + b\dot{x} + kx = b\dot{y} + ky$$

constitutes a simplified description of the motion of this model, where x is a displacement measured from equilibrium. Show that, if $z = x - y$,

$$m\ddot{z} + b\dot{z} + kz = -m\ddot{y},$$

where y is the road profile.

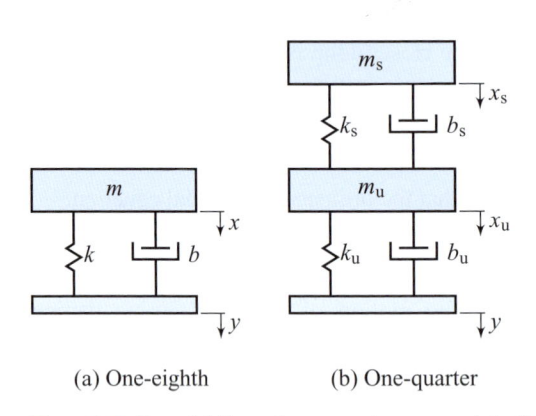

(a) One-eighth   (b) One-quarter

Figure 6.26 One-eighth- and one-quarter-car models, P6.19 and P6.24.

6.20 Consider the one-eighth-car model from P6.19. Calculate the transfer-function from the road profile, y, to the mass relative displacement, z. Calculate the value of the spring stiffness, k, and shock absorber damping coefficient, b, for a car with 1/4 mass equal to 640 kg to have a natural frequency $f_n = 2.5$ Hz and damping ratio $\zeta = 0.08$. Locate the roots in the complex plane.

6.21 Consider the one-eighth-car model from P6.19. Use MATLAB to plot the response of the car suspension with parameters as in P6.20 to a pothole with a profile as shown in Fig. 6.27, where $w = 1$ m and $d = 5$ cm for a car traveling at 10 km/h. Repeat for a car traveling at 100 km/h. Comment on your findings.

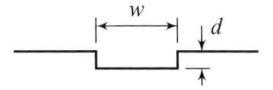

Figure 6.27 Road profile for P6.21.

6.22 Consider the one-eighth-car model from P6.19 and a road with profile $y(t) = \cos(vt/\lambda)$, where v is the car's velocity. What is the worst possible velocity a car with suspension parameters as in P6.20 could be traveling at as a function of the road wavelength λ?

6.23 Consider the one-eighth-car model from P6.19. Show that

$$m\ddot{z} = u - m\ddot{y}, \qquad u = -(kz + b\dot{z}),$$

from which u can be interpreted as the output of a PD controller. Use this fact to repeat P6.20 using the root-locus method.

6.24 The mass–spring–damper diagram in Figure 6.26(b) is an improved version of the model considered in P6.19 which is used to model the wheel–body dynamics of a car [Jaz08, Chapter 15]. It is known as the *one-quarter-car model*. The mass m_s represents 1/4 of the mass of the car without the wheels and m_u represents the mass of a single wheel. The constants k_s and b_s are the stiffness and damping coefficient of the spring and shock absorber. The constants k_u and b_u are the stiffness and damping coefficient of the tire. The ordinary differential equations

$$m_s\ddot{x}_s + b_s(\dot{x}_s - \dot{x}_u) + k_s(x_s - x_u) = 0,$$
$$m_u\ddot{x}_u + (b_u + b_s)\dot{x}_u + (k_u + k_s)x_u - b_s\dot{x}_s - k_s x_s = k_u y + b_u \dot{y}$$

constitute a simplified description of the motion of the one-quarter-car model, where x_s and x_u are displacements measured from equilibrium. If

$$x = x_s - x_u, \qquad z = x_u - y,$$

show that

$$m_s m_u \ddot{x} + b_s(m_s + m_u)\dot{x} + k_s(m_s + m_u)x - b_u m_s \dot{z} - k_u m_s z = 0,$$
$$m_u \ddot{z} + b_u \dot{z} + k_u z - b_s \dot{x} - k_s x = -m_u \ddot{y},$$

where y is the road profile.

6.25 Consider the one-quarter-car model from P6.24. If the tire stiffness is $k_u = 200,000$ N/m and the tire damping coefficient is negligible, i.e. $b_u = 0$, use MATLAB to calculate the transfer-function from the road profile, y, to the relative displacement $x + z$ and select values of the spring stiffness, k_s, and shock absorber damping coefficient, b_s, for a car with 1/4 mass $m_s = 600$ kg and wheel mass $m_u = 40$ kg to have its dominant poles display a natural frequency $f_n = 2.5$ Hz and damping ratio $\zeta = 0.08$. Locate all the roots in the complex plane.

6.26 Consider the one-quarter-car model from P6.24. Use MATLAB to plot the response of the car suspension with parameters as in P6.25 to a pothole with a profile as shown in Fig. 6.27, where $w = 1$ m and $d = 5$ cm for a car traveling at 10 km/h. Repeat for a car traveling at 100 km/h. Comment on your findings.

6.27 Consider the one-quarter-car model from P6.24 and a road with profile $y(t) = \cos(vt/\lambda)$, where v is the car's velocity. What is the worst possible velocity a car with suspension parameters as in P6.25 could be traveling at as a function of the road wavelength λ?

6.28 Consider the one-quarter-car model from P6.24. Show that

$$m_u\ddot{x} - b_u\dot{z} - k_u z = (1 + m_u/m_s)u,$$
$$m_u\ddot{z} + b_u\dot{z} + k_u z = -u - m_u\ddot{y},$$

where

$$u = -(k_s x + b_s \dot{x}),$$

can be interpreted as the output of a PD controller. Use this fact to repeat P6.25 using the root-locus method.

6.29 Compare the answers from P6.25–P6.28 with the answers from P6.20–P6.23.

6.30 You have shown in P2.41 that the ordinary differential equation

$$J\dot{\omega} + \left(b + \frac{K_e K_t}{R_a}\right)\omega = \frac{K_t}{R_a}v_a$$

is a simplified description of the motion of the rotor of the DC motor in Fig. 2.24. Let the voltage v_a be the control input and the rotor angular velocity ω be the measured output. Let $J = 227 \times 10^{-6}$ kg m^2, $K_t = 0.02$ N m/A, $K_e = 0.02$ V s/rad, $b = 289.4 \times 10^{-6}$ kg m^2/s, and $R_a = 7\,\Omega$. Use the root-locus method to design a dynamic feedback controller so that the closed-loop system is capable of asymptotically tracking a constant reference input $\bar{\omega}(t) = \bar{\omega}, t \geq 0$.

6.31 Repeat P6.30 to design a position controller that uses a measurement of the angular position

$$\theta(t) = \theta(0) + \int_0^t \omega(\tau)d\tau,$$

as output and is capable of asymptotically tracking a constant angular reference $\bar{\theta}(t) = \bar{\theta}, t \geq 0$.

6.32 What difference does it make using an integrator to integrate the angular velocity rather than a sensor to directly measure angular position in P6.31?

6.33 You have shown in P4.34 that the torque of a DC motor, τ, is related to the armature voltage, v_a, through the transfer-function

$$\frac{T(s)}{V_a(s)} = \frac{K_t}{R_a} \frac{(s + b/J)}{s + b/J + K_e K_t/(R_a J)}.$$

Use the data from P6.30 and the root-locus method to design a controller that uses the voltage v_a as the control input and the torque τ as the measured output so that the closed-loop system is capable of asymptotically tracking a constant reference input torque $\bar{\tau}(t) = \bar{\tau}, t \geq 0$.

6.34 You have shown in P2.49 that the temperature of a substance, T (in K or in °C), flowing in and out of a container kept at the ambient temperature, T_o, with an inflow temperature, T_i, and a heat source, q (in W), can be approximated by the differential equation

$$mc\dot{T} = q + wc\,(T_i - T) + \frac{1}{R}(T_o - T),$$

where m and c are the substance's mass and specific heat, and R is the overall system's thermal resistance. The input and output flow mass rates are assumed to be equal to w in kg/s. Assume that water's density and specific heat are 997.1 kg/m^3 and $c = 4186$ J/kg K. Use the root-locus method to design a dynamic feedback controller that uses the heat source q as the control input and the temperature T as the measured output for a 50 gal (≈ 0.19 m^3) water heater rated at $\bar{q} = 40{,}000$ BTU/h (≈ 12 kW) and thermal resistance $R = 0.27$ K/W at ambient temperature, $T_o = 77\,°$F ($\approx 25\,°$C). The controller should achieve asymptotic tracking of a reference temperature $\bar{T} = 140\,°$F ($\approx 60\,°$C) without any in/out flow, i.e. $w = 0$.

6.35 Repeat P6.34 with a constant in/out flow of 20 gal/h ($\approx 21 \times 10^{-6}$ m^3/s) at ambient temperature.

6.36 Repeat P6.34 with a sinusoidal in/out flow perturbation

$$w(t) = \frac{\bar{w}}{2}(1 + \cos(\omega t)),$$

where $\bar{w} = 20$ gal/h ($\approx 21 \times 10^{-6}$ m^3/s) at ambient temperature and $\omega = 2\pi/24\,\text{h}^{-1}$. Approximate $w(t)(T_i - T(t)) \approx w(t)(T_i - \bar{T})$.

6.37 You have shown in P5.42 that

$$\dot{x} = \begin{bmatrix} 0 & 1 & 0 \\ 3\Omega^2 & 0 & 2\Omega \\ 0 & -2\Omega & 0 \end{bmatrix} x + \begin{bmatrix} 0 \\ 0 \\ 1/m \end{bmatrix} u_t, \qquad x = \begin{pmatrix} r - R \\ \dot{r} \\ R(\omega - \Omega) \end{pmatrix},$$

$$y = \begin{bmatrix} 1 & 0 & 0 \end{bmatrix} x, \qquad\qquad\qquad\qquad\qquad y = r - R$$

is a simplified description of the motion of a satellite orbiting earth as in Fig. 5.18, where r is the satellite's radial distance from the center of the earth, ω is the satellite's angular velocity, m is the mass of the satellite, M is the mass of the earth, G is the universal gravitational constant, and u_t is a force applied by a thruster in the tangential direction. These equations were obtained by linearizing around the equilibrium orbit $u_t(t) = u_r(t) = \dot{r}(t) = 0$, $r(t) = R$, and $\omega(t) = \Omega$, where $\Omega^2 R^3 = GM$. Let $M \approx 6 \times 10^{24}$ kg be the mass of the earth, and $G \approx 6.7 \times 10^{-11}$ N m^2/kg^2. Use MATLAB to calculate the transfer-function from the tangential thrust, u_t, to the radial distance deviation, y, for a 1600 kg GPS satellite in medium earth orbit (MEO) with a period of 11 h. Use the root-locus method to design a dynamic feedback controller that uses tangential thrust u_t as the control input and the radial distance y as the measured output and that can regulate the radial distance of the satellite, y, in closed-loop.

6.38 You have shown in P5.52 that the linearized transfer-function from the insulin plasma release rate, u, to the glucose level, y, is

$$\frac{\tilde{Y}(s)}{\tilde{U}(s)} = \frac{-gb\bar{x}_2}{(s+f)(s+a)(s+c+\bar{x}_1)}.$$

The authors of [Ste+03] propose that glucose homeostasis is maintained by the following feedback mechanism:

$$u(t) = K_\mathrm{p}(y(t) - \bar{y}) + K_\mathrm{p} T_\mathrm{d} \frac{dy(t)}{dt} + \frac{K_\mathrm{p}}{T_\mathrm{i}} \int_0^t (y(\tau) - \bar{y}) d\tau, \qquad (6.26)$$

where y is the glucose level and u is the rate of release of insulin in the plasma. Draw a block diagram representing the complete closed-loop insulin homeostasis system, including the signals $\tilde{y}$, y, and u. What kind of "controller" is represented by (6.26)? Explain why the feedback controller (6.26) can be defined in terms of the actual glucose level, y, rather then its variation from equilibrium, $\tilde{y}$, when $\bar{y}$ is constant.

6.39 In [Ste+03], the authors have determined experimentally that the values $T_\mathrm{i} = 100$, $T_\mathrm{d} = 38$, and $K_\mathrm{p} = 0.17$ for the "controller" proposed in (6.26) in P6.38 seem to match the physiological response to glucose level response to insulin plasma delivery. Calculate the transfer-function corresponding to the controller (6.26). Use the values of T_i and T_d above and calculate the loop transfer-function, $L(s)$, that can be used for feedback analysis of the closed-loop glucose homeostasis system with respect to the proportional gain, $K_\mathrm{p} > 0$, and sketch the corresponding root-locus diagram. Is the closed-loop insulin homeostasis system asymptotically stable?

6.40 Can you determine $K_\mathrm{p} > 0$ in P6.39 with $T_\mathrm{d} = 0$ so that the closed-loop glucose homeostasis system is internally stable?

7 Frequency Domain

The frequency response of a linear time-invariant system is a complex-valued function that encodes the response of the system to a family of sinusoidal input functions parametrized by the frequency variable ω. The frequency response can be obtained experimentally or from a model in the form of a transfer-function. The study of the frequency response is a powerful source of insight into the behavior of feedback systems. It also plays a key role in many controller design methods to be introduced in this chapter. You will learn how to sketch Bode plots, polar plots, and Nyquist plots, with which you can analyze the stability of open- and closed-loop systems.

7.1 Bode Plots

The complex-valued frequency response function, $G(j\omega)$, was introduced in Section 3.8. One way to visualize the frequency response is to plot its magnitude and phase as a function of the frequency variable, ω. These plots are popular in systems and control. In fact, you have already encountered frequency response plots in Figs. 4.7, 4.14, 4.16, 6.18, and 6.23. It is customary to have the magnitude expressed in terms of the logarithm:

$$20 \log_{10} |G(j\omega)|,$$

in which case we say that the magnitude is expressed in *decibels* (dB), no matter the original units of G, such as in Figs. 6.18 and 6.23. Figures 4.7, 4.14, and 4.16 were plotted in linear scale, not in dB, taking the units of the associated transfer-function. The reason for the scaling factor "20" and the use of base 10 logarithms is mostly historic, dating back to the measurements of gain and attenuation on early communication systems. A pair of logarithmic plots of the magnitude in dB and the phase in degrees is known as a *Bode plot*. If G is a transfer-function with real coefficients then $|G(-j\omega)| = |G(j\omega)|$ and $\angle G(j\omega) = -\angle G(-j\omega)$. For this reason it is necessary only to plot the frequency response for $\omega > 0$. When the transfer-function is rational, it is possible to compute straight-line asymptotes to quickly sketch a Bode plot. The trick is to break up the frequency response into contributions from individual poles and zeros.

Consider the general[1,2] rational transfer-function with m zeros and n poles

$$G(s) = \frac{\beta(s-z_1)^{\ell_1}(s-z_2)^{\ell_2}\ldots(s-z_r)^{\ell_r}}{s^{k_0}(s-p_1)^{k_1}(s-p_2)^{k_2}\ldots(s-p_t)^{k_t}},$$

where $z_j \neq 0$, $j = 1, \ldots, r$, $\sum_{j=1}^{r} \ell_j = m$, and $p_i \neq 0$, $i = 1, \ldots, t$, $\sum_{i=0}^{t} k_i = n$. We *normalize* each term and rewrite

$$G(s) = \frac{\tilde{\beta}(1+\sigma_1 s)^{\ell_1}(1+\sigma_2 s)^{\ell_2}\ldots(1+\sigma_r s)^{\ell_r}}{s^{k_0}(1+\tau_1 s)^{k_1}(1+\tau_2 s)^{k_2}\ldots(1+\tau_t s)^{k_t}},$$

where

$$\sigma_j = -z_j^{-1}, \quad j = 1, \ldots, r, \qquad \tau_i = -p_i^{-1}, \quad i = 1, \ldots, t.$$

Normalization makes each term with a nonzero pole or nonzero zero equal to one at $s = 0$. The term

$$\tilde{\beta} = (-1)^{m-n-k_0} \frac{\beta z_1^{\ell_1} z_2^{\ell_2} \ldots z_r^{\ell_r}}{p_1^{k_1} p_2^{k_2} \ldots p_t^{k_t}}.$$

is a gain. When $k_0 = 0$, $\tilde{\beta}$ is the *DC gain* of the transfer-function, i.e. $\tilde{\beta} = G(0)$.

After applying logarithms, the magnitude of the frequency response is split into sums and differences of first-order terms:

$$\begin{aligned}20\log_{10}|G(j\omega)| = {} & 20\log_{10}|\tilde{\beta}| - 20k_0\log_{10}|j\omega| \\ & + 20\ell_1\log_{10}|1+j\sigma_1\omega| + \cdots + 20\ell_r\log_{10}|1+j\sigma_r\omega| \\ & - 20k_1\log_{10}|1+j\tau_1\omega| - \cdots - 20k_t\log_{10}|1+j\tau_t\omega|,\end{aligned}$$

each term involving a distinct root. The phase of the magnitude response naturally splits into sums and differences without the need to use logarithms:

$$\begin{aligned}\angle G(j\omega) = {} & \angle\tilde{\beta} - k_0\angle j\omega + \ell_1\angle(1+j\sigma_1\omega) + \cdots + \ell_r\angle(1+j\sigma_r\omega) \\ & - k_1\angle(1+j\tau_1\omega) - \cdots - k_t\angle(1+j\tau_t\omega).\end{aligned}$$

In the next paragraphs we will derive asymptotes for the low- and high-frequency behavior of first- and also second-order terms.

First-Order Real Poles and Zeros

Consider the transfer-function with a single real nonzero pole with multiplicity k:

$$G(s) = \frac{1}{(1+s\tau)^k},$$

---

[1] If the transfer-function has no poles or zeros at the origin $k_0 = 0$.
[2] This form can also accommodate a transfer-function with $\ell_0 > 0$ zeros at the origin instead of poles. In this case set $k_0 = -\ell_0$, $\sum_{j=0}^{r} \ell_j = m$, and $\sum_{i=1}^{t} k_i = n$.

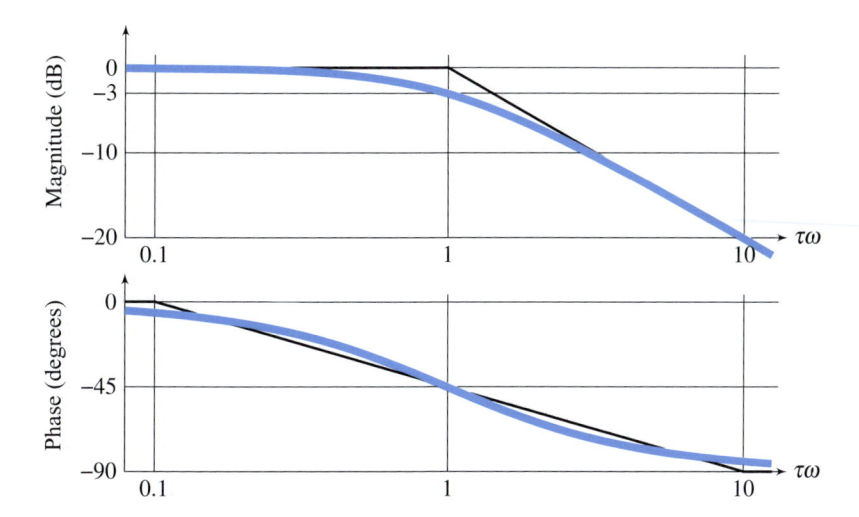

Figure 7.1 Normalized Bode plots for the first-order transfer-function $G(s) = (1 + \tau s)^{-1}$, $\tau > 0$; the thick solid curve is the exact frequency response and the thin solid curve is the straight-line approximation.

where τ is assumed to be real and the integer k is positive. The complex case will be analyzed later. When ω is small, that is when $|\omega| \ll |\tau|^{-1}$, the term $|1 + j\tau\omega| \approx 1$ and the magnitude of the frequency response in dB is

$$20\log_{10}|G(j\omega)| = -20k\log_{10}|1 + j\tau\omega| \approx 0, \quad |\omega| \ll |\tau|^{-1},$$

and when ω is large, that is, $|\omega| \gg |\tau|^{-1}$, the term $|1 + j\tau\omega| \approx |\tau\omega|$ and

$$20\log_{10}|G(j\omega)| = -20k\log_{10}|1 + j\tau\omega| \approx -20k\log_{10}|\tau\omega|, \quad \omega \gg |\tau|^{-1}.$$

When plotted on a logarithmic scale versus $\log_{10}|\omega|$, the function

$$-20k\log_{10}|\tau\omega| = -20k\log_{10}|\tau| - 20k\log_{10}|\omega|$$

is a straight line of slope $-20k$ that intersects zero at $\omega = |\tau|^{-1}$. These asymptotes can be combined into the straight-line approximation

$$20\log_{10}|G(j\omega)| \approx \begin{cases} 0, & |\omega| < |\tau|^{-1}, \\ -20k\log_{10}|\tau| - 20k\log_{10}|\omega|, & |\omega| \geq |\tau|^{-1}. \end{cases}$$

Figure 7.1 compares the exact response (thick solid) with the asymptotes (thin) when $k = 1$ over two decades. If more information is required at the *corner frequency* $\omega = |\tau|^{-1}$ we can use the fact that $|1 \pm j| = \sqrt{2}$ and

$$20\log_{10}|G(j|\tau|^{-1})| = 20\log_{10}|(1 \pm j)^{-k}| = -10k\log_{10}2 \approx -3k.$$

As for the phase,

$$\angle(1 + j\tau\omega)^k = k\angle(1 + j\tau\omega) = k\tan^{-1}(\tau\omega)$$

Frequency Domain

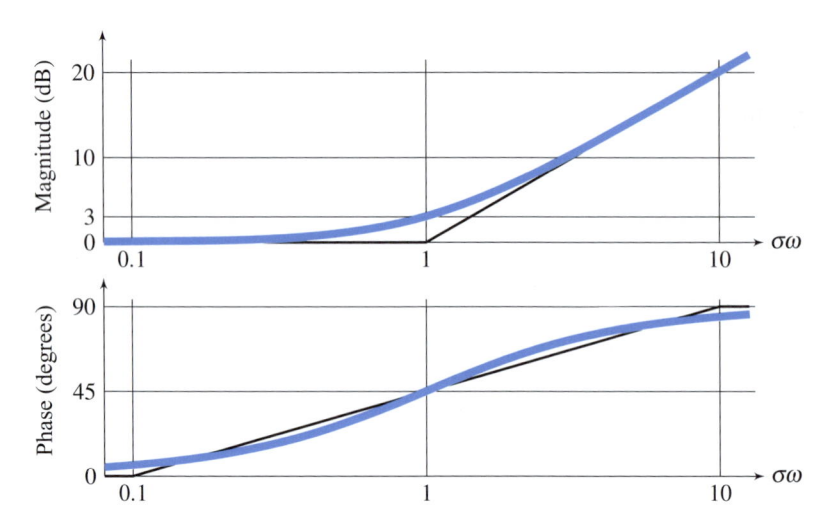

Figure 7.2 Normalized Bode plots for the first-order transfer-function $G(s) = (1 + \sigma s)$, $\sigma > 0$; the thick solid curve is the exact frequency response and the thin solid curve is the straight-line approximation.

leads to the approximation

$$\angle G(j\omega) = -k\angle(1 + j\tau\omega) \approx 0, \qquad |\omega| \ll \frac{1}{10}|\tau|^{-1},$$

when ω is small and[3]

$$\angle G(j\omega) = -k\angle(1 + j\tau\omega) \approx \frac{\tau}{|\tau|}\frac{-k\pi}{2}, \qquad \omega \gg 10|\tau|^{-1},$$

when ω is large. As Fig. 7.1 illustrates, it is reasonable to consider small and large to be about a decade above and below $\omega = |\tau|^{-1}$, and the straight-line approximation corresponds to

$$\angle G(j\omega) \approx \frac{\tau}{|\tau|} \begin{cases} 0, & |\omega| < \frac{1}{10}|\tau|^{-1}, \\ -\frac{k\pi}{4}\log_{10} 10|\tau|\omega, & \frac{1}{10}|\tau|^{-1} \leq \omega < 10|\tau|^{-1}, \\ -\frac{k\pi}{2}, & \omega \geq 10|\tau|^{-1}, \end{cases}$$

which are the thin lines plotted in Fig. 7.1.

For a transfer-function with a single nonzero zero with multiplicity ℓ

$$G(s) = (1 + s\sigma)^\ell,$$

where $\sigma \neq 0$ is real and the integer ℓ is positive, exactly the same analysis is possible upon substitution of $\tau = \sigma$ and $k = -\ell$. When $\sigma > 0$ this implies a change in slope in both the magnitude and phase plots, as shown in Fig. 7.2.

The sign of τ (or σ) does not affect the magnitude of $G(j\omega)$, which depends on $|\tau|$ (or $|\sigma|$) alone, but does change the sign of the slope in the phase plot. The diagrams in

---

[3] The function $\tau/|\tau|$ is equal to sign(τ) when $\tau \neq 0$.

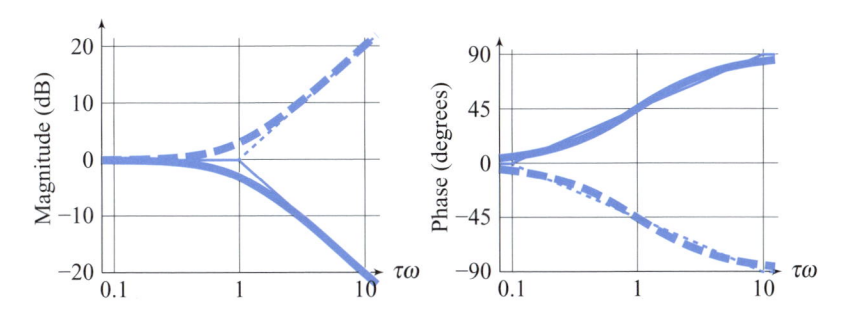

Figure 7.3 Exact (thick) and straight-line approximations (thin) of the normalized Bode plots for the first-order transfer-functions $G_1(s) = (1 - \tau s)^{-1}$, $\tau > 0$ (solid), and $G_2(s) = (1 - \sigma s)$, $\sigma = \tau > 0$ (dashed); the magnitude response is unaffected by the sign of τ or σ but the slope of the phase is reversed.

Fig. 7.3 illustrate the frequency response when both σ and τ are negative: the phase of a pole looks like the phase of a zero and vice versa. Poles with negative τ, i.e. positive real part, are associated with unstable systems, and therefore will appear frequently in control design. Zeros with negative σ, i.e. positive real part, appear in interesting and sometimes difficult-to-control systems. Systems with poles and zeros with positive real part introduce extra phase into the phase diagram and for this reasons are called *non-minimum-phase* systems. See Section 7.2 for details.

Second-Order Complex Poles and Zeros

The diagrams studied so far are for transfer-functions with real poles and zeros. When complex poles and zeros are present the best approach is to group these poles and zeros into complex-conjugate pairs and study the magnitude and phase of the pairs.

Consider the transfer-function with a pair of complex poles with multiplicity k:

$$G(s) = \frac{1}{(1 + 2\zeta s/\omega_n + s^2/\omega_n^2)^k}, \qquad \omega_n > 0, \qquad |\zeta| < 1,$$

which is the kth power of the canonical second-order system (6.5). As studied in detail in Section 6.1, when $|\zeta| < 1$ the roots are complex conjugate and

$$G(j\omega) = \frac{1}{(1 - \omega^2/\omega_n^2 + j 2\zeta\omega/\omega_n)^k},$$

from which

$$20 \log_{10} |G(j\omega)| = -10k \log_{10}\left((1 - \omega^2/\omega_n^2)^2 + 4\zeta^2 \omega^2/\omega_n^2\right).$$

As before, the magnitude of the frequency response in dB is

$$20 \log_{10} |G(j\omega)| \approx 0, \qquad |\omega| \ll \omega_n,$$

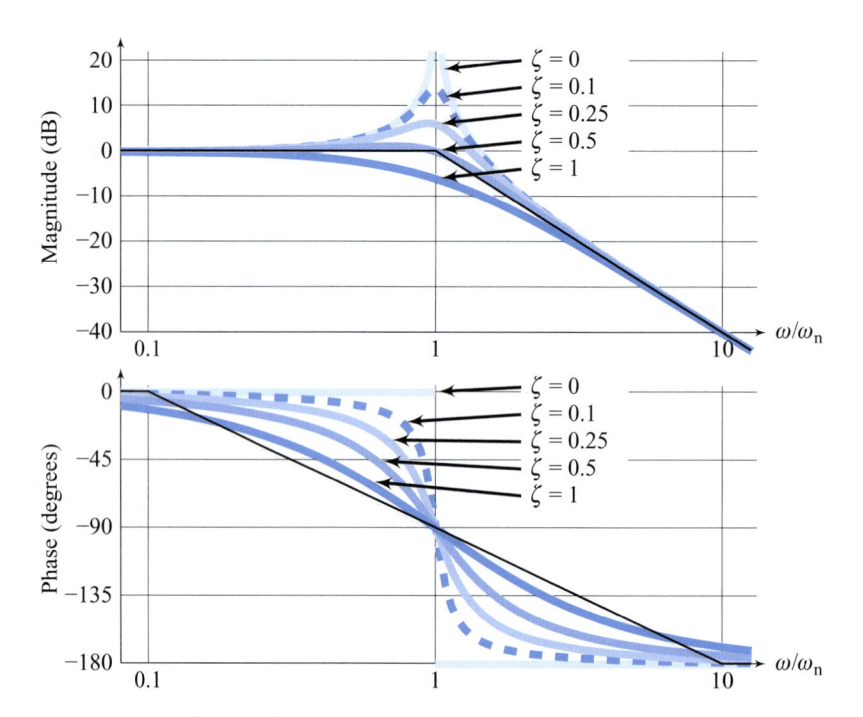

Figure 7.4 Normalized Bode plots for the second-order transfer-function $G(s) = (1 + 2\zeta s/\omega_n + s^2/\omega_n^2)^{-1}$, $\omega_n > 0$, $\zeta \in [0, 1]$. Thick curves show the exact frequency response; the thin solid curve is the straight-line approximation.

when $|\omega|$ is small. When $|\omega|$ is large,

$$20 \log_{10}|G(j\omega)| \approx -40k \log_{10}|\omega/\omega_n|, \qquad |\omega| \gg \omega_n.$$

These provide the straight-line approximation

$$20 \log_{10}|G(j\omega)| \approx \begin{cases} 0, & |\omega| < \omega_n, \\ 40k \log_{10}|\omega_n| - 40k \log_{10}|\omega|, & |\omega| \geq \omega_n, \end{cases}$$

whose two lines intersect at $\omega = \omega_n$. Figure 7.4 compares the exact responses (thick) obtained for various values of $|\zeta| \leq 1$ with the asymptotes (thin solid) when $k = 1$ over two decades. Note the impact of the damping ratio, ζ, on the magnitude of the response near $\omega = \omega_n$. When $\zeta = 1$ the poles are real and the response is similar to that of a transfer-function with a single real pole with multiplicity 2. Note the 40 dB/decade slope. When ζ approaches zero, the magnitude shows accentuated peaks near ω_n, which are characteristic of lightly damped second-order systems. In particular, when $\zeta = 0$ the magnitude response is unbounded at ω_n, since G has imaginary poles at $s = \pm j\omega_n$. When sketching Bode plots of second-order systems by hand it is therefore important to take into account the value of the damping ratio ζ near the natural frequency ω_n.

As seen in Fig. 7.4, the normalized magnitude of the frequency response of a pair of complex poles can have a maximum that exceeds 0 dB. On differentiating $|G(j\omega_n x)|^2$

with respect to $x = \omega/\omega_n$ and equating to zero we obtain

$$\frac{d}{dx}|G(j\omega_n x)|^2 = 4x(2\zeta^2 + x^2 - 1) = 0,$$

which indicates that the magnitude of the frequency response can be potentially maximized at

$$\omega = 0 \quad \text{or} \quad \omega = \omega_r = \omega_n\sqrt{1 - 2\zeta^2}.$$

Clearly, the maximum can occur at $\omega = \omega_r$ only when $1 - 2\zeta^2 > 0$ or $|\zeta| < \sqrt{2}/2 \approx 0.7$, in which case ω_r is known as the *resonance frequency*. As $\zeta \to 0$ the resonance frequency approaches the natural frequency, that is $\omega_r \to \omega_n$.

The phase of $G(j\omega)$ with $0 < |\zeta| < 1$ is

$$\angle G(j\omega) = -k \tan^{-1}\left(\frac{2\zeta\omega/\omega_n}{1 - \omega^2/\omega_n^2}\right).$$

Note that as $\omega \to \omega_n^-$ the ratio inside the inverse tangent approaches $+\infty$ when $\zeta > 0$ and $-\infty$ when $\zeta < 0$, from which we conclude that

$$\lim_{\omega \to \omega_n^-} \angle G(j\omega) = -k\frac{\zeta}{|\zeta|}\frac{\pi}{2}.$$

When $\omega \to \omega_n^+$ the ratio inside the tangent inverse approaches $-\infty$ when $\zeta > 0$ and $+\infty$ when $\zeta < 0$, and, if we use the second branch of the inverse tangent,

$$\lim_{\omega \to \omega_n^+} \angle G(j\omega) = -k\frac{\zeta}{|\zeta|}\frac{\pi}{2}.$$

This implies that the phase is actually continuous at $\omega = \omega_n$ when $\zeta \neq 0$. See Fig. 7.4. As before,

$$\angle G(j\omega) \approx 0, \qquad |\omega| \ll \frac{1}{10}\omega_n.$$

and

$$\angle G(j\omega) \approx -k\frac{\zeta}{|\zeta|}\pi, \qquad \omega \gg 10\omega_n.$$

As with real poles, it is reasonable to consider small and large to be about a decade above and below $\omega = \omega_n$ and to use the straight-line approximation

$$\angle G(j\omega) \approx \frac{\zeta}{|\zeta|}\begin{cases} 0, & |\omega| < \frac{1}{10}\omega_n, \\ -(k\pi/2)\log_{10}(10\omega/\omega_n), & \frac{1}{10}\omega_n \leq \omega < 10\omega_n, \\ -k\pi, & \omega \geq 10\omega_n, \end{cases}$$

which is shown by the thin lines in Fig. 7.4. As with the magnitude, the value of ζ has important effects on the phase, especially near $\omega = \omega_n$. The value of $\zeta = 0$ needs special attention. In this case

$$G(j\omega) = \frac{1}{(1 - \omega^2/\omega_n^2)^k},$$

is real. Therefore, for consistency[4] with the case $0 < |\zeta| < 1$ the phase of $G(j\omega)$ at $\omega = \omega_n$ must have a discontinuity, $-k\pi$ when $\zeta > 0$ or $k\pi$ when $\zeta < 0$, as shown in Fig. 7.4, even when k is even.

The case of complex zeros with multiplicity ℓ,

$$G(s) = (1 + 2\zeta s/\omega_n + s^2/\omega_n^2)^\ell, \qquad \omega_n > 0, \qquad |\zeta| < 1,$$

can be treated similarly by letting $\ell = -k$. The sign of $k = -\ell$ will flip the slopes in both the magnitude and phase diagrams. As with real poles, when $\zeta < 0$ the poles (or zeros) have positive real part, leaving the magnitude of $G(j\omega)$ unaltered but flipping the phase to produce non-minimum-phase systems.

Poles and Zeros at the Origin

The last case we need to discuss is the simplest: poles and zeros at the origin. First note that a SISO transfer-function can have either poles or zeros at the origin, but not both, due to cancellations. In either case, the frequency response for a pole at the origin with multiplicity k,

$$G(s) = \frac{1}{s^k},$$

is simply

$$20\log_{10}|G(j\omega)| = -20k\log_{10}|\omega|, \qquad \angle G(j\omega) = -k\angle j\omega = -k\frac{\pi}{2}.$$

Zeros at the origin with multiplicity ℓ can be handled by letting $k = -\ell$.

In summary, zeros and poles at the origin contribute straight lines to the magnitude of the frequency response when plotted on a logarithmic scale versus $\log_{10}|\omega|$. These lines intersect 0 dB at $\omega = 1$ and have slope $\pm 20k$ dB. One difference here is that this is not an approximation but the exact response. Insofar as the phase is concerned, poles at the origin contribute an additional constant $-k\pi/2$ and zeros contribute $k\pi/2$.

A Simple Example

As a first example consider the transfer-function of the lead controller from (6.22):

$$G(s) = C_{\text{lead}}(s) = 3.83\frac{s+7.5}{s+21}.$$

First normalize:

$$G(s) = \tilde{\beta}\frac{1+s/7.5}{1+s/21}, \qquad \tilde{\beta} = \frac{3.83 \times 7.5}{21} \approx 1.37.$$

We will sketch the magnitude of the frequency response first. The gain $\tilde{\beta}$ will offset the magnitude by

$$20\log_{10}\tilde{\beta} \approx 2.7\,\text{dB}, \quad \omega \leq 7.5. \tag{A}$$

---

[4] A better reason will be provided in Section 7.6, where you will learn how to handle imaginary poles in Nyquist plots.

7.1 Bode Plots

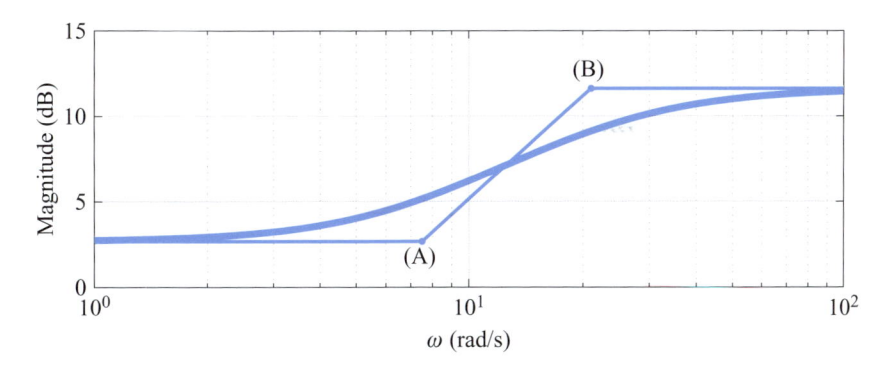

Figure 7.5 Exact plot (thick solid line) and straight-line approximation (thin solid line) of the magnitude of the frequency response of the lead controller (6.22).

At $\omega = 7.5$, which corresponds to a zero, the slope becomes $+20$ dB/decade. This zero is followed by a pole at $\omega = 21$. The approximate magnitude of the frequency response at $\omega = 21$ is therefore

$$2.7 \text{ dB} + 20 \log_{10}(21/7.5) \approx 11.6 \text{ dB}, \qquad \omega \geq 21, \qquad \text{(B)}$$

and it will remain at this level because the -20 dB/decade slope contributed by the pole cancels the $+20$ dB/decade of the earlier zero. Points (A) and (B) are joined to trace the straight-line approximation for the magnitude of the frequency response shown in Fig. 7.5 as thin solid lines.

We will now sketch the phase response in degrees. Start by compiling the slope contribution from each pole and zero. The gain $\tilde{\beta}$ contributes nothing since $\angle\tilde{\beta} = 0°$. The phase remains constant until one decade below the first zero at $\omega = 7.5$, which contributes a slope of

$$+45°/\text{decade}, \qquad 0.75 \leq \omega \leq 75.$$

The next pole at $\omega = 21$ contributes a slope of

$$-45°/\text{decade}, \qquad 2.1 \leq \omega \leq 210.$$

Add and compile all slope contributions per interval to obtain the slope profile:

$$0°/\text{decade}, \qquad 0 \leq \omega \leq 0.75,$$
$$+45°/\text{decade}, \qquad 0.75 \leq \omega \leq 2.1,$$
$$0°/\text{decade}, \qquad 2.1 \leq \omega \leq 75,$$
$$-45°/\text{decade}, \qquad 75 \leq \omega \leq 210,$$
$$0°/\text{decade}, \qquad \omega \geq 210,$$

which we use to compute the approximate phase at the points:

$$\approx 0°, \qquad \omega = 0.75, \qquad \text{(A)}$$
$$45° \log_{10}(2.1/0.75) \approx 20°, \qquad \omega = 2.1, \qquad \text{(B)}$$
$$45° \log_{10}(2.1/0.75) \approx 20°, \qquad \omega = 75, \qquad \text{(C)}$$
$$0° \log_{10}(2.1/0.75) \approx 0°, \qquad \omega = 210, \qquad \text{(D)}$$

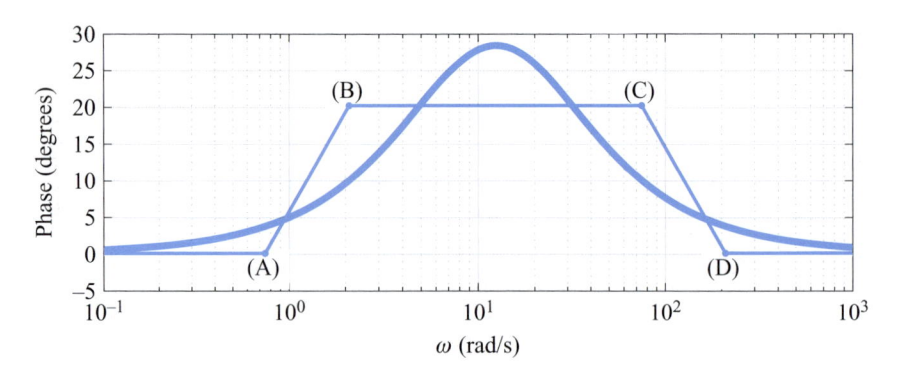

Figure 7.6 Exact plot (thick solid line) and straight-line approximation (thin solid line) of the phase of the frequency response of the lead controller (6.22).

Points (A)–(D) are joined to trace the straight-line approximation for the phase of the frequency response shown in Fig. 7.6 as thin solid lines. The name *lead* for this controller comes from the fact that the phase is positive for all $\omega \geq 0$. The steady-state response to a sinusoidal input will therefore appear to be *leading* the input. Conversely, a controller consisting of a real pole followed by a real zero is a *lag* controller, since the phase of the frequency response is now negative for all $\omega \geq 0$.

Putting It All Together

We will illustrate how you can use straight-line approximations to sketch both the magnitude and the phase of the frequency response of the more involved transfer-function

$$G(s) = \frac{28.1s^2 + 22.4s + 112.4}{s^6 + 5.7s^5 + 12.81s^4 + 47.6s^3 + 7.5s^2 + 11.25s}.$$

With the help of a calculator, rewrite $G(s)$ in terms of approximate products of first- and second-order terms:

$$G(s) = \frac{28.1(s^2 + 0.8s + 4)}{s(s^2 + 0.1s + 0.25)(s^2 + 0.6s + 9)(s + 5)} \quad (7.1)$$

and normalize to obtain

$$G(s) = \frac{\tilde{\beta}(s^2/2^2 + 0.4s/2 + 1)}{s(s^2/0.5^2 + 0.2s/0.5 + 1)(s^2/3^2 + 0.2s/3 + 1)(s/5 + 1)},$$

where

$$\tilde{\beta} = \frac{28.1 \times 4}{5 \times 0.25 \times 9} \approx 10.$$

Start by sketching the straight-line approximations for the magnitude of the frequency response. The gain $\tilde{\beta}$ offsets the magnitude of the frequency response by

$$20 \log_{10} \tilde{\beta} \approx 20 \text{ dB}.$$

Locate the first pole starting from $\omega = 0$, which in this case is the pole at zero, which contributes a slope of -20 dB. This pole is followed by a complex-conjugate pair of poles at

$$\omega_n = 0.5, \quad \zeta = 0.1.$$

The approximate magnitude of the frequency response at $\omega = 0.5$ is dictated by $\tilde{\beta}$ and the pole at zero:

$$20\,\text{dB} - 20\log_{10} 0.5 \approx 26\,\text{dB}, \quad \omega = 0.5. \tag{A}$$

After $\omega = 0.5$ the slope of the straight-line approximation of the magnitude of the frequency response is -60 dB/decade, which is the sum of the slope due to the pole at the origin, -20 dB, plus the slope due to the pair of complex-conjugate poles at $\omega = 0.5$, -40 dB. Next come the complex-conjugate zeros

$$\omega_n = 2, \quad \zeta = 0.2,$$

which intersect the current straight-line approximation at

$$26\,\text{dB} - 60\log_{10}(2/0.5) \approx -10.1\,\text{dB}, \quad \omega = 2. \tag{B}$$

The complex-conjugate pair of zeros will cancel the slope of the last pair of poles, which brings the slope of the magnitude back to -20 dB/decade. The next complex conjugate pair of poles is at

$$\omega_n = 3, \quad \zeta = 0.1,$$

with magnitude

$$-10.1\,\text{dB} - 20\log_{10}(3/2) \approx -13.6\,\text{dB}, \quad \omega = 3. \tag{C}$$

From $\omega = 3$ onward the magnitude decreases at a rate of -60 dB/decade. The last pole is a real pole at $\omega = 5$ at approximately

$$-13.6\,\text{dB} - 60\log_{10}(5/3) \approx -26.9\,\text{dB}, \quad \omega = 5. \tag{D}$$

The remaining part of the magnitude plot is approximated with a straight line with slope -80 dB/decade. The complete straight-line approximation to the magnitude of the frequency response is obtained by joining the points (A)–(D) and continuing with the appropriate slope. This is shown in Fig. 7.7 as thin solid lines. The thick solid curve is the exact magnitude of the frequency response. We could have also used the calculated ζs to improve on the straight-line approximation by raising or lowering the plot near the natural frequencies of the complex poles and zeros.

Let us now sketch the phase response. We start by compiling the slope contribution from each pole and zero. The pole at zero and the gain $\tilde{\beta}$ contribute a constant phase of

$$-90° + \angle\tilde{\beta} = -90° + 0° = -90°$$

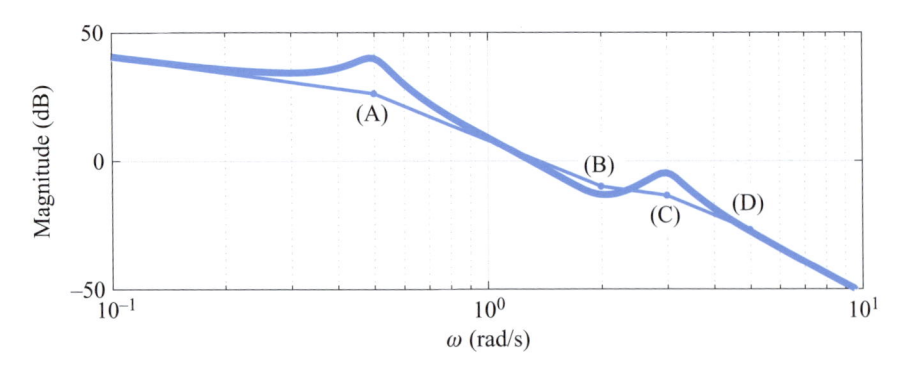

Figure 7.7 Exact plot (thick solid line) and straight-line approximation (thin solid line) of the magnitude of the frequency response of the transfer-function (7.1).

and no slope. Indeed, the phase is constant until one decade below the first pair of complex-conjugate poles at $\omega = 0.5$, which contributes a slope of

$$-90°/\text{decade}, \quad 0.05 \leq \omega \leq 5.$$

The next pair of zeros is at $\omega = 2$, which contributes a slope of

$$+90°/\text{decade}, \quad 0.2 \leq \omega \leq 20.$$

Finally, the pair of complex-conjugate poles at $\omega = 3$ contributes

$$-90°/\text{decade}, \quad 0.3 \leq \omega \leq 30,$$

and the real pole at $\omega = 5$ contributes

$$-45°/\text{decade}, \quad 0.5 \leq \omega \leq 50.$$

Add and compile all slope contributions per interval to obtain the slope profile:

$$\begin{aligned}
0°/\text{decade}, & \quad 0 \leq \omega \leq 0.05, \\
-90°/\text{decade}, & \quad 0.05 \leq \omega \leq 0.2, \\
0°/\text{decade}, & \quad 0.2 \leq \omega \leq 0.3, \\
-90°/\text{decade}, & \quad 0.3 \leq \omega \leq 0.5, \\
-135°/\text{decade}, & \quad 0.5 \leq \omega \leq 5, \\
-45°/\text{decade}, & \quad 5 \leq \omega \leq 20, \\
-135°/\text{decade}, & \quad 20 \leq \omega \leq 30, \\
-45°/\text{decade}, & \quad 30 \leq \omega \leq 50, \\
0°/\text{decade}, & \quad \omega \geq 50,
\end{aligned}$$

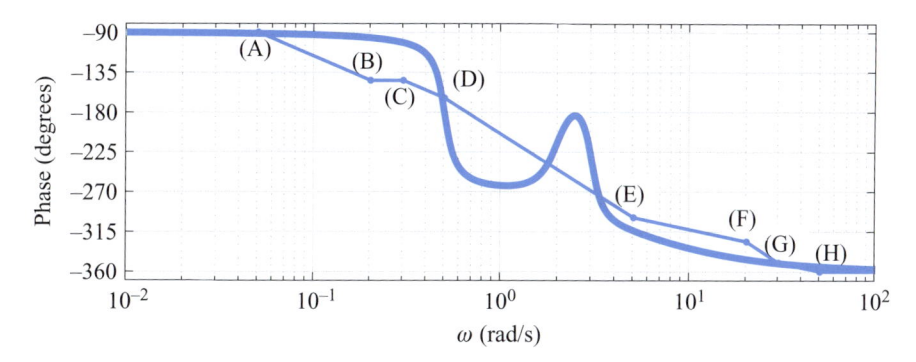

Figure 7.8 Exact plot (thick solid line) and straight-line approximation (thin solid line) of the phase of the frequency response of the transfer-function (7.1).

which we use to compute the approximate phase at the points:

$$-90°, \qquad \omega = 0.05, \qquad (A)$$
$$-90° - 90° \log_{10}(0.2/0.05) \approx -144°, \qquad \omega = 0.2, \qquad (B)$$
$$\approx -144°, \qquad \omega = 0.3, \qquad (C)$$
$$-144° - 90° \log_{10}(0.5/0.3) \approx -164°, \qquad \omega = 0.5, \qquad (D)$$
$$-164° - 135° \log_{10}(5/0.5) \approx -299°, \qquad \omega = 5, \qquad (E)$$
$$-299° - 45° \log_{10}(20/5) \approx -326°, \qquad \omega = 20, \qquad (F)$$
$$-326° - 135° \log_{10}(30/20) \approx -350°, \qquad \omega = 30, \qquad (G)$$
$$-350° - 45° \log_{10}(50/30) \approx -360°, \qquad \omega = 50. \qquad (H)$$

Points (A)–(H) are joined to trace the straight-line approximation for the phase of the frequency response shown in Fig. 7.8 as thin solid lines. Note how the ζs produce large errors in the straight-line approximation in the neighborhood of complex poles and zeros.

Of course, these days, modern computer software, such as MATLAB, can produce exact Bode plots without much effort. The point of learning about the straight-line approximations is to be able to anticipate changes in the diagram that come with the introduction of poles and zeros. This is a very useful skill to have when designing controllers and analyzing feedback systems.

7.2 Non-Minimum-Phase Systems

Consider the three distinct transfer-functions

$$G_1 = 1, \qquad G_2 = -1, \qquad G_3 = \frac{s-1}{s+1}.$$

Verify that

$$|G_1(j\omega)| = |G_2(j\omega)| = |G_3(j\omega)| = 1 \quad \text{for all } \omega.$$

In other words, their frequency responses have exactly the same magnitude. However, since they are not the same transfer-functions, their phases

$$\angle G_1(j\omega) = 0, \qquad \angle G_2(j\omega) = \pi, \qquad \angle G_3(j\omega) = \pi - 2\tan^{-1}\omega,$$

must differ. The transfer-function G_1 has the minimum possible phase among all of the transfer-functions that have $|G(j\omega)| = 1$, which include G_2 and G_3. For this reason, it is known as a *minimum-phase* transfer-function. By extension, a linear system modeled as a minimum-phase transfer-function is a minimum-phase system. All other transfer-functions and systems with the same frequency-response magnitude are said to be of *non-minimum-phase*.

It is not hard to be convinced that minimum-phase rational transfer-functions are well behaved: they have *all poles and zeros* with negative real part. Non-minimum-phase systems, on the other hand, can have intriguing behaviors. For example, when excited by a unit step input, $u(t) = 1(t)$, non-minimum-phase systems respond in the *opposite direction*. For example, the unit step response, $y(t)$, of a system with transfer-function G_1 is simply a unit step in the same direction as the input step, i.e. $y(t) = 1(t)$. However, the unit step response of a system with transfer-function G_2 is in the direction opposite to the input step, i.e. $y(t) = -1(t)$. The step response of the transfer-function G_3 is even more interesting. We calculate the step response using the inverse Laplace transform:

$$y(t) = \mathcal{L}^{-1}\{s^{-1}G_3(s)\} = \mathcal{L}^{-1}\left\{\frac{s-1}{s(s+1)}\right\} = 2e^{-t} - 1, \quad t \geq 0.$$

At $t = 0$ the response is equal to 1 and as t grows tends to

$$\lim_{t \to \infty} y(t) = \lim_{s \to 0} G_3(s) = -1,$$

which is again in a direction opposite to the input step.

A subtler example is

$$G_1 = \frac{1+s}{s^2+s+1}, \qquad G_2 = \frac{1-s}{s^2+s+1}, \qquad G_3 = \frac{(1-s)^2}{s^3+2s^2+2s+1}, \qquad (7.2)$$

for which $|G_1(j\omega)| = |G_2(j\omega)| = |G_3(j\omega)|$ and $G_1(s)$ is minimum-phase. The phases of G_1, G_2, and G_3 are quite different. See Fig. 7.9. When excited with a unit step, all three systems, which are asymptotically stable, converge to

$$\lim_{s \to 0} G_1(s) = \lim_{s \to 0} G_2(s) = \lim_{s \to 0} G_3(s) = 1$$

but the system with transfer-function $G_2(s)$ will first veer in the opposite direction, as shown in Fig. 7.10. Indeed, using the initial-value theorem to calculate the derivative of the step response, $\dot{y}(t)$, at $t = 0$,

$$\lim_{t \to 0^+} \dot{y}_1(t) = \lim_{s \to \infty} sG_1(s) = 1, \qquad \lim_{t \to 0^+} \dot{y}_2(t) = \lim_{s \to \infty} sG_2(s) = -1,$$

which explains why the trajectories for G_1 and G_2 diverge initially. Yet

$$\lim_{t \to 0^+} \dot{y}_1(t) = \lim_{t \to 0^+} \dot{y}_3(t) = \lim_{s \to \infty} sG_1(s) = \lim_{s \to \infty} sG_3(s) = 1$$

7.2 Non-Minimum-Phase Systems

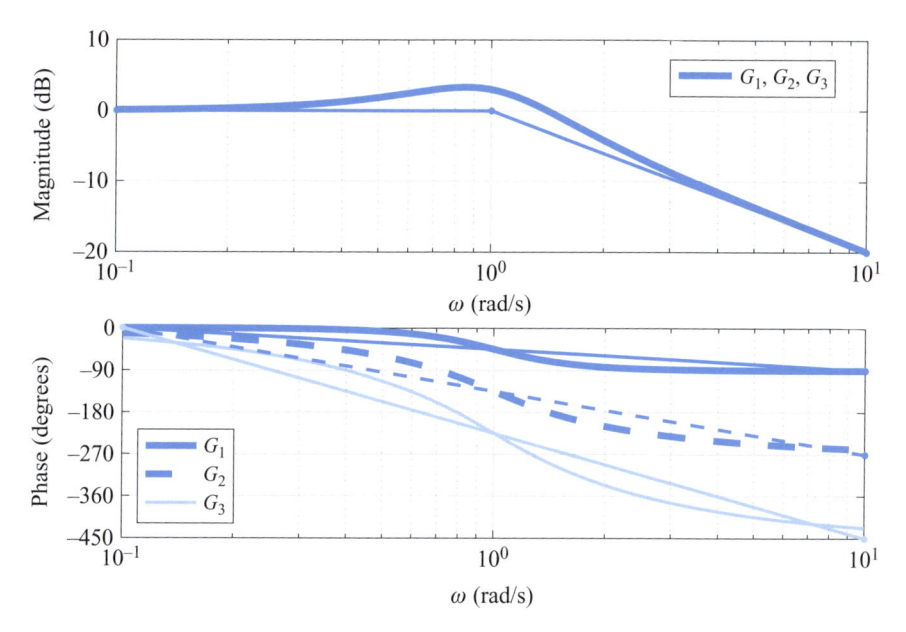

Figure 7.9 Bode plots for the transfer-functions G_1, G_2, and G_3 in (7.2); the magnitudes of their frequency response are the same but their phases differ; G_1 is minimum-phase; G_2 and G_3 are non-minimum-phase.

so G_1 and G_3 start with the same derivative, but the non-minimum-phase system G_3 ends up veering in the opposite direction after a while, as shown in Fig. 7.10. The odd behavior of non-minimum-phase systems often translates into challenges in control design.

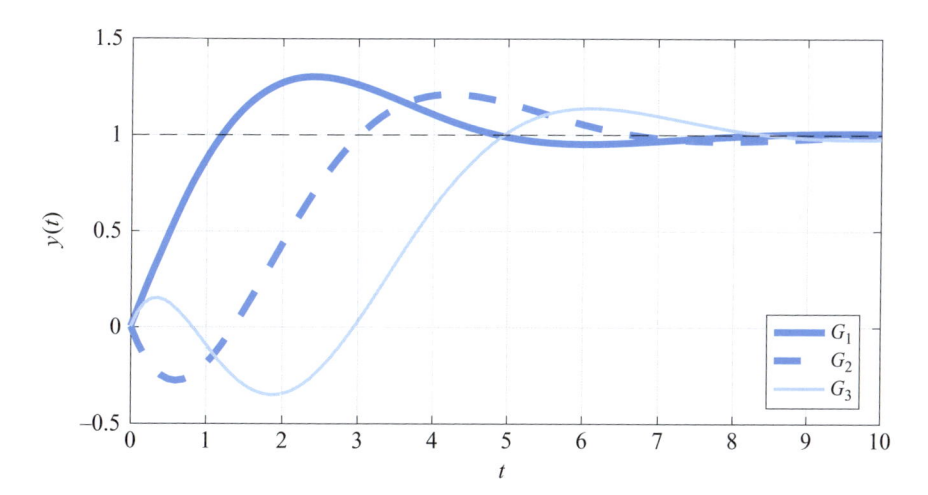

Figure 7.10 Step responses for the transfer-functions G_1, G_2, and G_3 in (7.2). G_1 is minimum-phase; G_2 and G_3 are non-minimum-phase. All responses converge to one. The responses of G_1 and G_3 start with the same derivative but G_2 and eventually G_3 veer toward negative values before converging to one.

The step responses of the non-minimum-phase systems G_2 and G_3 appear to be *delayed* when compared with G_1. Not surprisingly, linear systems with delays are always non-minimum-phase. Indeed, because $\mathcal{L}\{y(t-\tau)\} = e^{-s\tau}$ (see Table 3.2), a linear system with minimum-phase transfer-function $G(s)$ in which one requires $\tau > 0$ seconds to process the measurement is represented by a transfer-function $G_\tau(s) = e^{-s\tau} G(s)$, which is non-minimum-phase: $|G_\tau(j\omega)| = |G(j\omega)|$ but $\angle G_\tau(j\omega) = \angle G(j\omega) - \omega\tau$. When the delays are not very large, it is possible to approximate the transcendental $e^{-s\tau}$ by a rational function:

$$e^{-s\tau} \approx \frac{1 - a_1 s + a_2 s^2 - \cdots \pm a_n s^n}{1 + a_1 s + a_2 s^2 + \cdots + a_n s^n},$$

where the coefficients a_i, $i = 1, \ldots, n$ depend on the value of the delay, $\tau > 0$, and the desired degree, n. Such approximations are known as *Padé* approximations [FPE14, Section 5.6.3]. For example, $a_1 = \tau/2$ for $n = 1$ and $a_1 = \tau/2$, $a_2 = \tau^2/12$ for $n = 2$.

For a familiar example of a non-minimum-phase system consider the steering car from Section 5.7. Set the y-coordinate of the mid-point of the front axle to be the output:

$$y(t) = z_y(t) + \ell \sin \theta(t),$$

and linearize around the horizontal straight-line trajectory (5.22) at constant velocity, v, to obtain the time-invariant state-space model in the form (5.16) with matrices

$$A = \begin{bmatrix} 0 & 0 & 0 \\ 0 & 0 & v \\ 0 & 0 & 0 \end{bmatrix}, \quad B = \begin{bmatrix} 0 \\ 0 \\ v/\ell \end{bmatrix},$$

$$C = \begin{bmatrix} 0 & 1 & \ell \end{bmatrix}, \quad D = \begin{bmatrix} 0 \end{bmatrix},$$

from which the transfer-function from the steering input, $u = \tan \psi$, to the output, the y-coordinate of the mid-point of the front axle, is

$$G(s) = \frac{v(s + v/\ell)}{s^2}.$$

If $v > 0$ then G is minimum-phase. However, if $v < 0$, which corresponds to the car being driven backwards, G is non-minimum-phase because the zero, $-v/\ell$, moves to the right-hand side of the complex plane. Check out the y-coordinate changing maneuver depicted in Fig. 7.11, in which a backward-moving car has to first decrease its y-coordinate before that can be increased.

7.3 Polar Plots

As an alternative to a pair of Bode plots, it is possible to represent the magnitude and phase of the frequency response of a transfer-function in a single plot. The idea is to see $G(j\omega)$ as a curve on the complex plane parametrized by $\omega \in \mathbb{R}$. If a Bode plot is

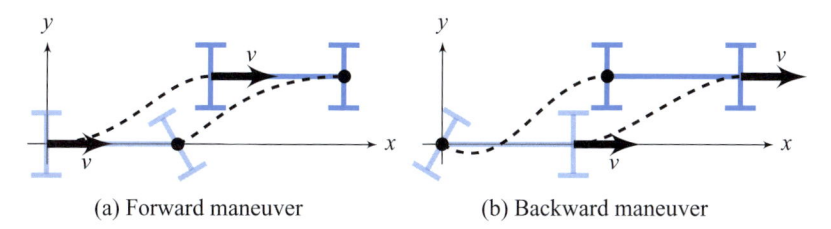

Figure 7.11 Trajectories of a four-wheeled vehicle changing its y-coordinate; the mid-point of the forward axle is marked with a circle. The forward maneuver is *minimum-phase*; backward motion is *non-minimum-phase*. The mid-point of the front axle has to first decrease its y-coordinate before that can be increased.

(a) Forward maneuver   (b) Backward maneuver

available, then sketching the polar plot is not a hard task. Consider for example the transfer-function

$$G(s) = \frac{1}{s+1}.$$

The polar plot of G is the parametric curve

$$s = G(j\omega), \quad \omega \in \mathbb{R}, \quad -\infty < \omega < \infty.$$

Information about the segment of the curve $0 \leq \omega < \infty$ is contained in the Bode plot of G. Because G has real coefficients and $|G(j\omega)| = |G(-j\omega)|$, $\angle G(j\omega) = -\angle G(-j\omega)$, the segment $-\infty < \omega \leq 0$ is obtained using symmetry: just reflect the plot about the real axis. From the Bode plot of G we can directly obtain some points, say

$$G(j0.1) \approx 1, \quad G(j10) \approx e^{j\pi/2}, \tag{7.3}$$

and calculate

$$G(j0) = 1, \quad G(j1) = \frac{\sqrt{2}}{2} e^{-j\pi/4}, \quad G(j\infty) = 0 e^{-j\pi/2}, \tag{7.4}$$

which can be used to sketch the polar plot by hand. The points (7.3) and (7.4), along with complete Bode and polar plots, are shown in Fig. 7.12.

In this simple case the exact curve

$$G(j\omega) = \frac{1}{j\omega+1} = \frac{1}{2} + \frac{1}{2}\frac{1-j\omega}{1+j\omega}, \quad -\infty < \omega < \infty,$$

can also be explicitly evaluated. Because

$$\frac{|1-j\omega|}{|1+j\omega|} = 1, \quad \angle \frac{1-j\omega}{1+j\omega} = -2\tan^{-1}\omega \in (-\pi, \pi),$$

the plot of $G(j\omega)$, $-\infty < \omega < \infty$ is a circle of radius $1/2$ centered at $1/2$, which is the curve shown in Fig. 7.12(c). The arrows indicate the direction of traversal of the curve as ω goes from $-\infty$ to ∞. The thick solid segment is the one obtained directly from the Bode plot, i.e. $0 < \omega < \infty$, and the thick dashed segment is obtained by reflecting the thick solid segment about the real axis.

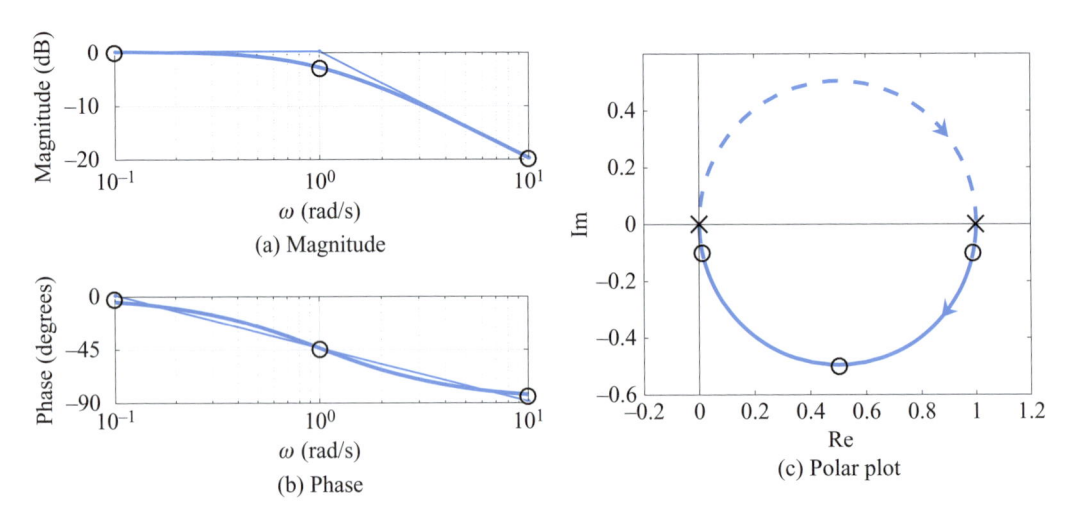

Figure 7.12 Bode and polar plot of $G(s) = 1/(s+1)$: crosses mark $G(0)$ and $\lim_{\omega \to \infty} G(j\omega)$; circles mark $G(j0.1)$, $G(j)$, and $G(j10)$.

We will encounter more examples of polar plots later when we learn how to use these beautiful figures to make inferences about the stability of open- and closed-loop systems in Sections 7.5 and 7.6.

7.4 The Argument Principle

Before we talk about stability in the frequency domain it is necessary to introduce the *argument principle*. The principle itself is very simple and many readers may choose to skip the latter parts of this section that are dedicated to its proof on a first reading. Indeed, many standard books do not provide a proof, but rather work out the principle using graphical arguments, e.g. [FPE14, Section 6.3.1] and [DB10, Section 9.2]. If you have endured (perhaps even secretly enjoyed) the most technical parts of Chapter 3, then the proof will be enlightening. The argument principle can be stated as follows.

THEOREM 7.1 (Argument principle) *If a function f is analytic inside and on the positively oriented simple closed contour C except at a finite[5] number of poles[6] inside C and f has no zeros on C then*

$$\frac{1}{2\pi} \Delta_C^0 \arg f(s) = \frac{1}{2\pi j} \int_C \frac{f'(s)}{f(s)} ds = Z_C - P_C, \qquad (7.5)$$

where Z_C is the number of zeros and P_C is the number of poles of f that lie inside the contour C counting their multiplicities.

---

[5] This implies that all singularities are isolated.
[6] More precisely, f is *meromorphic* in a domain containing C [BC14, Section 93].

7.4 The Argument Principle

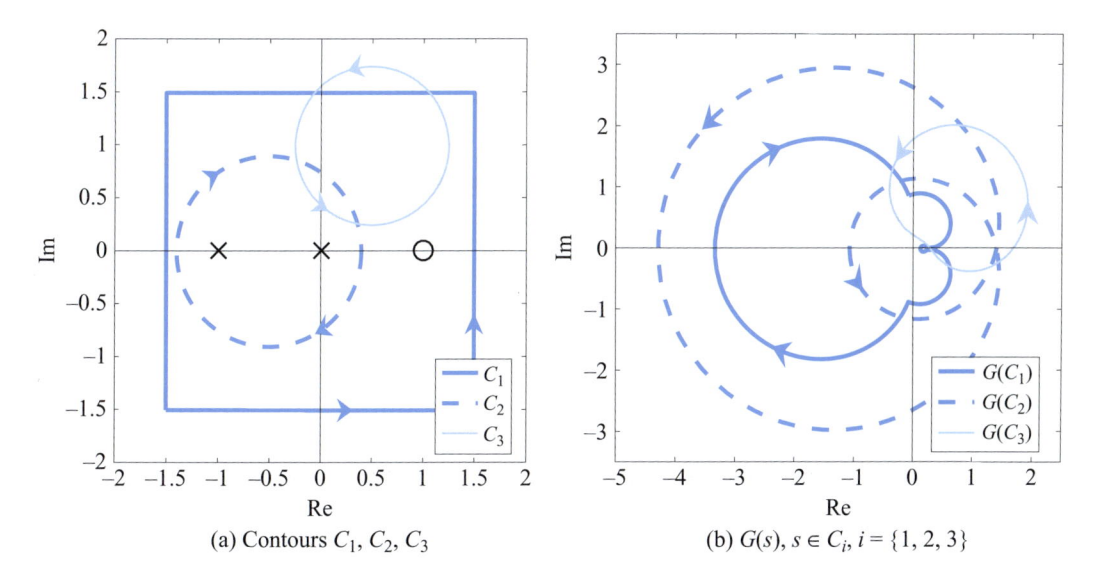

Figure 7.13 Mapping $G(s) = (s-1)/(s^2+s)$: crosses mark poles and circles mark zeros; arrows indicate direction of travel.

(a) Contours C_1, C_2, C_3

(b) $G(s), s \in C_i, i = \{1, 2, 3\}$

Because C is a closed contour and f has no singularities on C, the quantity

$$\Delta_C^0 \arg f(s),$$

which is the total argument variation recorded as we traverse the image of C under f, must be an integer multiple of 2π. Indeed, the quantity on the left-hand side of (7.5) is an integer that indicates how many times the image of the contour encircles the origin. The notation $\Delta_C^0 \arg f(s)$ reflects the fact that encirclements should be counted around the origin.

For example, let f be the rational function

$$G(s) = \frac{s-1}{s(s+1)} \tag{7.6}$$

and recall the positively oriented simple closed contours C_1, C_2, and C_3 introduced earlier in Fig. 3.1 and reproduced in Fig. 7.13(a). The images of C_1, C_2, and C_3 under the mapping G are plotted in Fig. 7.13(b). The direction of traversal of the contours and their images is indicated by arrows along the paths. The total argument variation can be obtained directly from Fig. 7.13(b) by simply counting the net number of times the image of C_1, C_2, or C_3 under G encircles the origin, taking into account their direction of traversal. From Fig. 7.13(b)

$$\frac{1}{2\pi} \Delta_{C_1}^0 \arg G(s) = -1$$

because the image of the positively oriented simple closed contour C_1 encircles the origin once in the clockwise (negative) direction. Because the image of the contour C_3

never encircles the origin,

$$\frac{1}{2\pi}\Delta^0_{C_3} \arg G(s) = 0.$$

The case C_2 is a bit trickier because the closed contour C_2 is negatively oriented. This case can be accounted for after observing that reversing the direction of travel along a negatively oriented contour simply reverses the direction of travel of its image, that is,

$$\Delta^0_C \arg G(s) = -\Delta^0_{-C} \arg G(s), \qquad (7.7)$$

where $-C$ denotes the positively oriented contour obtained by reversing the direction of travel of a negatively oriented contour C. Hence

$$\frac{1}{2\pi}\Delta^0_{C_2} \arg G(s) = 2 \quad \text{and} \quad \frac{1}{2\pi}\Delta^0_{-C_2} \arg G(s) = -2,$$

because the image of C_2 under G encircles the origin twice in the counter-clockwise (positive) direction.

The right-hand side of (7.5) can be evaluated by locating the poles and zeros relative to the contours. In Fig. 7.13(a) poles are marked with crosses and zeros with circles. $G(s)$ has two simple poles, at $s = 0$ and $s = -1$, and one simple zero, at $s = 1$. From Fig. 7.13(a)

$$Z_{C_1} = 1, \qquad P_{C_1} = P_{-C_2} = 2, \qquad Z_{-C_2} = Z_{C_3} = P_{C_3} = 0,$$

and

$$Z_{C_1} - P_{C_1} = -1, \qquad Z_{-C_2} - P_{-C_2} = -2, \qquad Z_{C_3} - P_{C_3} = 0,$$

which agree with our previous argument calculations obtained graphically from Fig. 7.13(b).

It is also possible to restate the argument principle for negatively oriented contours. If C is a negatively oriented contour satisfying the assumptions of Theorem 7.1 then

$$\frac{1}{2\pi}\Delta^0_C \arg G(s) = P_C - Z_C, \qquad (7.8)$$

which follows directly from (7.7). Indeed, $P_{C_2} = 2$, $Z_{C_2} = 0$, and

$$\frac{1}{2\pi}\Delta^0_{C_2} \arg G(s) = P_{C_2} - Z_{C_2} = 2,$$

which once again agrees with our previous calculations. This form of the argument principle will be used in Sections 7.5 and 7.6. The rest of this section is a proof of the argument principle.

Proof of Theorem 7.1

As seen in our simple example, application of the argument principle usually involves evaluating the quantities on the far left and far right of (7.5). The proof will rely on the contour integral in the middle.

7.4 The Argument Principle

Integration around simple closed contours should immediately bring to mind Cauchy's residue theorem (Theorem 3.1) and its calculus of residues. Before applying Theorem 3.1 we need to enumerate the singular points of the integrand function f'/f. By assumption f is analytic inside C except at its poles. When f is analytic at s_0 and $f(s_0) \neq 0$ then f' and hence f'/f are also analytic at s_0 [BC14, Section 57]. This means that the only candidate points at which f'/f can be singular are the poles and zeros of f.

If s_0 is a zero of f with multiplicity $m > 0$ then f is analytic at s_0 and

$$f(s) = (s - s_0)^m h(s), \qquad h(s_0) \neq 0,$$

where h is analytic at s_0. If f is polynomial or rational, then h is obtained by factoring the roots of the numerator of f.[7] Therefore

$$f'(s) = m(s - s_0)^{m-1} h(s) + (s - s_0)^m h'(s)$$

and

$$\frac{f'(s)}{f(s)} = \frac{m(s - s_0)^{m-1} h(s) + (s - s_0)^m h'(s)}{(s - s_0)^m h(s)} = \frac{m}{s - s_0} + \frac{h'(s)}{h(s)},$$

where h' and also h'/h are analytic[8] at s_0. Consequently

$$\operatorname*{Res}_{s=s_0} \frac{f'(s)}{f(s)} = \operatorname*{Res}_{s=s_0} \frac{m}{s - s_0} + \operatorname*{Res}_{s=s_0} \frac{h'(s)}{h(s)} = \operatorname*{Res}_{s=s_0} \frac{m}{s - s_0} = m > 0,$$

which means that a zero of f is always a pole of f'/f at which the residue is equal to the multiplicity m.

The rationale for poles is similar. If s_0 is a pole of f with multiplicity n,

$$f(s) = (s - s_0)^{-n} g(s), \qquad g(s_0) \neq 0,$$

with g analytic at s_0. This fact has been used before in Section 3.4. We perform the same analysis as in the case of zeros, replacing h for g and the multiplicity m for $-n$, to obtain

$$\operatorname*{Res}_{s=s_0} \frac{f'(s)}{f(s)} = \operatorname*{Res}_{s=s_0} \frac{-n}{s - s_0} + \operatorname*{Res}_{s=s_0} \frac{g'(s)}{g(s)} = \operatorname*{Res}_{s=s_0} \frac{-n}{s - s_0} = -n,$$

concluding that a pole of f is always a pole of f'/f at which the residue is equal to the negative of the multiplicity n, i.e. $-n$.

This property of f'/f is so incredible that we cannot resist the urge to show an example. Let f be equal to G from (7.6). Compute

$$\frac{G'(s)}{G(s)} = \frac{1 + 2s - s^2}{s^2(s+1)^2} \times \frac{s(s+1)}{s-1} = \frac{1 + 2s - s^2}{s(s+1)(s-1)},$$

which does not seem very enlightening until we expand in partial fractions

$$\frac{G'(s)}{G(s)} = \frac{1}{s-1} - \frac{1}{s+1} - \frac{1}{s}.$$

---

[7] In general h is obtained from the power series expansion of f about s_0 [BC14, Section 83].
[8] Because h and h' are analytic at s_0 and $h(s_0) \neq 0$.

to reveal the poles of G'/G at the zeros and poles of G and the corresponding residues of 1 or -1 depending on whether it concerns what was originally a zero or a pole in G.

Summing the residues at the poles of f'/f, that is, the poles and zeros of f, inside the positively oriented contour C we obtain from Theorem 3.1

$$\frac{1}{2\pi j}\int_C \frac{f'(s)}{f(s)}\,ds = Z_C - P_C, \tag{7.9}$$

where Z_C is the number of zeros and P_C is the number of poles inside C, counting their multiplicities. This is the right-hand side of (7.5).

The left-hand side of (7.5) is obtained through direct evaluation of the integral. For that we introduce a parametric representation of the contour C in terms of the real variable $r \in [a, b] \subseteq \mathbb{R}$:

$$s = \eta(r) \in C, \qquad a \leq r \leq b.$$

From the assumptions in Theorem 7.1, $f(s) \neq 0$ whenever $s \in C$, which allows the complex integral to be rewritten in terms of the parametric integral on the real variable r:

$$\int_C \frac{f'(s)}{f(s)}\,ds = \int_a^b \frac{f'(\eta(r))\eta'(r)}{f(\eta(r))}\,dr.$$

The simplest case is when η is smooth in $[a, b]$, which we assume next. More complicated contours can be handled by letting η be piecewise smooth.[9] Because $f(s) \neq 0$ for all $s \in C$, the image of any point $s = \eta(r) \in C$ can be written in polar form:

$$f(\eta(r)) = \rho(r)e^{j\theta(r)}, \qquad \rho(r) > 0 \qquad a \leq r \leq b,$$

in which ρ and θ are smooth functions of the parameter r. Because

$$f'(\eta(r))\eta'(r) = \frac{d}{dr}f(\eta(r)) = \rho'(r)e^{j\theta(r)} + j\rho(r)e^{j\theta(r)}\theta'(r)$$

the integral becomes

$$\int_a^b \frac{f'(\eta(r))\eta'(r)}{f(\eta(r))}\,dr = \int_a^b \frac{\rho'(r)e^{j\theta(r)} + j\rho(r)e^{j\theta(r)}\theta'(r)}{\rho(r)e^{j\theta(r)}}\,dr$$

$$= \int_a^b \frac{\rho'(r)}{\rho(r)}\,dr + j\int_a^b \theta'(r)\,dr$$

$$= \ln \rho(r)\big|_a^b + j\theta(r)\big|_a^b.$$

Because C is closed we have that $\rho(a) = \rho(b)$ and the first term is zero. The second term is related to the desired total[10] angular variation:

$$\frac{1}{2\pi j}\int_C \frac{f'(s)}{f(s)}\,ds = \frac{1}{2\pi}[\theta(b) - \theta(a)] = \frac{1}{2\pi}\Delta_C^0 \arg f(s),$$

which is the left-hand side of (7.5).

---

[9] See [BC14, Section 93].
[10] This is a consequence of the fact that the contour is parametrized by a piecewise continuous function.

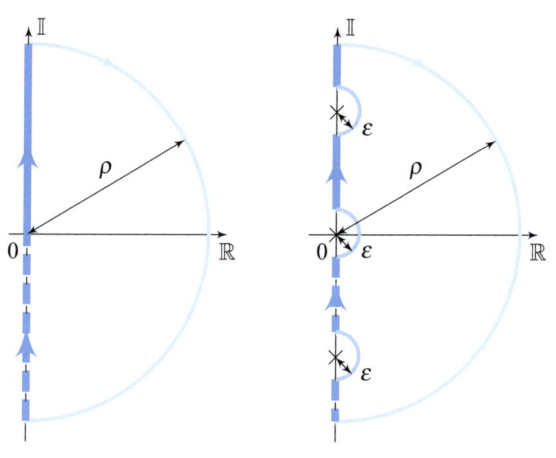

(a) No imaginary poles   (b) Three imaginary poles

Figure 7.14 Simple closed contour, Γ, for assessing stability; imaginary poles need to be excluded from the contour as shown in (b); as ρ is made large and ϵ is made small these contours cover the entire right-hand side of the complex plane; the image of the thick solid and thick dashed segments under a mapping G can be obtained directly from the polar or Bode plot of $G(j\omega)$.

7.5 Stability in the Frequency Domain

In order to use the argument principle, Theorem 7.1, to check for stability of a linear time-invariant system with transfer-function $G(s)$ we need to define a suitable closed contour. We shall use a special case of the contour Γ_-^α, introduced earlier in Fig. 3.2(a), in which $\alpha = 0$. The resulting contour, which we refer to in the rest of this book simply as Γ, is reproduced in Fig. 7.14(a). As in Section 3.4, it can cover the entire right-hand side of the complex plane by taking the limit $\rho \to \infty$. It is convenient to split the contour into three smooth segments:

$$\Gamma = \begin{cases} s = j\omega, & -\rho \leq \omega \leq 0 \quad \text{(dashed line)}, \\ s = j\omega, & 0 \leq \omega \leq \rho \quad \text{(thick line)}, \\ s = \rho e^{-j\theta}, & -\pi/2 \leq \theta \leq \pi/2 \quad \text{(thin semicircle)}. \end{cases}$$

The thick- and dashed-line segments lie on the imaginary axis and their image under the mapping G coincides exactly with the polar plot of G, Section 7.3, as ρ is made larger. The thin semi-circular part of the path closes the contour to allow the application of the argument principle. Note, however, that transfer-functions that satisfy the convergence condition (3.23) are such that

$$\lim_{\rho \to \infty} G(\rho e^{-j\theta}) = 0, \qquad (7.10)$$

consequently the radius, ρ, cannot be made infinite without violating the assumption of Theorem 7.1 that no zeros should lie on the contour. We will handle this complication later. For now, assume that G does not have any finite or infinite poles or zeros on Γ. In this case, the total argument of the image of Γ under G offers an assessment of the

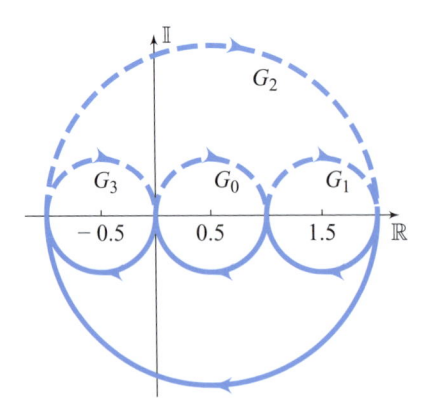

Figure 7.15 Images of the contour Γ under the mappings $G_0(s) = 1/(s+1)$, $G_1(s) = (s+2)/(s+1)$, $G_2(s) = (2-s)/(s+1)$, and $G_3(s) = -s/(s+1)$.

number of poles and zeros of G on the right-hand side of the complex plane. Because the contour Γ is negatively oriented,[11,12] see Fig. 7.14(a), the argument principle, Theorem 7.1, is best applied in the form given by formula (7.8):

$$\frac{1}{2\pi} \Delta_\Gamma^0 \arg G(s) = P_\Gamma - Z_\Gamma.$$

In practice, the reversal of the direction of travel of the contour Γ means that encirclements of the origin should be counted as positive if they are clockwise and as negative if they are counter-clockwise. See Section 7.4.

It is easy to compute the total argument variation, $\Delta_\Gamma^0 \arg G(s)$, from the polar plot of G. If Z_Γ is known or is easy to compute, then G is asymptotically stable if and only if

$$P_\Gamma = Z_\Gamma + \frac{1}{2\pi} \Delta_\Gamma^0 \arg G(s) = 0. \tag{7.11}$$

Otherwise, G has at least one pole on the right-hand side of the complex plane and therefore is not asymptotically stable.

Consider for example the transfer-function

$$G_1(s) = \frac{s+2}{s+1} = 1 + G_0(s), \qquad G_0(s) = \frac{1}{s+1}.$$

The function G_1 has no poles or zeros on the contour Γ. It has a zero, $s = -2$, on the left-hand side of the complex plane so that $Z_\Gamma = 0$. Because $G_1 = 1 + G_0$, the polar plot of G_1 is the polar plot of G_0, the clockwise circle of radius $1/2$ centered at $1/2$ shown in Fig. 7.12, translated by 1. See Fig. 7.15. That is, the polar plot of G_1 is the circle of radius $1/2$ centered at $1 + 1/2 = 3/2$, which never encircles the origin. Consequently

$$P_\Gamma = Z_\Gamma + \frac{1}{2\pi} \Delta_\Gamma^0 \arg G_1(s) = 0 + 0 = 0$$

---

[11] See discussion at the end of Section 7.4.
[12] This is for convenience, as it allows one to *read* Bode plots from left to right as we traverse the imaginary axis from $-j\infty$ to $+j\infty$ to produce polar and Nyquist plots.

and G_1 is asymptotically stable. As a second example, consider

$$G_2(s) = \frac{2-s}{s+1} = 3G_0(s) - 1.$$

The polar plot of $3G_0$ is a clockwise circle of radius $3/2$ centered at $3/2$. The polar plot of $G_2 = 3G_0 - 1$ is therefore a clockwise (negative) circle of radius $3/2$ centered at $3/2 - 1 = 1/2$, Fig. 7.15, which encircles the origin once in the clockwise direction. Because the zero of G_2, i.e. $s = 2$, is on the right-hand side of the complex plane, $Z_\Gamma = 1$, and

$$P_\Gamma = Z_\Gamma + \frac{1}{2\pi} \Delta_\Gamma^0 \arg G_2(s) = 1 + (-1) = 0,$$

from which we conclude that G_2 is asymptotically stable. As mentioned earlier, clockwise encirclements of the origin obtained from the polar plot are counted as positive because Γ is negatively oriented.

When the transfer-function G has poles or zeros on the contour Γ, the assumptions of the argument principle, Theorem 7.1, are violated. The case of poles is simple: if G satisfies (3.23) or the weaker (3.25), then it can only have finite poles; therefore, any poles on Γ must be on the imaginary axis and G is not asymptotically stable. The case of zeros on the contour requires some more thought because a transfer-function can have zeros on the imaginary axis and still be asymptotically stable. Moreover, as mentioned earlier, if G satisfies (3.23) then it necessarily has *zeros at infinity*. In either case, the symptom is the same: if $s_0 \in \Gamma$ is a zero then $G(s_0) = 0$, therefore the image of Γ under G contains the origin, making it difficult to analyze encirclements. This should be no surprise. In fact, the polar plot of the transfer-function

$$G_0(s) = \frac{1}{s+1}$$

shown in Fig. 7.12 contains the origin as $s = j\rho, \rho \to \infty$. Finite zeros on Γ also produce crossings of the origin. For example, the transfer-function

$$G_3(s) = \frac{-s}{s+1} = G_0(s) - 1$$

has a zero at $s = 0 \in \Gamma$ and its polar plot is a clockwise circle of radius $1/2$ centered at $1/2 - 1 = -1/2$, Fig. 7.15, which contains the origin. Both G_0 and G_3 are asymptotically stable but it is not clear how many encirclements of the origin have occurred. In the case of infinite zeros, a formal workaround is possible if all poles of G on the right-hand side of the complex plane are finite so that there exists a large enough ρ for which Γ contains all such poles. For example, by plotting the image of the contour Γ under the mapping G for such a large yet finite radius ρ in the case of the transfer-function $G_0(s)$, we obtain the graphic in Fig. 7.16. It is now possible to see that no encirclements of the origin occur and, since $Z_\Gamma = 0$ and

$$P_\Gamma = Z_\Gamma + \frac{1}{2\pi} \Delta_\Gamma^0 \arg G_0(s) = 0 + 0 = 0,$$

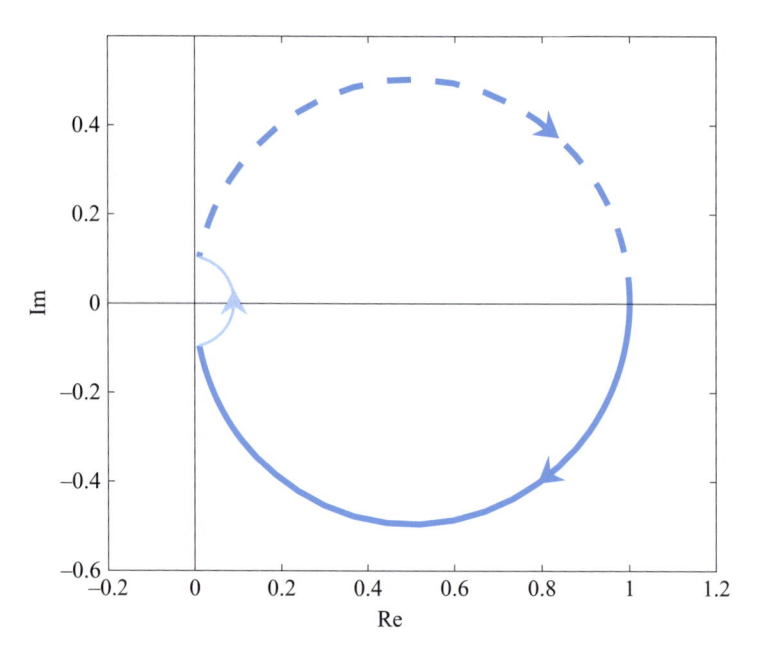

Figure 7.16 Image of the contour Γ under the mapping $G_0(s) = 1/(s+1)$ showing the behavior near the origin as ρ is made large.

to correctly conclude that G_0 is asymptotically stable. In the case of finite zeros, a solution is to *indent* the contour around imaginary zeros, as shown in Fig. 7.14(b). We will develop this technique in Section 7.6 when we study the Nyquist stability criterion.

Do not let these difficulties at the origin obfuscate the power of the argument principle applied to stability analysis. The real virtue of the test is that one does not need to explicitly calculate the poles of a transfer-function, in other words the roots of the characteristic equation, to make inferences about the presence of poles on the right-hand side of the complex plane. This makes it useful even if the characteristic equation is not polynomial or G is not rational. For example, a linear system in closed-loop with a delay has a characteristic equation which is not polynomial. A simple example is

$$\phi(s) = s + 1 + e^{-s} = 0.$$

It is possible to use the argument principle to assess the location of the roots of $\phi(s)$ even in this case in which we do not know the exact number[13] of roots. Define the function

$$G(s) = \frac{s+1}{\phi(s)} = \frac{s+1}{s+1+e^{-s}}. \tag{7.12}$$

Since $|e^{-s}| \leq 1$ for all s such that $\text{Re}(s) \geq 0$,

$$|G(s)| \geq \frac{|s+1|}{1+|s+1|} > 0, \qquad \text{whenever } \text{Re}(s) \geq 0,$$

---

[13] The characteristic equations of linear systems with delay have an infinite number of roots [GKC03].

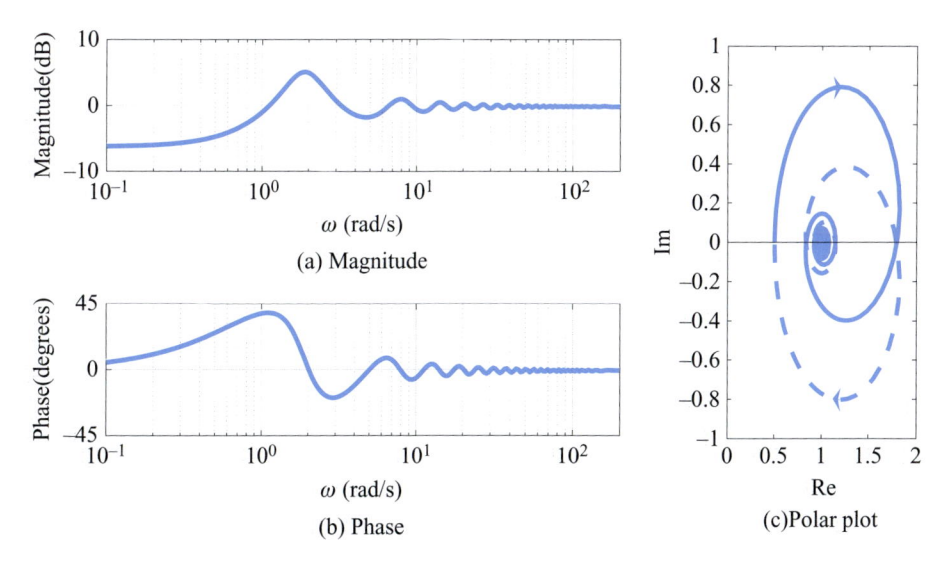

Figure 7.17 Bode and polar plots of $G(s) = (s+1)/(s+1+e^{-s})$.

which means that G does not have any zeros on the right-hand side of the complex plane, $Z_\Gamma = 0$. All poles on the right-hand side of the complex plane must also be finite because

$$\lim_{\substack{\text{Re}(s) \geq 0, \\ |s| \to \infty}} G(s) = \frac{s+1}{s+1+e^{-s}} = 1.$$

The transfer-function G is asymptotically stable because

$$P_\Gamma = Z_\Gamma + \frac{1}{2\pi} \Delta_\Gamma^0 \arg G(s) = 0,$$

that is the image of Γ under G does not encircle the origin and G has no zeros or poles on the imaginary axis. This condition is easily checked graphically in Fig. 7.17. Note how the term e^{-s} significantly affects the phase of $G(j\omega)$, which oscillates and crosses the positive real axis an infinite number of times. Yet, as the polar plot of G in Fig. 7.17(c) shows, the image of Γ under G never encircles the origin, and G does not have any poles (the plot is bounded) or zeros (does not contain the origin) on the imaginary axis. The conclusion is that G is asymptotically stable, that is, the characteristic equation $\phi(s) = 0$ has no roots on the right-hand side of the complex plane.

7.6 Nyquist Stability Criterion

We are finally ready to study the celebrated Nyquist stability criterion. The setup is the same as for root-locus analysis (Section 6.4): the *loop transfer-function*, L, is placed in feedback with a static gain, α, as shown in Fig. 7.18. The Nyquist criterion is an indirect

Frequency Domain

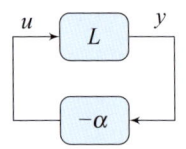

Figure 7.18 Closed-loop feedback configuration for Nyquist criterion; $\alpha \geq 0$.

graphical method to asses the location of the zeros of the characteristic equation

$$1 + \alpha L(s) = 0, \quad \alpha > 0.$$

By properly constructing L one can represent the various feedback configurations studied earlier in Chapters 4 and 6. When $\alpha > 0$, the poles and zeros of $1 + \alpha L(s)$ are the same as the poles and zeros of

$$L_\alpha(s) = \frac{1}{\alpha} + L(s), \quad \alpha > 0. \tag{7.13}$$

Furthermore, the *poles* of L_α are the same as the (open-loop) poles of L, which we assume are known by the designer. The information we are after is the locations of the *zeros* of L_α, which correspond to the *closed-loop poles* of the feedback system in Fig. 7.18.

The idea behind the Nyquist criterion is that the image of the simple closed contour Γ, Fig. 7.14(a), under the (closed-loop) mapping L_α is readily obtained from the image of Γ under the (open-loop) mapping L: the image of Γ under L_α is simply the image of Γ shifted by $1/\alpha$, as illustrated in Fig. 7.19. Such plots are known as *Nyquist plots*.

(a) Image of Γ under L

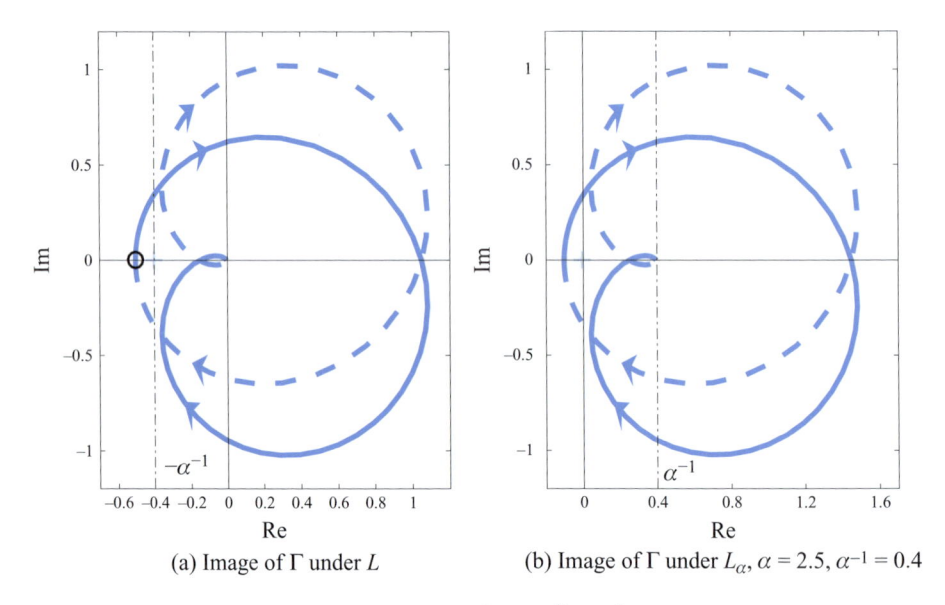

(b) Image of Γ under L_α, $\alpha = 2.5$, $\alpha^{-1} = 0.4$

Figure 7.19 Nyquist plots for $L(s) = (s - 0.5)/(s^4 + 2.5s^3 + 3s^2 + 2.5s + 1)$. The circle in (a) marks $-\frac{1}{\alpha} = -0.5$, which corresponds to the largest positive value of $\alpha = 2$ for which no encirclements occur.

7.6 Nyquist Stability Criterion

Moreover, clockwise encirclements of the origin by the image of Γ under L_α are the same as clockwise encirclements of the image of Γ under L around

$$-\frac{1}{\alpha}, \quad \alpha > 0.$$

See Fig. 7.19. Recalling the argument principle, Theorem 7.1, and formula (7.8), the number of encirclements is equal to

$$\frac{1}{2\pi}\Delta_\Gamma^{-1/\alpha}\arg L(s) = \frac{1}{2\pi}\Delta_\Gamma^0 \arg L_\alpha(s) = P_\Gamma - Z_\Gamma.$$

Because the poles of L_α on the right-hand side of the complex plane are the same as the poles of L, that is P_Γ, the zeros of L_α on the right-hand side of the complex plane are

$$Z_\Gamma = P_\Gamma - \frac{1}{2\pi}\Delta_\Gamma^{-1/\alpha}\arg L(s),$$

which are the right-hand-side closed-loop poles of the feedback system in Fig. 7.18. Closed-loop asymptotic stability ensues if and only if $Z_\Gamma = 0$, that is,

$$P_\Gamma = \frac{1}{2\pi}\Delta_\Gamma^{-1/\alpha}\arg L(s).$$

This is the *Nyquist stability criterion*:

THEOREM 7.2 (Nyquist) *Assume that the transfer-function L satisfies* (3.25) *and has no poles on the imaginary axis. For any given $\alpha > 0$ the closed-loop connection in Fig. 7.18 is asymptotically stable if and only if the number of counter-clockwise encirclements of the image of the contour Γ, Fig. 7.14(a), under the mapping L around the point $-1/\alpha$ is equal to the number of poles of L on the right-hand side of the complex plane.*

Note that the Nyquist stability criterion can be applied even if L is not rational, in contrast with the root-locus method studied in Chapter 6, which is better suited to rational transfer-functions. The assumption that L has no poles on the imaginary axis will be removed later in this section.

For example, we trace the Nyquist plot for

$$L(s) = \frac{s - 0.5}{s^4 + 2.5s^3 + 3s^2 + 2.5s + 1}, \quad \alpha = 2.5,$$

in Fig. 7.19. Verify that L is asymptotically stable, that is $P_\Gamma = 0$. The image of Γ under L encircles the point $-1/\alpha = -0.4$ once in the clockwise direction. Therefore

$$0 = P_\Gamma \neq \frac{1}{2\pi}\Delta_\Gamma^{-0.4}\arg L(s) = -1,$$

and the closed-loop system is not asymptotically stable. Because $P_\Gamma = 0$, no net encirclements of the point $-1/\alpha$ should occur if the closed-loop system is to be asymptotically stable. Note also that

$$Z_\Gamma = P_\Gamma - \frac{1}{2\pi}\Delta_\Gamma^{-0.4}\arg L(s) = 1,$$

indicating that exactly one closed-loop pole is on the right-hand side of the complex plane.

Frequency Domain

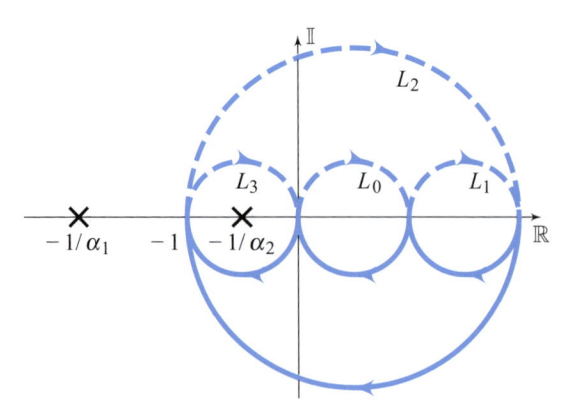

Figure 7.20 Nyquist plots for $L_0(s) = 1/(s+1)$, $L_1(s) = (s+2)/(s+1)$, $L_2(s) = (2-s)/(s+1)$, and $L_3(s) = -s/(s+1)$; L_2 and L_3 intersect the negative real axis at -1. $P_\Gamma = 0$ for all transfer-functions. L_0 and L_1 are asymptotically stable in closed-loop for any feedback gain $\alpha > 0$ because their Nyquist plots do not encircle any negative real number. L_2 and L_3 are asymptotically stable in closed-loop for any feedback gain $\alpha < 1$ because their Nyquist plots do not encircle $-1/\alpha < -1$; e.g. α_1 does and α_2 does not stabilize L_2 and L_3 in closed-loop.

The largest value of α for which no encirclements occur is $\alpha = 2$, from $-1/\alpha = -0.5$, which is marked with a circle in Fig. 7.19(a). This point is the farthest point at which the Nyquist plot crosses the negative real axis, and it determines a limit (margin) on possible gains that lead to asymptotically stable closed-loop systems, in this case $0 < \alpha < 2$. We will have more to say about crossings of the negative real axis in Section 7.7.

For another simple example consider the first-order transfer-functions

$$L_0(s) = \frac{1}{s+1}, \quad L_1(s) = \frac{s+2}{s+1}, \quad L_2(s) = \frac{2-s}{s+1}, \quad L_3(s) = \frac{-s}{s+1}.$$

We have traced their polar plots in Fig. 7.15, which we reproduce in Fig. 7.20. All these transfer-functions have one pole at $s = -1$, therefore $P_\Gamma = 0$. For closed-loop stability, there should be no net encirclement of the point $-1/\alpha$, $\alpha > 0$. For L_0 and L_1, this is always the case, as their Nyquist plots never intersect the negative real axis. The Nyquist plots of L_2 and L_3 intersect the negative real axis at -1 and produce no encirclement if $-1/\alpha < -1$. That is, L_2 and L_3 are asymptotically stable in closed-loop if $\alpha < 1$, e.g. $1/\alpha_1$ in Fig. 7.20. However, when $0 < \alpha < 1$, the Nyquist plots of G_2 and G_3 display one clockwise (negative) encirclement, hence the associated closed-loop systems are not stable and exhibit

$$Z_\Gamma = P_\Gamma - \frac{1}{2\pi} \Delta_\Gamma^{1/\alpha} \arg L_k(s) = 0 - (-1) = 1, \quad k = \{2, 3\},$$

poles on the right-hand side of the complex plane. Verify that these conclusions match exactly what is expected from the corresponding root-locus plots.[14] In the case of G_2 and G_3, the point where the Nyquist plot intersects the negative real axis also provides

---

[14] Note that G_2 and G_3 have $\beta < 0$ when their numerators and denominators are made monic, so be careful with their root-locus plots!

7.6 Nyquist Stability Criterion

the gain, $\alpha = 1$, at which the closed-loop system ceases to be asymptotically stable, that is, the value of the gain at which the single branch of the root-locus crosses toward the right-hand side of the complex plane. There is more to come in Section 7.7.

As in Section 7.5, difficulties arise when a pole or zero of L_α lies on the contour Γ. We will study poles on the contour first. If L satisfies (3.25) then

$$\lim_{|s|\to\infty} |L_\alpha(s)| \leq \frac{1}{\alpha} + \lim_{|s|\to\infty} |L(s)| \leq \frac{1}{\alpha} + M < \infty,$$

which means that L_α has only finite poles. Since the poles of L_α are the same as the poles of L, any finite pole of L_α or L on Γ must be imaginary. The solution is to *indent* the contour Γ by drawing a small semicircle around each imaginary pole of L, thus removing the pole from the indented contour Γ, as illustrated in Fig. 7.14(b) in the case of three poles on the imaginary axis. The image of the indented contour Γ is then traced as we make ρ large and ϵ, the radii of the indented semicircles, small. This covers the entire open right-hand side of the complex plane. Because L is singular at the indented poles, one should expect that the image of Γ under L becomes unbounded as $\epsilon \to 0$. Yet stability can still be assessed if we are careful about keeping track of the direction and number of encirclements.

Let $s_0 = j\omega_0$ be a pole of L with multiplicity $n \geq 1$ (poles at the origin can be handled by setting $\omega_0 = 0$). Then

$$L(s) = \frac{F(s)}{(s - j\omega_0)^n}, \qquad F(j\omega_0) \neq 0,$$

where F is analytic at $s = j\omega_0$. Parametrize the indented semicircle around $s = j\omega_0$ shown in Fig. 7.14(b) by

$$s = j\omega_0 + \epsilon e^{j\theta}, \qquad -\frac{\pi}{2} \leq \theta \leq \frac{\pi}{2}, \qquad \epsilon > 0,$$

and substitute this curve into L to obtain

$$L(j\omega_0 + \epsilon e^{j\theta}) = \epsilon^{-n} e^{-jn\theta} F(j\omega_0 + \epsilon e^{j\theta}).$$

When ϵ is small[15]

$$|L(j\omega_0 + \epsilon e^{j\theta})| \approx \epsilon^{-n} |F(j\omega_0)|, \qquad |F(j\omega_0)| \neq 0, \qquad \epsilon \approx 0,$$

which becomes unbounded as $\epsilon \to 0$. As for the phase,

$$\angle L(j\omega_0 + \epsilon e^{j\theta}) \approx \angle F(j\omega_0) - n\theta, \qquad -\frac{\pi}{2} \leq \theta \leq \frac{\pi}{2}, \qquad \epsilon \approx 0.$$

This formula shows that, starting at

$$\lim_{\omega \to \omega_0^-} \angle L(j\omega) = \angle F(j\omega_0) + \frac{n\pi}{2},$$

the phase of the image of the indented contour Γ under the mapping L decreases by a total of $n\pi$ radians. That is, it is an arc with very large radius that spins $n\pi$ radians in the clockwise direction. A negative value of ω_0 does not change this behavior and the direction of the spin is independent of the segment within which the pole is located (thick solid or thick dashed line in Fig. 7.14(b)).

---

[15] This argument can be made precise by expanding F in a power series.

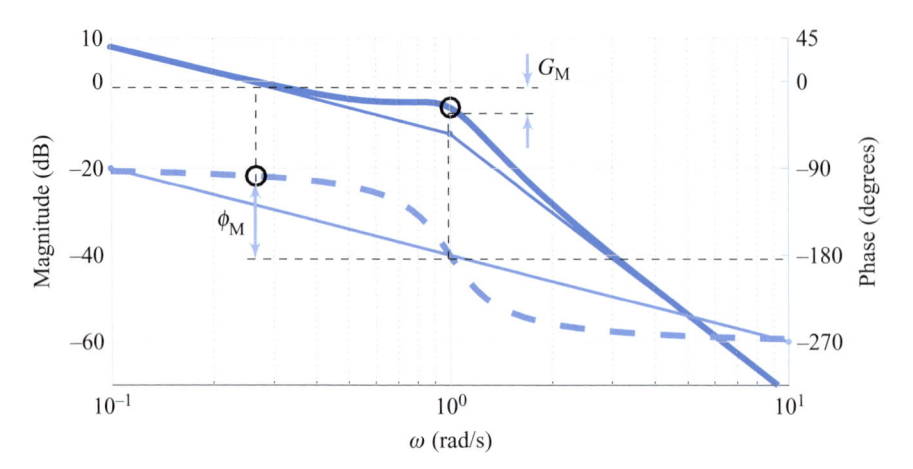

Figure 7.21 Combined Bode plot of $L(s) = (4s(s^2 + s/2 + 1))^{-1}$; G_M is the gain margin and ϕ_M is the phase margin.

For example, the transfer-function

$$L(s) = \frac{1}{4} \frac{1}{s(s^2 + s/2 + 1)} \qquad (7.14)$$

has a pole at zero, and its frequency response is unbounded as $\omega \to 0$. This can be seen in the Bode plot of L shown in Fig. 7.21, where the straight-line approximation to the magnitude of the frequency response has a constant negative slope of -20 dB/decade for $0 < \omega \leq 1$. The behavior of the Nyquist plot as it approaches the origin can be inferred by analyzing

$$L(\epsilon e^{j\theta}) = -\epsilon^{-1} e^{-j\theta} F(\epsilon e^{j\theta}), \qquad F(s) = \frac{1}{4} \frac{1}{s^2 + s/2 + 1},$$

from which, for $\epsilon \approx 0$,

$$|L(\epsilon e^{j\theta})| \approx \frac{|F(0)|}{\epsilon} = \frac{1}{4\epsilon},$$

and

$$\angle L(\epsilon e^{j\theta}) \approx \angle F(0) - \theta = -\theta, \qquad -\frac{\pi}{2} \leq \theta \leq \frac{\pi}{2},$$

so that the Nyquist plot has a semicircle with large radius that swings from $\pi/2$ to $-\pi/2$, as shown in Fig. 7.22(a). The asymptote $s = -1/8$ is obtained from

$$L(j\omega) = \frac{1}{4} \frac{1}{j\omega[j\omega/2 + (1 - \omega^2)]}$$

$$= \frac{-1}{4(1-\omega^2)^2 + \omega^2} \left(\frac{1}{2} + j\frac{1-\omega^2}{\omega} \right)$$

$$\approx -\frac{1}{8} - j\frac{1}{4\omega}, \qquad \omega \approx 0.$$

7.6 Nyquist Stability Criterion

(a) Nyquist plot

(b) Gain and phase margins

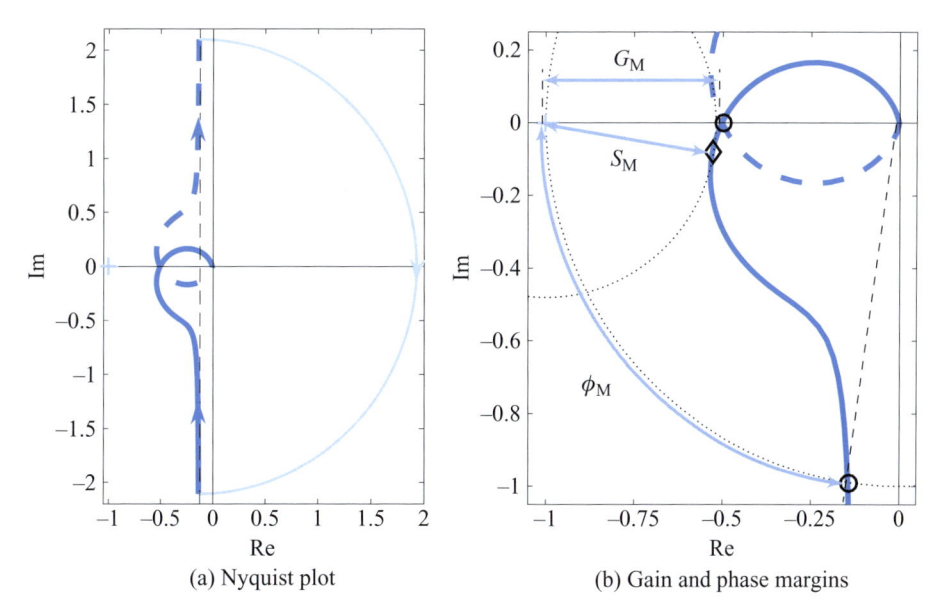

Figure 7.22 Nyquist plot of the function $L(s) = (4s(s^2 + s/2 + 1))^{-1}$. Plot (b) is zoomed in to show the gain margin, G_M, and the phase margin, ϕ_M, and stability margin, S_M.

Unfortunately, this asymptote is not easily computed from the Bode plot. In this example, L has no poles inside the indented contour Γ, that is $P_\Gamma = 0$. If $\alpha = 1$ then no encirclements happen and

$$P_\Gamma = \frac{1}{2\pi} \Delta_\Gamma^{-1} \arg L(s) = 0.$$

That is, there are no closed-loop poles on the right-hand side of the complex plane and the closed-loop system is asymptotically stable. Note that the hypothesis that $\alpha > 0$ is important. In this example, the closed-loop system with $\alpha = 0$ is not asymptotically stable because of the pole at the origin.

We shall now analyze what happens if zeros of L_α are on the contour Γ. As in Section 7.5, a finite or infinite zero on the path Γ means that the origin is part of the image of Γ under the mapping L_α. If s_0 is a finite or infinite zero of L_α then

$$0 = L_\alpha(s_0) = \frac{1}{\alpha} + L(s_0), \qquad \alpha > 0.$$

In other words,

$$L(s_0) = -\frac{1}{\alpha}, \qquad \angle L(s_0) = \pi, \qquad \alpha > 0,$$

and the Nyquist plot of L crosses the negative real axis at exactly $-1/\alpha < 0$ when s_0 is a zero of L_α. Because the location of the zeros of L_α, that is, the closed-loop poles, is precisely the information we are looking for, it is not practical to indent the contour at these (unknown) zeros. Luckily, it often suffices simply to *shift* the point $-1/\alpha$.

For example, in Fig. 7.22(a), the function L_α has a zero on the contour Γ only if $-1/\alpha = -1/2$. On the one hand, no encirclements of the point $-1/\alpha$ happen if $0 < \alpha < 2$, that is, the closed-loop system is asymptotically stable if $0 < \alpha < 2$. On the other hand, if $\alpha > 2$ then one clockwise encirclement occurs and the closed-loop system is not asymptotically stable for any $\alpha > 2$.

More generally, the exact behavior at the crossing of the point $-1/\alpha$ will depend on whether the associated zero is finite or infinite. For example, if

$$L(s) = \frac{2-s}{s+1} = 3\frac{1}{s+1} - 1$$

the Nyquist plot of L is a clockwise circle of radius $3/2$ centered at $1/2$ that crosses the negative real axis at -1 (L_2 in Fig. 7.20). The loop transfer-function, L, is asymptotically stable, $P_\Gamma = 0$, and the closed-loop system is asymptotically stable if $0 \leq \alpha < 1$, because $Z_\Gamma = 0$; the closed-loop system is not asymptotically stable if $\alpha > 1$, because $Z_\Gamma = 1$. At $\alpha = 1$, the Nyquist plot crosses the point $-1/\alpha = -1$, which means that L_α has a zero on the contour Γ. In this case, this is a zero at infinity, and an analysis similar to the one used in Section 7.5 to obtain Fig. 7.16 reveals that no crossings of the point -1 occur if we let ρ be large yet finite, and the corresponding closed-loop system is asymptotically stable. Indeed,

$$1 + L(s) = 3\frac{1}{s+1},$$

which has no finite zeros. In this very special case the numerator has order zero, which indicates that an exact closed-loop pole–zero cancellation happened.

Most commonly, however, crossings of the negative real axis by the Nyquist plot happen due to finite zeros of L_α in Γ. This will be the case, for example, whenever L satisfies the convergence condition (3.23), e.g. L is strictly proper. In this case, zeros on Γ must be imaginary. In fact, any point $s_0 = j\omega_0$ for which $L(s_0) = L(j\omega_0)$ is real and negative is such that

$$L(j\omega_0) = -\frac{1}{\alpha_0}, \qquad \angle L(j\omega_0) = \pi,$$

for some $\alpha_0 > 0$. On comparing this with (6.19) we conclude that these very special points are also part of the root-locus of $L(s)$. Indeed, when the Nyquist plot crosses the negative real axis at $s = j\omega_0$, a root must cross the imaginary axis in the root-locus plot for a corresponding $\alpha = \alpha_0$. If $L(s)$ has n poles on the right-hand side of the complex plane, all n poles will eventually have to cross over the imaginary axis on the root-locus plot if the closed-loop system is to become asymptotically stable. We illustrate this relationship with a simple yet somewhat puzzling example. Consider the loop transfer-function

$$L(s) = \frac{1}{(s+1)(s-1)}.$$

Its root-locus is shown in Fig. 7.23(a). There exists no $\alpha > 0$ for which the closed-loop system is asymptotically stable. When $0 \leq \alpha < 1$ the closed-loop system has two real poles, one of which is always on the right-hand side of the complex plane, and when

(a) Root-locus plot   (b) Nyquist plot

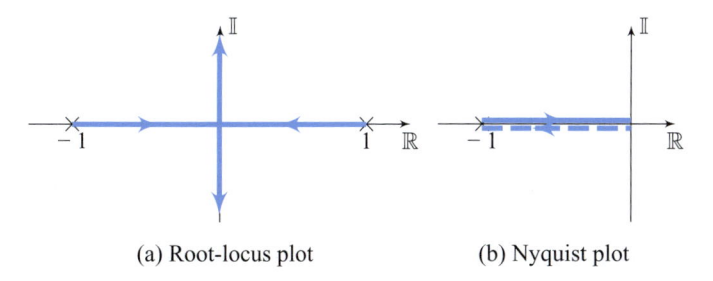

Figure 7.23 Root-locus and Nyquist plot for $L(s) = [(s+1)(s-1)]^{-1}$; both lines on the Nyquist plot lie on the negative real axis but are shown separated for the sake of illustration.

$\alpha > 1$ the closed-loop system has a pair of imaginary poles. At $\alpha = 1$ the closed-loop poles are both at the origin. Because

$$L(j\omega) = \frac{-1}{1+\omega^2}, \qquad |L(j\omega)| = \frac{1}{1+\omega^2} \leq 1, \qquad \angle L(j\omega) = \pi,$$

the Nyquist plot is simply the segment of line on the negative real axis from -1 to 0 shown in Fig. 7.23(b). Since L has one pole with positive real part, there must be at least one encirclement of the point $-1/\alpha$ for the closed-loop system to be asymptotically stable. No encirclements happen when $\alpha < 1$, indicating that one pole remains on the right-hand side of the complex plane. When $\alpha \geq 1$ no encirclements happen[16] but the point $-1/\alpha$ is always *on* the Nyquist plot. This corresponds to a pair of closed-loop poles, i.e. zeros of L_α, which are on the imaginary axis.

7.7 Stability Margins

Crossings of the negative real axis in the Nyquist plot and crossings of the imaginary axis in the root-locus plot play a critical role in the stability of closed-loop systems. By measuring how much additional gain or phase is needed to make an asymptotically stable closed-loop system become unstable we obtain a measure of the available stability *margin*. In the Nyquist plot, we define *gain* and *phase margins* after normalizing the closed-loop gain to $\alpha = 1$. The *gain margin* is defined as

$$G_M = \frac{1}{|L(j\omega_G)|}, \qquad \text{for } \omega_G \text{ such that } \angle L(j\omega_G) = \pi, \qquad (7.15)$$

and the *phase margin* as

$$\phi_M = \angle L(j\omega_\phi) - \pi, \qquad \text{for } \omega_\phi \text{ such that } |L(j\omega_\phi)| = 1, \qquad (7.16)$$

in which the argument $\angle L(j\omega)$ is always calculated in the interval $[0, 2\pi)$. If there exists no ω_G such that $\angle L(j\omega_G) = \pi$, then $G_M = \infty$. Likewise, $\phi_M = \infty$ if the magnitude of the frequency response of L is always less than one.

---

[16] Do not get fooled by the drawing in Fig. 7.23(b)! The two line segments coincide.

The margins (7.15) and (7.16) can be computed in the Bode plot or the Nyquist diagram. For G given in (7.14), the margins[17] are

$$\omega_G = 1, \quad G_M = \frac{1}{|G(j1)|} = 4|\,j/2\,| = 2,$$
$$\omega_\phi \approx 0.27, \quad \phi_M = \angle G(j0.27) - \pi \approx 82°,$$

which are indicated in the Bode plot in Fig. 7.21 and in the Nyquist plot in Fig. 7.22(b).

As noted earlier, the Nyquist plot often crosses the negative real axis at points that are part of the root-locus. This means that the gain margin can also be calculated in the root-locus diagram, corresponding to the smallest distance (smallest additional gain) needed for a root to cross the imaginary axis toward the right-hand side of the complex plane. Indeed, when the Nyquist plot crosses the negative real axis at some $s = j\omega_0$, a root must cross the imaginary axis in the root-locus plot for some $\alpha = \alpha_0 > 0$.

Gain and phase margins can be fully understood only in the broader context of the Nyquist stability criterion. For example, if the closed-loop systems is unstable, crossings of the negative real axis and the unit circle can be computed but they do not imply any margins. Even when closed-loop stability is possible, the Nyquist plot may intersect the negative real axis many times, so one needs to be careful about calculating the appropriate margins. If the loop transfer-function, L, is unstable and has n poles on the right-hand side of the complex plane, then it will be necessary to encircle the point -1 exactly n times in the counter-clockwise direction in order to have closed-loop stability. This implies crossing the negative real axis at least n times! In this case, the gain margin may be less than one (negative in dB), which means that a *reduction* in the gain will lead to instability. If L is asymptotically stable then the gain margin is greater than one, and is often obtained at the first crossing. This is the case, for example, in Fig. 7.19(a).

The margins can also be interpreted in terms of *robustness* to changes in the gain and phase of the system being controlled. For example, gain margin provides guarantees that the closed-loop system will remain asymptotically stable if the overall gain changes. With respect to the closed-loop diagram of Fig. 7.18 with $\alpha > 0$,

$$|\alpha L(j\omega)| = \alpha |L(j\omega)|, \quad \angle \alpha L(j\omega) = \angle L(j\omega).$$

If the closed-loop system in Fig. 7.18 is asymptotically stable for $\alpha = 1$ and $G_M > 1$ then it remains asymptotically stable when

$$|\alpha L(j\omega_G)| = \frac{\alpha}{G_M} < 1 \quad \Longrightarrow \quad 1 \le \alpha < G_M.$$

When $0 < G_M < 1$, the closed-loop system remains asymptotically stable if

$$|\alpha L(j\omega_G)| = \frac{\alpha}{G_M} > 1 \quad \Longrightarrow \quad G_M < \alpha \le 1.$$

Those estimates are conservative, but it is sometimes possible to extend the guarantees of closed-loop asymptotic stability beyond these ranges with just a little more information on L and its Nyquist plot. For example, when $G_M > 1$, L is asymptotically

---

[17] $\angle G(j0.27) \approx -98°$, which is $262°$ in $[0°, 360°)$; hence $\phi_M \approx 262° - 180° = 82°$.

(a) Complex phase   (b) Delay $\tau \geq 0$

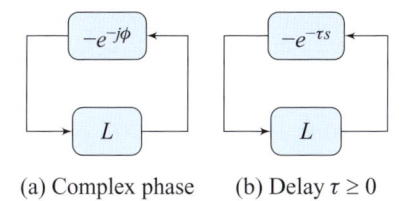

Figure 7.24 Feedback configurations with phase disturbance in the loop.

stable, and there are no crossings of the negative real axis that happen on the left of the point -1, then the closed-loop system is asymptotically stable for all $0 \leq \alpha < G_M$. This is the case, for example, for L given in (7.14) and its Nyquist plot in Fig. 7.22, for which $G_M = 2$ and the closed-loop system in Fig. 7.18 is asymptotically stable for all $0 \leq \alpha < 2$; also for the transfer-function $L(s) = (s+1)^{-1}$ and its Nyquist plot in Fig. 7.16, which has infinite gain margin and the closed-loop system in Fig. 7.18 is asymptotically stable for any value of $\alpha \geq 0$. Compare these statements with the corresponding root-locus plots.

Phase margin offers guarantees in case the overall phase changes. A pure change in phase requires the introduction of a complex disturbance, shown in Fig. 7.24(a). In this case

$$|e^{-j\phi}L(j\omega)| = |L(j\omega)|, \qquad \angle e^{-j\phi}L(j\omega) = \angle L(j\omega) - \phi.$$

When $\phi_M > 0$ and L is asymptotically stable under unit feedback then the closed-loop system in Fig. 7.24(a) remains asymptotically stable as long as

$$\angle e^{-j\phi}L(j\omega_\phi) = \phi_M + \pi - \phi > \pi \quad \Longrightarrow \quad 0 \leq \phi < \phi_M,$$

because keeping the phase away from π will prevent changes in the number of encirclements of the point -1. For example, L given in (7.14) is such that the closed-loop system in Fig. 7.24(a) remains asymptotically stable for all $0 \leq \phi < \phi_M \approx 82°$. When $\phi_M < 0$ a similar reasoning applies.

The pure complex phase disturbance in the block-diagram in Fig. 7.24(a) is a bit artificial, as it involves a complex element. In practical systems, additional phase is often introduced by delays, as shown in the diagram in Fig. 7.24(b), for which

$$|e^{-j\tau\omega}L(j\omega)| = |L(j\omega)|, \qquad \angle e^{-j\tau\omega}L(j\omega) = \angle L(j\omega) - \tau\omega.$$

Changes in phase due to delay are frequency-dependent and one needs to be careful when interpreting the phase margin as robustness against delays. When $\phi_M > 0$, L is asymptotically stable, and $L(j\omega)$ intercepts the unit circle only once for $0 \leq \omega < \infty$, then the closed-loop system in Fig. 7.24 remains asymptotically stable if

$$\phi_M + \pi - \tau\omega_\phi > \pi \quad \Longrightarrow \quad 0 \leq \tau < \phi_M/\omega_\phi, \quad \omega_\phi \neq 0.$$

For example, L given in (7.14) intercepts the unit circle only once and $\phi_M \approx 82° \approx 1.43$ at $\omega_\phi \approx 0.27$, therefore the closed-loop system in Fig. 7.24(b) is asymptotically stable for all $0 \leq \tau < \phi_M/\omega_\phi \approx 5.36$ s. If $\phi_M = \infty$ then stability of the closed-loop system is guaranteed for all values of delay. If L intercepts the unit circle more than once, one needs to be careful and evaluate $\angle L(j\omega)/\omega$ at all ω such that $|L(j\omega)| = 1$ and select the smallest value. When $\phi_M < 0$ a similar reasoning applies but this time

$$\angle L_\tau(j\omega_\phi) > -\pi \quad \Longrightarrow \quad 0 \leq \tau < (2\pi + \phi_M)/\omega_\phi, \quad \omega_\phi \neq 0.$$

Recall that because $\tau \geq 0$, delays can only add negative phase. We will have more to say about robustness and stability margins in Section 8.2.

Gain and phase margins are complementary measures of how close the Nyquist plot is from the point -1. An alternative stability margin is obtained by directly measuring the smallest distance[18] from the Nyquist plot to -1:

$$S_M = \inf_{\omega \in [0,\infty)} |L(j\omega) - (-1)| = \inf_\omega |1 + L(j\omega)| = \frac{1}{\sup_{\omega \in [0,\infty)} |1 + L(j\omega)|^{-1}}.$$

When the loop transfer-function, L, is obtained from the standard feedback connection diagram in Figs. 1.8 and 4.2 then

$$S_M = \frac{1}{\sup_{\omega \in [0,\infty)} |S(j\omega)|} = \|S\|_\infty^{-1}, \qquad (7.17)$$

that is, the inverse of the H_∞ norm of the sensitivity transfer-function S (see Section 3.9). When L satisfies the convergence condition (3.23) then $\lim_{|s| \to \infty} S(s) = \lim_{|s| \to \infty}(1 + L(s))^{-1} = 1$, in which case we can conclude that

$$\|S\|_\infty \geq 1 \quad \Longrightarrow \quad S_M \leq 1.$$

The smaller S_M the closer the Nyquist plot is to the point -1. Note also that, if $|S(j\omega)|$ has a peak, for example S has low-damped complex poles, then we expect that S_M will be small. We will have much more to say about the role of the sensitivity function in stability and performance in Section 8.1.

7.8 Control of the Simple Pendulum – Part II

We now revisit the control of the simple pendulum from Section 6.5. Recall that

$$G_0(s) = \frac{1}{J_r s^2 + bs + mgr}, \qquad G_\pi(s) = \frac{1}{J_r s^2 + bs - mgr},$$

which are models linearized around the stable equilibrium point, $\theta = 0$, and around the unstable equilibrium point, $\theta = \pi$, respectively. We use the same set of parameters as in (6.20) and (6.21):

$$m = 0.5 \text{ kg}, \quad \ell = 0.3 \text{ m}, \quad r = \ell/2 = 0.15 \text{ m},$$
$$b = 0 \text{ kg/s}, \quad g = 9.8 \text{ m/s}^2, \quad J = 3.75 \times 10^{-3} \text{ kg m}^2, \quad J_r = J + mr^2.$$

---

[18] inf is a fancy replacement for min as sup is for max. See footnote 31 on p. 75.

7.8 Control of the Simple Pendulum – Part II

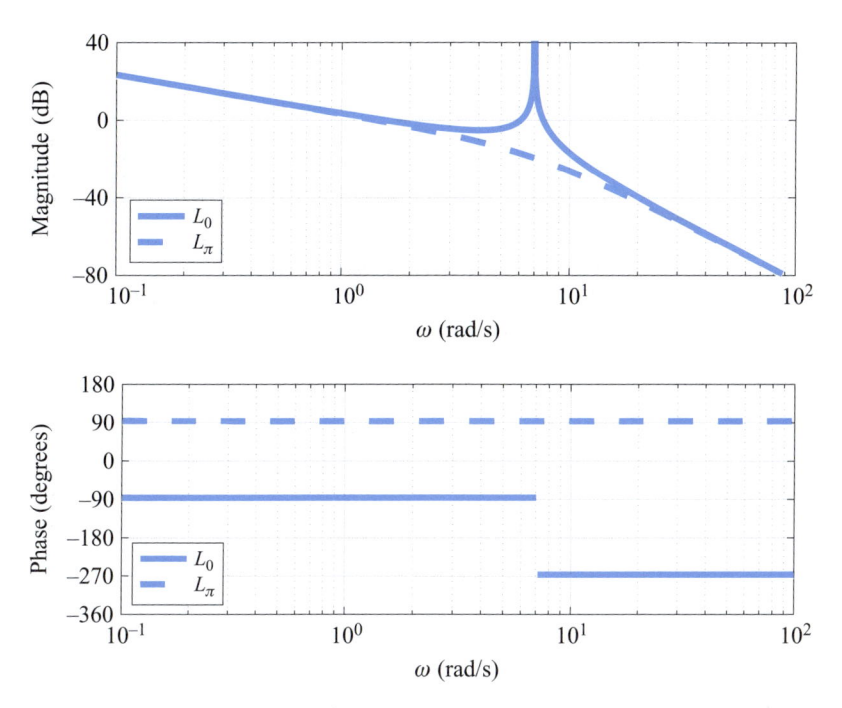

Figure 7.25 Bode plots of $L_0 = s^{-1}G_0$ (solid), and $L_\pi = s^{-1}G_\pi$ (dashed).

As in Section 6.5, we assume during design that there is no friction and that the controller is responsible for providing all necessary damping. We require integral action, which means that the controller must have a pole at the origin. For this reason we trace in Fig. 7.25 the Bode plots of the loop transfer-functions:

$$L_0(s) = \frac{G_0(s)}{s} = \frac{1/J_r}{s(s + j\omega_{n_{ol}})(s - j\omega_{n_{ol}})},$$

$$L_\pi(s) = \frac{G_\pi(s)}{s} = \frac{1/J_r}{s(s + \omega_{n_{ol}})(s - \omega_{n_{ol}})}, \qquad \omega_{n_{ol}} = \sqrt{\frac{mgr}{J_r}} = 7.$$

We analyze L_π first because its frequency response looks less intimidating. We start by normalizing:

$$L_\pi(s) = \frac{-1/(mgr)}{s(1 + s/\omega_{n_{ol}})(1 - s/\omega_{n_{ol}})} \approx \frac{-1.36}{s(1 + s/7)(1 - s/7)}.$$

Because of the pole at the origin, the magnitude of the frequency response is unbounded as $\omega \to 0$. The phase of the frequency response at $\omega \to 0^+$ is 90°, with the negative numerator contributing 180° and the pole at the origin contributing $-90°$. The phases of the symmetric pair of real poles cancel each other and the phase of L_π remains constant at 90° for any frequency $\omega > 0$, as seen in Fig. 7.25. The polar plot of L_π is therefore the entire imaginary axis. In addition to the imaginary axis, the pole at zero creates a 180° arc of infinite radius in the clockwise direction starting at $\lim_{\omega \to 0^-} \angle L_\pi(j\omega) \to -90°$ and ending at $\lim_{\omega \to 0^+} \angle L_\pi(j\omega) \to 90°$ in the Nyquist plot of L_π. Verify that this

Nyquist plot coincides with the contour Γ_-^0, from Fig. 3.2(a), traversed in the reverse direction, which encircles the point -1 once in the clockwise (negative) direction. Hence there are

$$Z_\Gamma = P_\Gamma - \frac{1}{2\pi}\Delta_\Gamma^{-1}\arg L_\pi(s) = 1 - (-1) = 2$$

closed-loop poles on the right-hand side of the complex plane. For closed-loop stability it is necessary that the Nyquist plot encircles the point -1 once but in the counter-clockwise (positive) direction. For that to happen we need to add zeros to the controller. We prefer[19] non-minimum-phase poles and zeros, which do not affect the behavior of the phase of L_π near the origin.

If $L_\pi = G_\pi C$ is the loop transfer-function computed with a controller, C, two things must happen if a counter-clockwise encirclement of the point -1 is to occur: (a) the phase of $L(j\omega)$ has to reach $180°$; (b) the magnitude of $L(j\omega)$ has to be greater than one when that happens.

Starting from $90°$ at $\omega \to 0^+$, it is necessary to add phase in excess of $90°$ if the phase of the frequency response of L is to reach $180°$. This is possible only if (at least) two (minimum-phase) zeros are added to the controller. Because the controller already has a pole at the origin, we add a real pole beyond those two zeros to keep the controller proper. This reasoning suggests a controller with two zeros, one pole at the origin, and one pole beyond the two zeros:

$$C(s) = K\frac{(s+z_1)(s+z_2)}{s(s+p_1)}, \quad \text{Re}(z_2) \geq \text{Re}(z_1) > 0, \quad p_1 > \text{Re}(z_2). \quad (7.18)$$

Note how quickly the frequency response leads to a general location of the controller poles and zeros that matches the general form of the controller C_6, from (6.25), obtained only at the end of Section 6.5. Further analysis will reveal even more information about the locations of the zeros.

Now substitute (7.18) into L to obtain

$$L_\pi(s) = G_\pi(s)C(s) = \frac{K}{J_r}\frac{(s+z_1)(s+z_2)}{s(s+p_1)(s+\omega_{n_{ol}})(s-\omega_{n_{ol}})},$$

$$\approx \frac{-1.36Kz_1z_2}{p_1}\frac{(1+s/z_1)(1+s/z_2)}{s(1+s/p_1)(1+s/7)(1-s/7)}.$$

The simplest way to produce a phase that exceeds $180°$ is to have z_1, z_2, and p_1 be real and positive, satisfying

$$p_1 > z_2 \geq z_1 > 0. \quad (7.19)$$

We expect the phase to grow from $90°$ at $\omega \to 0^+$ then exceed $180°$ momentarily somewhere in the interval $z_2 \leq \omega \leq p_1$, to finally fall back to $180°$ as $\omega \to \infty$. This accomplishes the first objective of producing a crossing of the negative real axis. As discussed earlier, it is necessary that the magnitude of L be greater than one at the crossing if a counter-clockwise encirclement is to be produced in the Nyquist plot. We plug in the

---

[19] Minimum-phase poles and zeros also result in controllers that are easier to implement. How would you "start" a feedback loop with an unstable controller?

7.8 Control of the Simple Pendulum – Part II

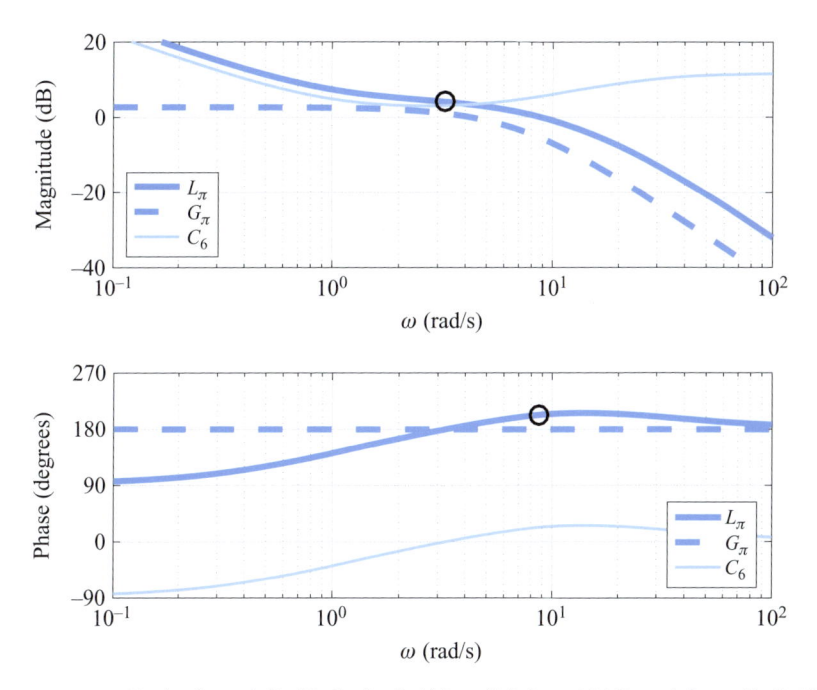

Figure 7.26 Bode plots of G_π (dashed), C_6 (thin solid) from (6.25), and $L_\pi = G_\pi C_6$ (thick solid).

zeros and poles of the controller C_6, from (6.25):

$$z_1 = 0.96, \quad z_2 = 6.86, \quad p_1 = 21.$$

We showed in Section 6.5 that (6.25) leads to an asymptotically stable closed-loop system. The locations of the poles and zeros already satisfy (7.19). Let us now verify whether they also achieve the second goal, i.e. having a magnitude of L greater than one (0 dB) at the crossing. Indeed, the Bode diagrams of G_π, C_6, and $L_\pi = G_\pi C_6$ plotted in Fig. 7.26 display phase exceeding 180° then falling to 180°, and magnitude greater than one (> 0 dB) at the crossing of the negative real axis. The associated gain, phase, and stability margins are

$$G_M \approx 0.63 \approx -4\,\text{dB}, \quad \phi_M \approx 23°, \quad S_M = 0.40.$$

The frequencies ω_G and ω_ϕ are marked in Fig. 7.26 by circles. The fact that the gain margin is less than one (negative in dB), indicates that the closed-loop will become unstable if the gain is reduced. This can be visualized in the associated Nyquist diagram shown in Fig. 7.27, which displays a large arc on the left-hand side of the complex plane due to the pole at the origin and one counter-clockwise (positive) encirclement of -1. Consequently,

$$P_\Gamma = \frac{1}{2\pi}\Delta_\Gamma^{-1}\arg L_\pi(s) = 1,$$

and the closed-loop system is asymptotically stable. The key to achieving a magnitude greater than one at the crossing of the negative real axis is the location of the zeros: by placing the first zero well before $\omega_{n_{ol}} = 7$, at $z_1 = 0.96$, we stop the decrease in magnitude due to the pole at the origin; by adding the second zero near $\omega_{n_{ol}} = 7$, at

Frequency Domain

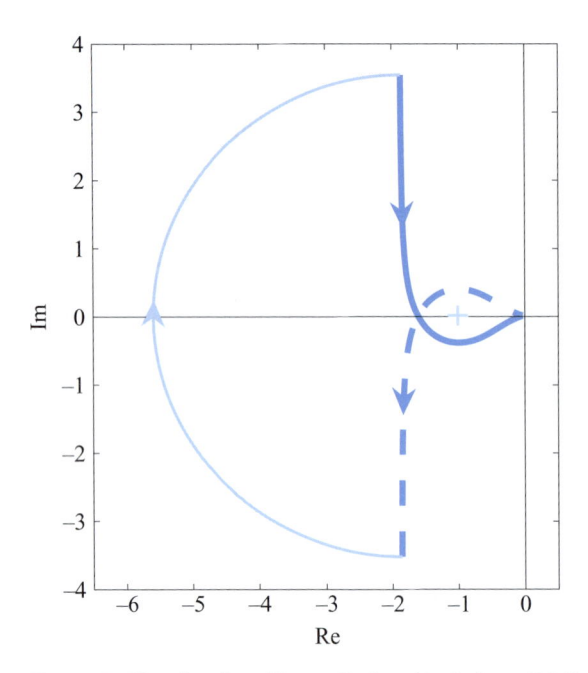

Figure 7.27 Nyquist plot of $L_\pi = G_\pi C_6$, with C_6 from (6.25).

$z_2 = 6.86$, we compensate for the magnitude drop in G_π and add the necessary phase to produce the crossing at 180°; finally, a controller pole added beyond both zeros, at $p_1 = 21$, keeps the controller transfer-function proper without affecting the behavior near $\omega_{n_{ol}} = 7$. Locate these features in the Bode plot in Fig. 7.26.

In the control of the simple pendulum, the ordering of poles and zeros established in (7.19) seems to be more important than the actual locations of the poles and zeros. With that in mind, we will attempt to shift the poles and zeros in the controller to suit other design requirements. For example, we would like to move the zero z_2 and the pole p_1 in order to increase the loop gain near $\omega_{n_{ol}} = 7$. Since the role of the first controller zero, z_1, is to interrupt the gain decrease due to the integrator, we set the first controller zero at $z_1 = 1$, which is close to the original $z_1 = 0.96$. By shifting the second controller zero, z_2, and pole, p_1, to

$$z_2 = 5, \quad p_1 = 11,$$

we should be able to raise the controller gain, and hence the loop gain, near $\omega_{n_{ol}} = 7$. We also adopt the following guideline: the controller must compensate for magnitude losses introduced by G_π at $\omega_{n_{ol}} = 7$. Because $|G_\pi(j\omega_{n_{ol}})| = |G_\pi(j7)| = 1/2$ (double real poles at ω_n) we select the controller gain, K, so that $|C_7(j7)| = 2$. The result is the controller

$$C_7(s) \approx 3.00 \frac{(s+1)(s+5)}{s(s+11)}. \tag{7.20}$$

For comparison, the Bode plots of the controller C_6 and C_7 are plotted in Fig. 7.28, from which $|C_6(j7)| \approx 1.7$. The controller C_7 is slightly more aggressive than the controller C_6 in the bandwidth of the system G_π but has a smaller high-frequency gain. Because the

7.8 Control of the Simple Pendulum – Part II

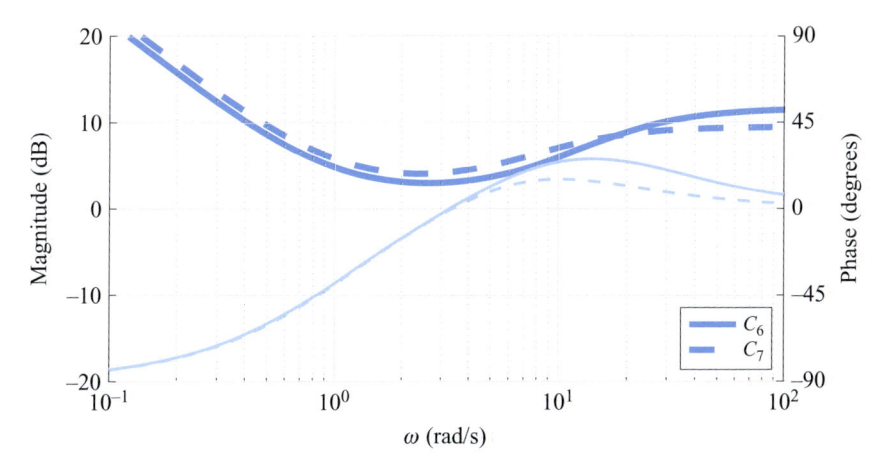

Figure 7.28 Combined Bode plots of C_6 (solid), from (6.25), and C_7 (dashed), from (7.20); thick curves are magnitude and thin curves are phase.

crossing of the negative real axis happens near $\omega_{n_{ol}} = 7$ we expect a slightly increased gain margin. Controller C_7 also has smaller phase near $\omega = 10$, which reduces the phase margin when compared with C_6. Indeed,

$$G_M \approx 0.55 \approx -5.2 \text{ dB}, \qquad \phi_M \approx 15°, \qquad S_M = 0.27.$$

The gain and phase margins are indicated in the frequency response of the controller C_7, from (7.20), along with the frequency response of G_π and the loop transfer-function, $L_\pi = G_\pi C_7$, in Fig. 7.29. The overall behavior is very close to the one obtained with controller C_6, from (6.25). Of course, the corresponding Nyquist diagram is also very close to the one plotted in Fig. 7.27, and can be easily sketched if wanted. Note that there is a somewhat significant reduction in stability margin: the value of S_M is 35% smaller than before, which may stand as a red flag for a possible deterioration in closed-loop performance and robustness. We will investigate these issues in more detail in Sections 8.1 and 8.2.

Finally, as was done in Section 6.5, we verify that the controller C_7 designed to stabilize G_π also works when placed in feedback with the model linearized around the equilibrium $\theta = 0$. On substituting (7.18) into L with G_0 we obtain

$$L_0(s) = G_0(s)C(s) \approx \frac{K}{J_r} \frac{(s+z_1)(s+z_2)}{s(s+p_1)(s^2+7^2)},$$

$$\approx \frac{1.36 K z_1 z_2}{p_1} \frac{(1+s/z_1)(1+s/z_2)}{s(1+s/p_1)(1+s^2/7^2)},$$

and trace the Bode plots of G_0, C, and the loop transfer-function, $L_0 = G_0 C_7$, with C_7 from (7.20) in Fig. 7.30. Because of the pair of imaginary poles at $s = \pm j\omega_{n_{ol}} = \pm j7$, the magnitude of the frequency response is singular and the phase is discontinuous at $\omega_{n_{ol}} = 7$.

The Nyquist diagram, Fig. 7.31, is obtained after indenting the contour, Γ, to exclude the three imaginary poles. The loop transfer-function, $L_0 = G_0 C$, has no poles or zeros on the right-hand side of the complex plane, that is, $P_\Gamma = 0$, therefore the Nyquist plot

Frequency Domain

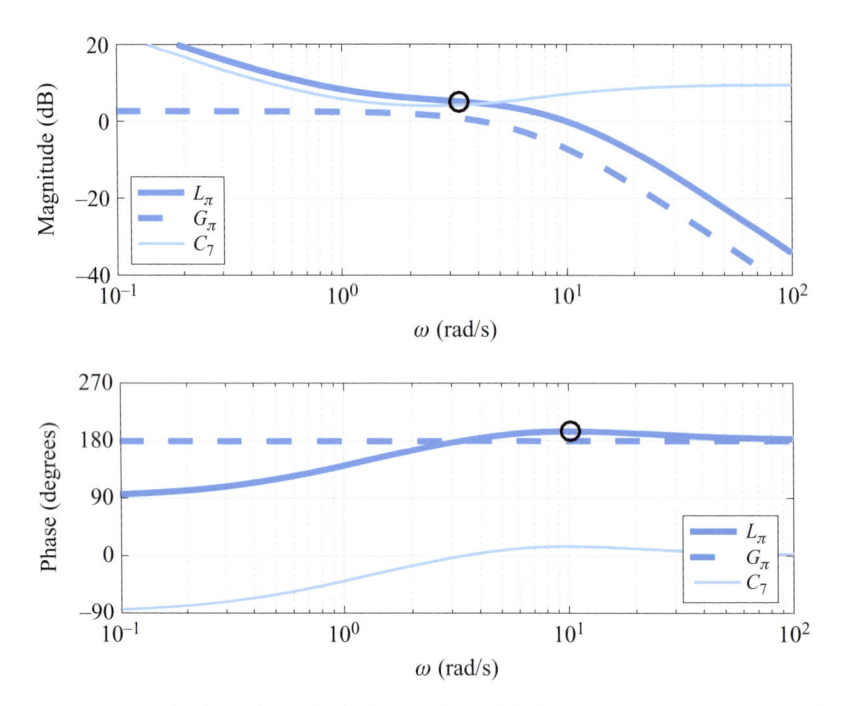

Figure 7.29 Bode plots of G_π (dashed), C_7 (thin solid) from (7.20), and $L_\pi = G_\pi C_7$ (thick solid).

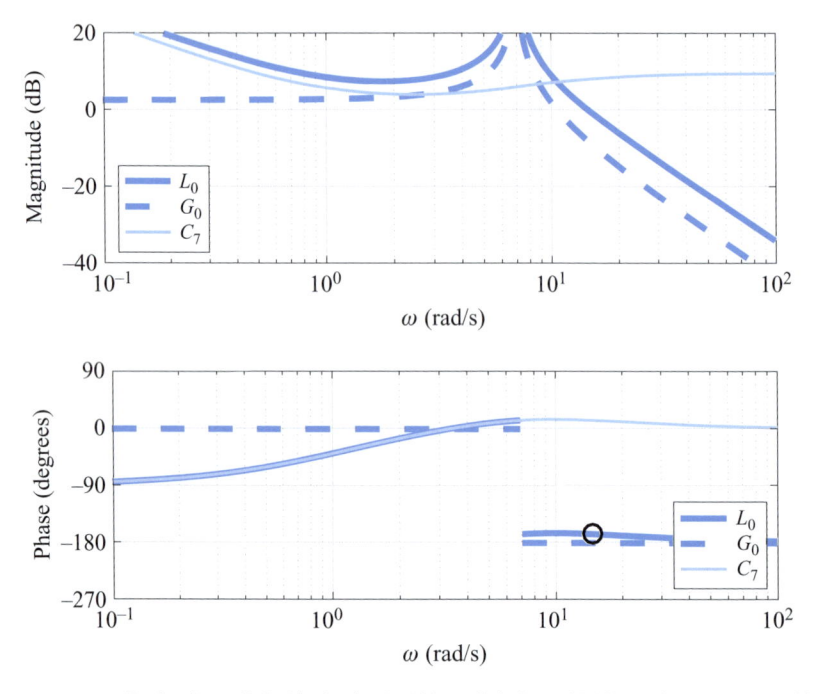

Figure 7.30 Bode plots of G_0 (dashed), C_7 (thin solid) from (7.20), and $L_0 = G_0 C_7$ (thick solid).

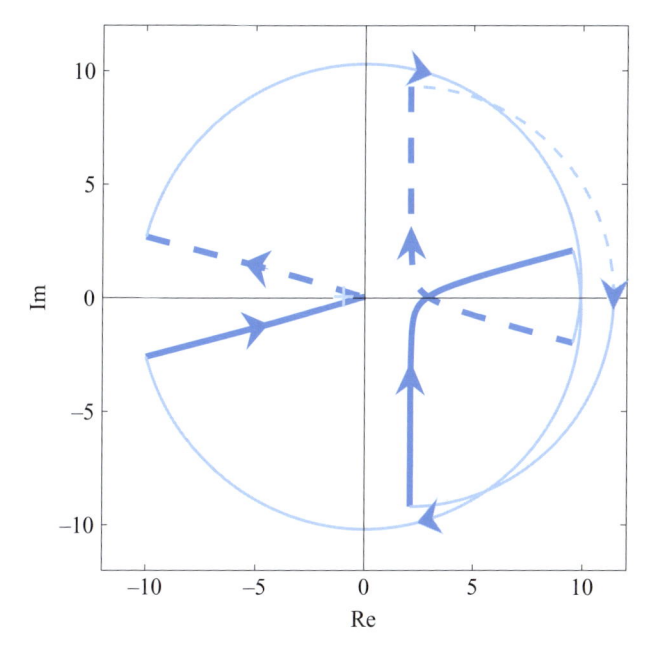

Figure 7.31 Nyquist plot of $L_0 = G_0 C_7$, with C_7 from (7.20).

no longer needs to encircle -1 in order for the closed-loop system to be asymptotically stable. This seems completely different from the design for G_π, which required a counter-clockwise encirclement. Yet, the requirements on the form and locations of the zeros and poles are very similar. Starting from $90°$ at $\omega \to 0^-$, the phase of the loop transfer-function swings $180°$ in the clockwise direction to reach $-90°$ at $\omega \to 0^+$ because of the integrator in the controller. As ω approaches $\omega_n = 7$ the phase swings $180°$ in the clockwise direction, connecting the discontinuity in the phase diagram, because of the pair of imaginary poles of G_0. This time it is necessary to raise the phase above $0°$ before reaching ω_n in order to avoid crossing the negative real axis and producing an encirclement of -1. This can be done by adding two zeros and a pole as in (7.18) and (7.19) to introduce phase in excess of $90°$ and keep the controller proper. Remarkably, in spite of the significant differences in the frequency responses of G_π and G_0, both designs require controllers that have the same structure (7.18)! Because $P_\Gamma = 0$ the Nyquist plot in Fig. 7.31, with three large $180°$ arc swings, neither crosses the negative real axis nor encircles the point -1, that is,

$$P_\Gamma = \frac{1}{2\pi} \Delta_\Gamma^{-1} \arg L_0(s) = 0,$$

and the closed-loop is asymptotically stable. The margins are

$$G_M = \infty, \quad \phi_M \approx 14°, \quad S_M = 0.24,$$

and the frequency ω_ϕ is marked in Fig. 7.30 by a circle. The gain margin is infinite due to the absence of crossings of the negative real axis. The phase and stability margins are slightly smaller than the ones obtained in closed-loop around the unstable equilibrium, i.e. with G_π.

Problems

7.1 Draw the straight-line approximation and sketch the Bode magnitude and phase diagrams for each of the following transfer-functions:

(a) $G = \dfrac{1}{(s+1)(s+10)}$;

(b) $G = \dfrac{s+0.1}{(s+1)(s+10)}$;

(c) $G = \dfrac{s+10}{s^2+0.1s+1}$;

(d) $G = \dfrac{s-0.1}{(s-1)(s+10)}$;

(e) $G = \dfrac{s+10}{s^2-0.1s+1}$;

(f) $G = \dfrac{s}{(s+1)(s+10)}$;

(g) $G = \dfrac{s+10}{(s+1)^2}$;

(h) $G = \dfrac{s+1}{s(s+10)}$;

(i) $G = \dfrac{s+1}{s^2(s+10)}$;

(j) $G = \dfrac{s+10}{(s^2+0.1s+1)^2}$.

(k) $G = \dfrac{s+1}{s^2(s+10)^2}$;

(l) $G = \dfrac{(s+1)^2}{s(s+10)}$.

Compare your sketch with the one produced by MATLAB.

7.2 Draw the polar plots associated with the rational transfer-functions in P7.1.

7.3 If necessary, modify the polar plots associated with the rational transfer-functions in P7.1 to make them suitable for the application of the Nyquist stability criterion.

7.4 Use the Nyquist stability criterion to decide whether the rational transfer-functions in P7.1 are stable under negative unit feedback. If not, is there a gain for which the closed-loop system can be made asymptotically stable? Draw the corresponding root-locus diagrams and compare them.

7.5 Draw the Bode plots associated with the pole–zero diagrams in Fig. 7.32 assuming that the straight-line approximations of the magnitude of the frequency-response have unit gain at $\omega = 0$ (or at $\omega = 1$ if there is a pole or zero at zero).

7.6 Draw the Nyquist plots associated with the pole–zero diagrams in Fig. 7.32 assuming that the straight-line approximations of the magnitude of the frequency-response have unit gain at $\omega = 0$ (or at $\omega = 1$ if there is a pole or zero at zero).

7.7 Use the Nyquist stability criterion to decide whether the rational transfer-functions in P7.5 and P7.6 are stable under negative unit feedback. If not, is there a gain for which the closed-loop system can be made asymptotically stable? Draw the corresponding root-locus diagram and compare.

7.8 Find a minimum-phase rational transfer-function that matches the Bode magnitude diagrams in Fig. 7.33. The straight-line approximations are plotted as thin lines.

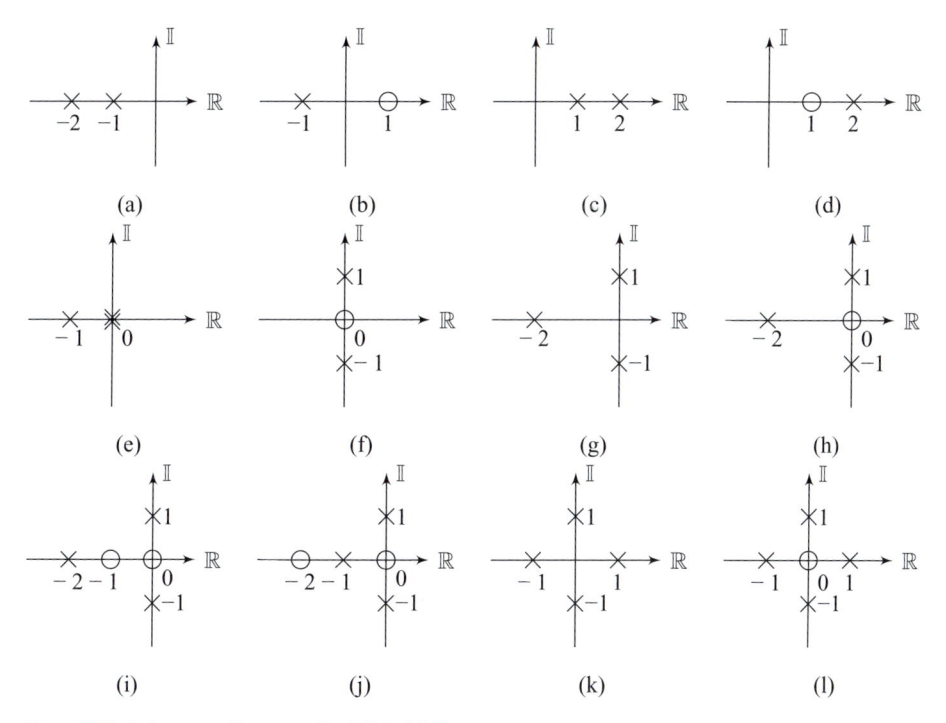

Figure 7.32 Pole–zero diagrams for P7.5–P7.7.

7.9 Find a minimum-phase rational transfer-function that matches the Bode phase diagrams in Fig. 7.34. The straight-line approximations are plotted as thin lines.

7.10 Calculate the rational transfer-function that simultaneously matches the Bode magnitude diagrams in Fig. 7.33 and the corresponding phase diagrams in Fig. 7.34.

7.11 Draw the polar plot associated with the Bode diagrams in Figs. 7.33 and 7.34. Use the Nyquist stability criterion to decide whether the corresponding rational transfer-functions are stable under negative unit feedback. If not, is there a gain for which the closed-loop system can be made asymptotically stable?

7.12 You have shown in P2.10 and P2.12 that the ordinary differential equation

$$\left(J_1 r_2^2 + J_2 r_1^2\right)\dot{\omega}_1 + \left(b_1 r_2^2 + b_2 r_1^2\right)\omega_1 = r_2^2 \tau, \qquad \omega_2 = (r_1/r_2)\omega_1$$

is a simplified description of the motion of a rotating machine driven by a belt without slip as in Fig. 2.18(a), where ω_1 is the angular velocity of the driving shaft and ω_2 is the machine's angular velocity. Let $r_1 = 25$ mm, $r_2 = 500$ mm, $b_1 = 0.01$ kg m²/s, $b_2 = 0.1$ kg m²/s, $J_1 = 0.0031$ kg m², and $J_2 = 25$ kg m². Use Bode plots and the Nyquist

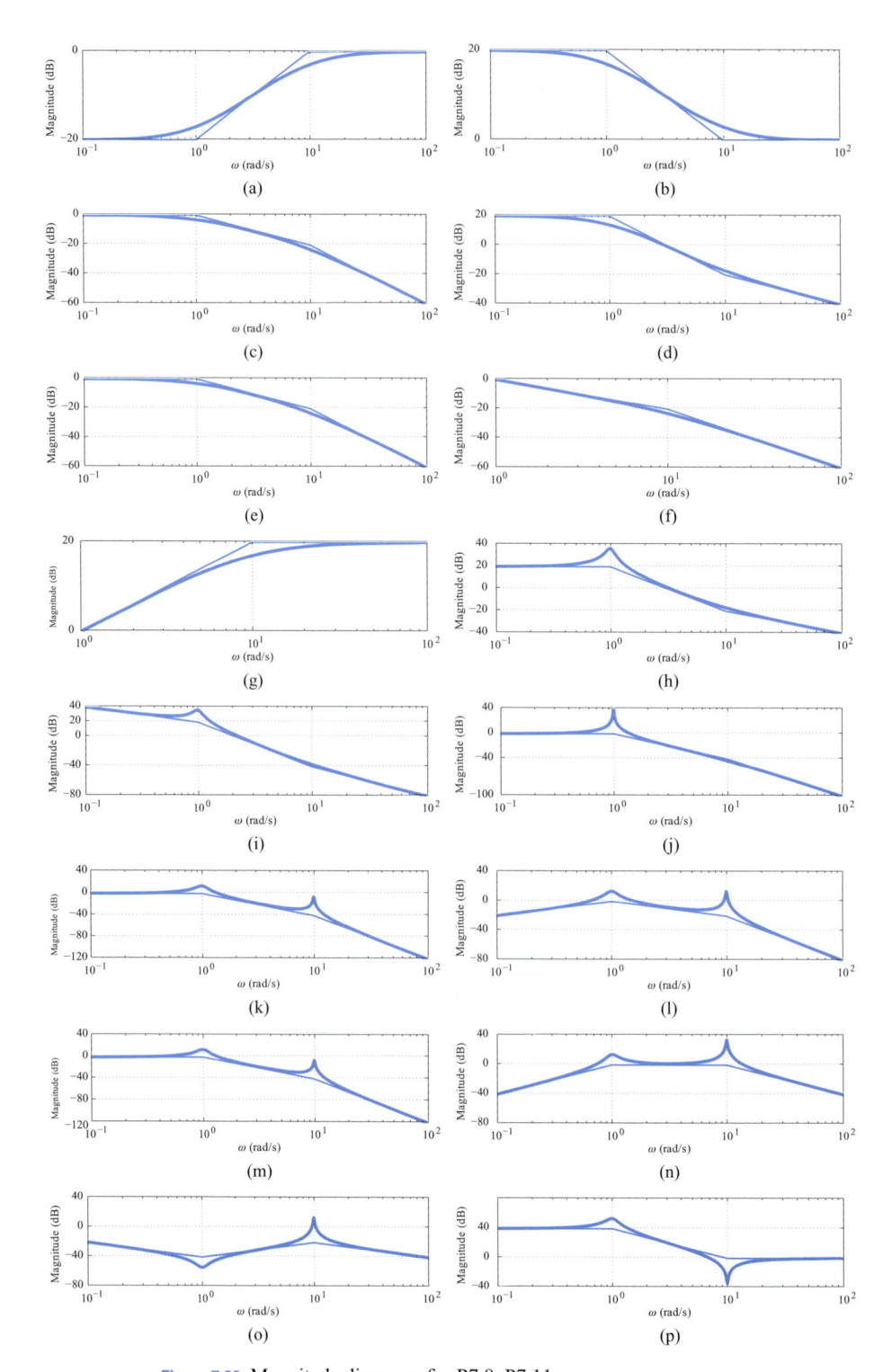

Figure 7.33 Magnitude diagrams for P7.8–P7.11.

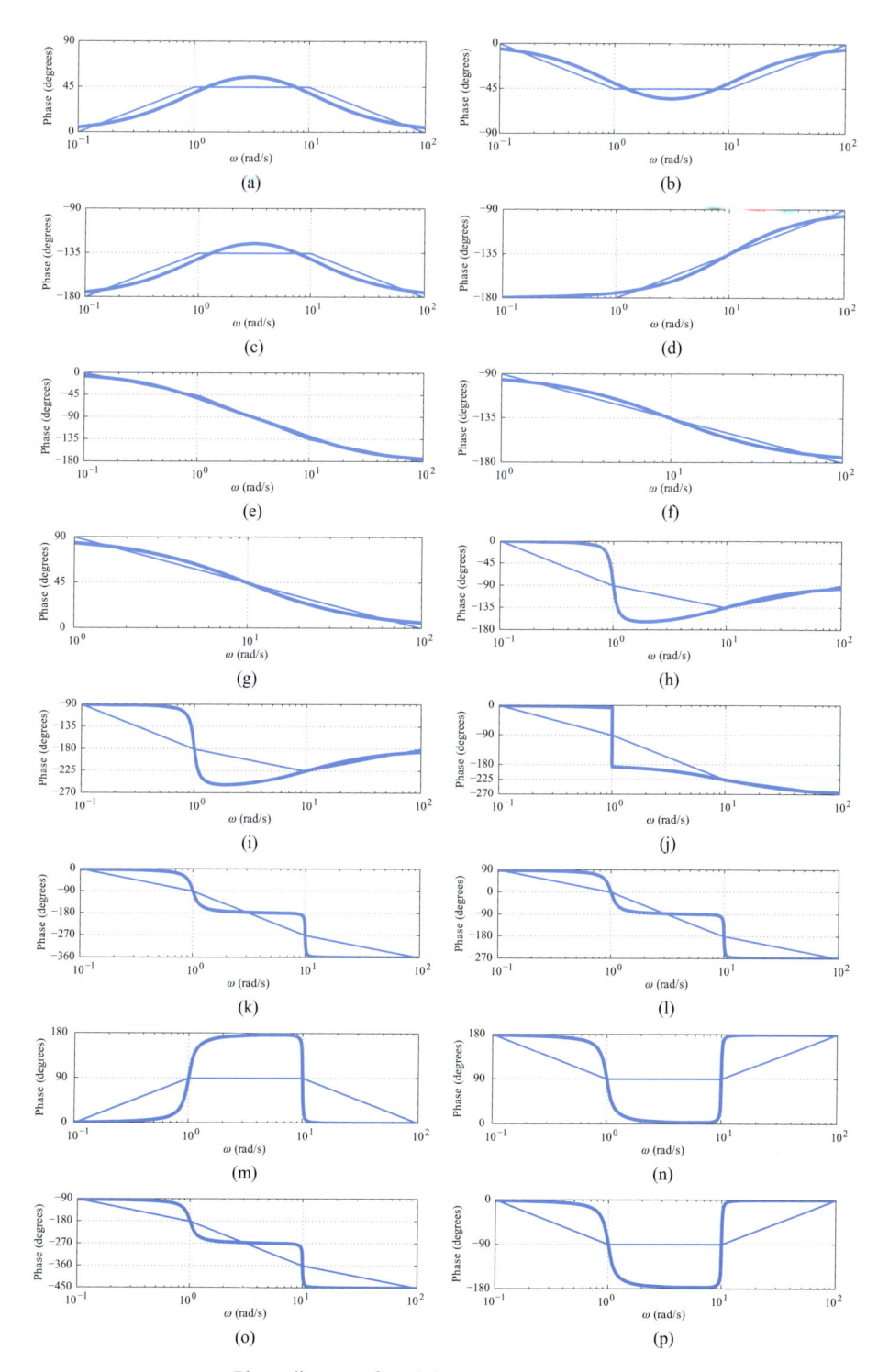

Figure 7.34 Phase diagrams for P7.8–P7.11.

stability criterion to design an I controller

$$\tau(t) = K_i \int_0^t e(\sigma)d\sigma, \qquad e = \bar{\omega}_2 - \omega_2,$$

and select K_i so that the closed-loop system is asymptotically stable. Calculate the corresponding gain and phase margin. Is the closed-loop capable of asymptotically tracking a constant reference input $\bar{\omega}_2$? Is the closed-loop capable of asymptotically rejecting a constant input torque disturbance?

7.13 Repeat P7.12 for the PI controller

$$\tau(t) = K_p e(t) + K_i \int_0^t e(\sigma)d\sigma, \qquad e = \bar{\omega}_2 - \omega_2.$$

7.14 The rotating machine in P6.11 is connected to a piston that applies a periodic torque that can be approximated by $\tau_2(t) = h\cos(\sigma t)$, where the angular frequency σ is equal to the angular velocity ω_2. The modified equation including this additional torque is given by

$$(J_1 r_2^2 + J_2 r_1^2)\dot{\omega}_1 + (b_1 r_2^2 + b_2 r_1^2)\omega_1 = r_2(r_2\tau + r_1\tau_2), \qquad \omega_2 = (r_1/r_2)\omega_1.$$

Use Bode plots and the Nyquist stability criterion to design a dynamic feedback controller that uses τ as control input and ω_2 as the measured output so that the closed-loop system is capable of asymptotically tracking a constant reference input $\bar{\omega}_2(t) = \bar{\omega}_2 = 4\pi$, $t \geq 0$, and asymptotically rejecting the torque perturbation $\tau_2(t) = h\cos(\sigma t)$ when $\sigma = \bar{\omega}_2$. Calculate the corresponding gain and phase margins.

7.15 You have shown in P2.18 that the ordinary differential equation

$$(J_1 + J_2 + r^2(m_1 + m_2))\dot{\omega} + (b_1 + b_2)\omega = \tau + gr(m_1 - m_2), \qquad v_1 = r\omega,$$

is a simplified description of the motion of the elevator in Fig. 2.18(b), where ω is the angular velocity of the driving shaft and v_1 is the elevator's load linear velocity. Let $r = 1$ m, $m_1 = m_2 = 1000$ kg, $b_1 = b_2 = 120$ kg m^2/s, $J_1 = J_2 = 20$ kg m^2, and $g = 10$ m/s^2. Use Bode plots and the Nyquist stability criterion to design a dynamic feedback controller that uses τ as control input and the elevator's load vertical position

$$x_1(t) = x_1(0) + \int_0^t v_1(\tau)d\tau$$

as measured output so that the closed loop is capable of asymptotically tracking a constant position reference $\bar{x}_1(t) = \bar{x}_1$, $t \geq 0$. Calculate the corresponding gain and phase margins.

7.16 Repeat P6.14 with $m_2 = 800$ kg.

7.17 You have shown in P2.28 that the ordinary differential equation

$$m\ddot{x} + b\dot{x} + kx = f + mg\sin\theta$$

is a simplified description of the motion of the mass–spring–damper system in Fig. 2.19(b), where g is the gravitational acceleration and x_0 is equal to the spring rest length ℓ_0. The additional force, f, will be used as a control input. Let $g = 10$ m/s^2,

$m = 1$ kg, $k = 1$ N/m, and $b = 0.1$ kg/s, and use Bode plots and the Nyquist stability criterion to design a dynamic feedback controller that uses f as control input and x as the measured output and that can regulate the position, x, at zero for any constant possible value of inclination $\theta \in (-\pi/2, \pi/2)$. Calculate the corresponding gain and phase margins.   *Hint: Treat the inclination as a disturbance.*

7.18 You have shown in P2.32 that the ordinary differential equations

$$m_1\ddot{x}_1 + (b_1 + b_2)\dot{x}_1 + (k_1 + k_2)x_1 - b_2\dot{x}_2 - k_2 x_2 = f_1,$$
$$m_2\ddot{x}_2 + b_2(\dot{x}_2 - \dot{x}_1) + k_2(x_2 - x_1) = f_2$$

constitute a simplified description of the motion of the mass–spring–damper system in Fig. 2.20(b), where x_1 and x_2 are displacements, and f_1 and f_2 are forces applied on the masses m_1 and m_2. Let the force, f_2, be the control input and let the displacement, x_2, be the measured output. Let $m_1 = m_2 = 1$ kg, $b_1 = b_2 = 0.1$ kg/s, $k_1 = 1$ N/m, and $k_2 = 2$ N/m. Use Bode plots and the Nyquist stability criterion to design a dynamic feedback controller that uses f_2 as control input and x_2 as the measured output and that can regulate the position, x_2, at zero for any constant possible value of force f_1. Calculate the corresponding gain and phase margins.   *Hint: Treat the force f_1 as a disturbance.*

7.19 Repeat problem P7.18 for a force $f_1 = \cos(2\pi t)$.

7.20 In P6.20 you have designed the spring and damper on the one-eighth-car model from P6.19 for a car with 1/4 mass equal to 640 kg to have a natural frequency $f_n = 2.5$ Hz and damping ratio $\zeta = 0.08$. Draw the Bode magnitude and phase diagrams corresponding to the design in P6.20. Interpret your result using Bode plots and the frequency-response method.

7.21 In P6.21 you calculated the response of the one-eighth-car model you designed in P6.20 to a pothole with a profile as shown in Fig. 6.27, where $w = 1$m and $d = 5$ cm for a car traveling first at 10 km/h and then at 100 km/h. Interpret your results using Bode plots and the frequency-response method.

7.22 In P6.22 you calculated the worst possible velocity a car modeled by the one-eighth-car model you designed in P6.20 could have when traveling on a road with profile $y(t) = \cos(vt/\lambda)$, where v is the car's velocity. Interpret your results using Bode plots and the frequency-response method.

7.23 You showed in P6.23 that the design of the spring and damper for the one-eighth-car model from P6.19 can be interpreted as a PD control design problem. Use this reformulation to evaluate the solution obtained in P6.20 using Bode plots and the Nyquist stability criterion. Calculate the corresponding gain and phase margins.

7.24 In P6.25 you designed the spring and damper on the one-quarter-car model from P6.24 for a car with 1/4 mass $m_s = 600$ kg, wheel mass $m_u = 40$ kg, tire stiffness equal

Frequency Domain

to $k_u = 200{,}000$ N/m, and negligible tire damping coefficient $b_u = 0$, to have its dominant poles display a natural frequency $f_n = 2.5$ Hz and damping ratio $\zeta = 0.08$. Interpret your result using Bode plots and the frequency-response method.

7.25 In P6.26 you calculated the response of the one-quarter-car model you designed in P6.25 to a pothole with a profile as shown in Fig. 6.27, where $w = 1$ m and $d = 5$ cm for a car traveling first at 10 km/h and then at 100 km/h. Interpret your results using Bode plots and the frequency-response method.

7.26 In P6.27 you have calculated the worst possible velocity a car modeled by the one-quarter-car model you designed in P6.25 could have when traveling on a road with profile $y(t) = \cos(vt/\lambda)$, where v is the car's velocity. Interpret your results using Bode plots and the frequency-response method.

7.27 You have shown in P6.28 that the design of the spring and damper for the one-quarter-car model from P6.24 can be interpreted as a PD control design problem. Use this reformulation to evaluate the solution obtained in P6.25 using Bode plots and the Nyquist stability criterion. Calculate the corresponding gain and phase margins.

7.28 Compare the answers from P7.24–P7.27 with the answers from P7.20–P7.23.

7.29 You have shown in P2.41 that the ordinary differential equation

$$J\dot{\omega} + \left(b + \frac{K_e K_t}{R_a}\right)\omega = \frac{K_t}{R_a} v_a$$

is a simplified description of the motion of the rotor of the DC motor in Fig. 2.24. Let the voltage, v_a, be the control input and the rotor angular velocity, ω, be the measured output. Let $J = 227 \times 10^{-6}$ kgm^2, $K_t = 0.02$ N m/A, $K_e = 0.02$ V s/rad, $b = 289.4 \times 10^{-6}$ kg m^2/s, and $R_a = 7$ Ω. Use Bode plots and the Nyquist stability criterion to design a dynamic feedback controller so that the closed-loop system is capable of asymptotically tracking a constant reference input $\bar{\omega}(t) = \bar{\omega}, t \geq 0$. Calculate the corresponding gain and phase margins.

7.30 Repeat P7.29 to design a position controller that uses a measurement of the angular position

$$\theta(t) = \theta(0) + \int_0^t \omega(\tau)d\tau$$

as output and is capable of asymptotically tracking a constant angular reference $\bar{\theta}(t) = \bar{\theta}, t \geq 0$.

7.31 You showed in P4.34 that the torque of a DC motor, τ, is related to the armature voltage, v_a, through the transfer-function

$$\frac{T(s)}{V_a(s)} = \frac{K_t}{R_a} \frac{(s + b/J)}{s + b/J + K_e K_t/(R_a J)}.$$

Use the data from P7.29 and Bode and Nyquist plots to design a controller that uses the voltage v_a as the control input and the torque τ as the measured output so that the closed-loop system is capable of asymptotically tracking a constant reference input torque $\bar{\tau}(t) = \bar{\tau}, t \geq 0$. Calculate the corresponding gain and phase margins.

7.32 You showed in P2.49 that the temperature of a substance, T (in K or in °C), flowing in and out of a container kept at the ambient temperature, T_o, with an inflow temperature, T_i, and a heat source, q (in W), can be approximated by the differential equation

$$mc\dot{T} = q + wc(T_i - T) + \frac{1}{R}(T_o - T),$$

where m and c are the substance's mass and specific heat, and R is the overall system's thermal resistance. The input and output flow mass rates are assumed to be equal to w (in kg/s). Assume that water's density and specific heat are 997.1 kg/m³ and $c = 4186$ J/kg K. Use Bode plots and the Nyquist stability criterion to design a dynamic feedback controller that uses the heat source q as the control input and the temperature T as the measured output for a 50 gal (≈ 0.19 m³) water heater rated at $\bar{q} = 40{,}000$ BTU/h (≈ 12 kW) and thermal resistance $R = 0.27$ K/W at ambient temperature, $T_o = 77\,°F$ ($\approx 25\,°C$). The controller should achieve asymptotic tracking of a reference temperature $\bar{T} = 140\,°F$ ($\approx 60\,°C$) without any in/out flow, i.e. $w = 0$. Calculate the corresponding gain and phase margins.

7.33 Repeat P7.32 with a constant in/out flow of 20 gal/h ($\approx 21 \times 10^{-6}$ m³/s) at ambient temperature.

7.34 Repeat P7.32 with a sinusoidal in/out flow perturbation:

$$w(t) = \frac{\bar{w}}{2}(1 + \cos(\omega t)),$$

where $\bar{w} = 20$ gal/h ($\approx 21 \times 10^{-6}$ m³/s) at ambient temperature and $\omega = 2\pi/24$ h^{-1}. Approximate $w(t)(T_i - T(t)) \approx w(t)(T_i - \bar{T})$.

7.35 You showed in P5.42 that

$$\dot{x} = \begin{bmatrix} 0 & 1 & 0 \\ 3\Omega^2 & 0 & 2\Omega \\ 0 & -2\Omega & 0 \end{bmatrix} x + \begin{bmatrix} 0 \\ 0 \\ 1/m \end{bmatrix} u_t, \qquad x = \begin{pmatrix} r - R \\ \dot{r} \\ R(\omega - \Omega) \end{pmatrix},$$

$$y = \begin{bmatrix} 1 & 0 & 0 \end{bmatrix} x, \qquad\qquad y = r - R$$

is a simplified description of the motion of a satellite orbiting earth as in Fig. 5.18, where r is the satellite's radial distance from the center of the earth, ω is the satellite's angular velocity, m is the mass of the satellite, M is the mass of the earth, G is the universal gravitational constant, and u_t is a force applied by a thruster in the tangential direction. These equations were obtained by linearizing around the equilibrium orbit $u_t(t) = u_r(t) = \dot{r}(t) = 0$, $r(t) = R$, and $\omega(t) = \Omega$, where $\Omega^2 R^3 = GM$. Let $M \approx 6 \times 10^{24}$ kg be the mass of the earth, and let $G \approx 6.7 \times 10^{-11}$ N m²/kg². Use MATLAB to calculate the transfer-function from the tangential thrust, u_t, to the radial distance deviation, y, for a 1600 kg GPS satellite in medium earth orbit (MEO) with a period of 11 h. Use Bode plots and the Nyquist stability criterion to design a dynamic feedback controller that uses tangential thrust u_t as the control input and the radial distance y as the measured output and that can regulate the radial distance of the satellite, y, in closed-loop. Calculate the corresponding gain and phase margins.

7.36 In P6.38 and P6.39 you reproduced the results of [Ste+03] by verifying using the root-locus method that the PID controller (6.26) is capable of stabilizing the insulin homeostasis system in closed-loop. Use the values $T_i = 100$ and $T_d = 38$, and calculate the loop transfer-function, $L(s)$, that can be used for feedback analysis of the closed-loop glucose homeostasis system with respect to the proportional gain, $K_p > 0$, and sketch the corresponding Bode and polar plots. Use the Nyquist stability criterion to show that the closed-loop insulin homeostasis system is asymptotically stable.

7.37 Can you determine $K_p > 0$ in P7.36 with $T_d = 0$ so that the closed-loop glucose homeostasis system is internally stable?

8 Performance and Robustness

In earlier chapters you have learned the basic techniques and methods that are used in classic control design. One main goal was to achieve closed-loop stability. In this chapter we take our discussion on some aspects of performance and robustness of feedback systems further. We also introduce the concepts of filtering and feedforward.

8.1 Closed-Loop Stability and Performance

In Chapter 7 we saw how frequency-domain methods can be used for control design. We studied the frequency response of the loop transfer-function and the associated Bode and Nyquist diagrams to obtain clues on the structure and location of the poles and zeros of a controller needed for closed-loop stabilization. We introduced gain, phase, and stability margins, in Section 7.7, as measures of closed-loop robustness. However, it is not yet clear how closed-loop robustness relates to performance. We have also not established whether closed-loop performance specifications can be translated in terms of the (open) loop transfer-function.

For a concrete scenario, consider the performance of the standard closed-loop diagram of Figs. 1.8 and 4.2, which we reproduce once again in Fig. 8.1. As discussed in detail in Chapter 4, tracking performance depends on the sensitivity transfer-function, S, that is the transfer-function from the reference input, $\bar{y}$, to the error signal, e:

$$S = \frac{1}{1+L}, \qquad L = GK.$$

Often the energy of the reference signal, $\bar{y}$, is concentrated on a limited region of the spectrum and the job of the control system designer is to keep $|S(j\omega)|$ as small as possible in this region in order to achieve good tracking. A number of obstacles can make this a difficult task. First, one should not expect to be able to set

$$|S(j\omega)| = 0, \qquad \text{for all } \omega \in \mathbb{R}. \tag{8.1}$$

Indeed, the poles and zeros of L dictate the behavior of $|S(j\omega)|$, which makes this task impossible. If p is a pole of L and z is a zero of L then

$$S(p) = 0, \qquad S(z) = 1.$$

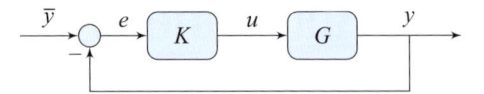

Figure 8.1 Closed-loop feedback configuration for tracking.

For instance, if L satisfies (3.23), e.g. L is rational and strictly proper, then

$$\lim_{|s| \to \infty} |L(s)| = 0 \quad \Longrightarrow \quad \lim_{|s| \to \infty} |S(s)| = 1.$$

This is the same as thinking of L as having a zero at infinity. Loosely speaking, achieving (8.1) would amount to having infinitely large gains, $|L(j\omega)| \to \infty$, throughout the entire spectrum $\omega \in \mathbb{R}$! Nevertheless, as detailed in Chapter 4, it is possible and even desirable to make $S(j\omega) = 0$ *at select frequencies* by cleverly placing the poles of the controller. The resulting closed-loop system achieves asymptotic tracking (Lemma 4.1).

Even if it were possible to make the sensitivity function equal to zero at all frequencies, there would still be plenty of reasons why it would not be a good idea to do that. As discussed in Section 4.6, by making $S(j\omega)$ small at a certain frequency, ω, we automatically make $H(j\omega) = 1 - S(j\omega)$ large at that frequency, in which case we should expect a deterioration in performance in the presence of measurement noise. Most control systems get around this limitation by making $|S(j\omega)|$ small at the low end of the spectrum, where most practical reference signals are concentrated, while making $|S(j\omega)| \approx 1$ at the upper end of the spectrum, which is permeated by noise. One usually makes $S(j\omega)$ small by making $L(j\omega)$ large. Indeed, if for some $\omega \in \mathbb{R}$

$$|L(j\omega)| > M \gg 1,$$

then[1]

$$|S(j\omega)| = \frac{1}{|1 + L(j\omega)|} < \frac{1}{M-1} \approx 0.$$

The larger M the smaller $|S(j\omega)|$ will be at $s = j\omega$. Conversely, when $|L(j\omega)|$ is small,

$$|L(j\omega)| \approx 0,$$

then

$$|S(j\omega)| = \frac{1}{|1 + L(j\omega)|} \approx 1.$$

This parallels the discussion in Chapters 1 and 4: large gains mean better tracking. But the real problem is not when gains are small or large but rather occurs when the gain is somewhere *in the middle*. If $|L(j\omega)| \approx 1$ then attention shifts from the magnitude to the phase. If $|L(j\omega)| \approx 1$ and $\angle L(j\omega)$ is close to $180°$ then a crossing of the negative real axis might occur, which, as seen in Chapter 7, has far-reaching implications for

---

[1] Because of the triangle inequality $|L(j\omega)| = |1 + L(j\omega) - 1| \leq |1 + L(j\omega)| + 1$, and hence $|1 + L(j\omega)| \geq |L(j\omega)| - 1$.

8.1 Closed-Loop Stability and Performance

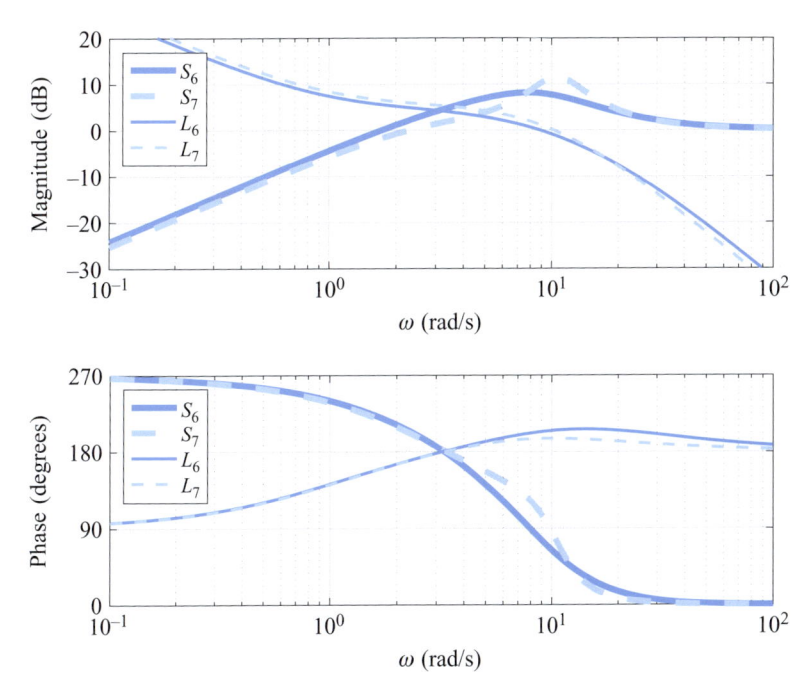

Figure 8.2 Bode plots of the sensitivity functions S_6 (thick solid) and S_7 (thick dashed) calculated for the simple pendulum model linearized around $\theta = \pi$, G_π, for the controllers C_6, from (6.25), and C_7, from (7.20). The corresponding loop transfer-functions, $L_6 = G_\pi C_6$ (thin solid) and $L_7 = G_\pi C_7$ (thin dashed), are plotted for comparison.

closed-loop stability; in terms of tracking performance, $|S(j\omega)|$ might become large, sometimes much larger than 1 if $|1 - L(j\omega)| \approx 0$, compromising the overall stability margin, as measured by (7.17). We illustrate this situation with an example.

Recall the two controllers for the simple pendulum, C_6, from (6.25), and C_7, from (7.20), which were analyzed in Sections 6.5 and 7.8. The Bode plots of the sensitivity functions, S_6 and S_7, calculated for the simple pendulum model linearized around the unstable equilibrium $\theta = \pi$, i.e. G_π, are shown in Fig. 8.2. The corresponding loop transfer-functions, L_6 and L_7, are plotted for comparison. As expected, the value of $|S(j\omega)|$ reaches its peak, which in both cases is much higher than one (>0 dB), precisely when $|L(j\omega)|$ is near 0 dB, that is $|L(j\omega)| \approx 1$. Note how much more pronounced $|S_7(j\omega)|$ is near $\omega = 10$ than $|S_6(j\omega)|$ due to the difference in phase in L_6 and L_7: the phase of L_7 is closer to 180°. Because the poles of the sensitivity function are the closed-loop poles, the magnitude of the sensitivity is often indicative of low-damped poles. Indeed, after calculating

$$S_7 = \frac{1}{1 + L_7} = \frac{s(s-7)(s+7)(s+11)}{(s^2 + 5.47s + 8.942)(s^2 + 5.53s + 111.9)},$$

we observe that the second pair of complex-conjugate poles has $\omega_n \approx 10.6$ with a relatively low damping ratio, $\zeta \approx 0.26$. The higher sensitivity near $\omega = 10$ coincides with the presence of the low-damped closed-loop pole. More oscillations are expected as a result of the low-damped closed-loop pole; robustness is also compromised as the stability margin, S_M, from (7.17), is reduced. We will analyze robustness in more detail in Section 8.2.

The peaking of the closed-loop sensitivity function in the above example may not be completely obvious to an inexperienced designer who is looking at the open-loop transfer-functions L_6 and L_7. The plots in Fig. 8.2 suggest, however, that something else may be at work: it seems as if the reduction in sensitivity achieved by the controller C_7 at frequencies $|\omega| < 7$ is *balanced* by an increase in sensitivity at $\omega > 7$. This phenomenon is similar to what is sometimes known in the literature as the *waterbed effect*[2] [DFT09, Section 6.2]: reducing the sensitivity at one frequency seems to raise the sensitivity at another frequency. In fact, the controller C_7 was designed precisely to increase the loop transfer-function gain for $|\omega| < 7$. This was accomplished but at the expense of raising the sensitivity for $\omega > 7$.

A quantitative account of the waterbed-type phenomenon can be obtained using the following result, which is similar to a conservation law in the frequency domain. The next theorem is a slightly more general version of Theorem 11.1 in [ÅM08].

THEOREM 8.1 (Bode's sensitivity integral) *Consider a loop transfer-function L having poles p_i, $\text{Re}(p_i) > 0$, $i = 1, \ldots, k$, on the right-hand side of the complex plane. Assume that*

$$\lim_{|s| \to \infty} s L(s) = \kappa, \qquad |\kappa| < \infty, \tag{8.2}$$

and that $S = (1 + L)^{-1}$ is asymptotically stable. Then

$$\int_0^\infty \ln|S(j\omega)|d\omega = \int_0^\infty \ln\left(\frac{1}{|1 + L(j\omega)|}\right) d\omega = \pi \sum_{i=1}^k p_i - \frac{\pi \kappa}{2}. \tag{8.3}$$

When L is rational, condition (8.2) is satisfied only if L is strictly proper. The value of κ is zero if the difference between the degree of the denominator, n, and the degree of the numerator, m, is at least two, i.e. $n - m > 1$. The integer $n - m$ is known as the *relative degree* of the transfer-function L.

Without getting distracted by the right-hand side of (8.3), notice the rather strong implication of Bode's sensitivity integral: the total variation of the magnitude of the closed-loop sensitivity function about 1 (see the ln?) is bounded. This means that if a controller is able to reduce the closed-loop sensitivity at a certain frequency range this must be at the expense of the sensitivity increasing somewhere else. Poles of L on the right-hand side further compromise the performance by raising the overall integral of the log of the magnitude of the sensitivity function. If the unstable poles of L come from the system, G, as they often do, then the right-hand side of (8.3) is the price to be paid to make these poles stable in closed-loop.

---

[2] The waterbed effect, as discussed in detail in [DFT09, Section 6.2] applies only to systems with non-minimum-phase zeros. A peaking phenomenon akin to the one displayed in Fig. 8.2 is also discussed in [DFT09, p. 98] in connection with a constraint on the bandwidth of the loop transfer-function.

8.1 Closed-Loop Stability and Performance

The balance imposed by (8.3) does not necessarily mean that the sensitivity function will have to peak. In practice, however, it often does. For instance, if the overall bandwidth of the loop transfer-function is bounded[3] then peaking will occur. In Fig. 8.2 the magnitude of $L(j\omega)$ rolls off quickly after $\omega > 11$ and the decreased sensitivity in the range $|\omega| < 7$ must be balanced in the relatively narrow region $7 < \omega < 11$. But peaking of the sensitivity need not occur if the controller is allowed to arbitrarily, and very often unrealistically, increase the bandwidth of the loop transfer-function. For example, let

$$G(s) = \frac{1}{s+1}, \qquad K(s) = K > 0, \qquad L(s) = G(s)K(s) = \frac{K}{s+1},$$

so that

$$S = \frac{1}{1+GK} = \frac{s+1}{s+1+K}, \qquad |S(j\omega)| \leq \|S\|_\infty = \lim_{\omega \to \infty} |S(j\omega)| = 1.$$

In this simple example one can obtain from the Bode plot of G that the range of frequencies for which $|G(j\omega)| < \sqrt{2}/2$, that is, the bandwidth of the open-loop system, is $|\omega| < 1$. However, the range[4] of frequencies for which $|L(j\omega)| < \sqrt{2}/2$ is $|\omega| < \sqrt{2K^2 - 1}$. In other words, the bandwidth of the loop transfer-function L can be made arbitrarily large by increasing the control gain K. In terms of the Bode integral (8.3),

$$\kappa = \lim_{|s| \to \infty} sL(s) = \lim_{|s| \to \infty} \frac{Ks}{s+1} = K$$

and

$$\int_0^\infty \ln|S(j\omega)|d\omega = -\frac{\pi K}{2}.$$

In this case raising the gain K lowers the overall sensitivity function. Interestingly, for a system in which $\kappa = 0$, raising the loop gain will not reduce the overall sensitivity and generally leads to peaking: recall from the root-locus method, Section 6.4, that if the relative degree of L is two or more, that is, if L has at least two more poles than zeros, then the root-locus will have at least two asymptotes and raising the gain will necessarily lead to low-damped closed-loop poles and even instability when the relative degree is greater than or equal to three.

From (8.3) alone, it is not clear how the zeros of L impact the closed-loop sensitivity function. Indeed, it does not seem to be possible to provide a simple account of the impact of zeros on the closed-loop performance. However, in the following simplified case, a quantitative analysis is possible. Let L have a single pole $p > 0$ and a zero $z > 0$, $z \neq p$, on the right-side of the complex plane. The asymptotically stable sensitivity function, S, will have a zero at $p > 0$. Define

$$\hat{S}(s) = S(s)\frac{s+p}{s-p},$$

---

[3] See the discussion in [DFT09, Chapter 6].   [4] With $K > \sqrt{2}/2$.

where $\hat{S}$ is now not only asymptotically stable but also minimum-phase since all remaining zeros are on the left-hand side of the complex plane. In P8.7 you will show that

$$|\hat{S}(j\omega)| = |S(j\omega)| \left| \frac{j\omega + p}{j\omega - p} \right| = |S(j\omega)|.$$

Because $z > 0$ is a zero of L we have that $S(z) = 1$ and

$$\hat{S}(z) = S(z)\frac{z+p}{z-p} = \frac{z+p}{z-p}.$$

Application of the *maximum modulus principle* [BC14, Section 59] (see also P8.1) yields the bound

$$\|S\|_\infty = \sup_{\omega \in \mathbb{R}} |S(j\omega)| = \sup_{\omega \in \mathbb{R}} |\hat{S}(j\omega)| \geq |\hat{S}(z)| = \frac{|z+p|}{|z-p|} > 1,$$

which shows that an unstable zero can *amplify* the already detrimental impact of an unstable pole: if z and p are close then $\|S\|_\infty$ can be much larger than one. See [GGS01] and [DFT09] for much more.

Proof of Theorem 8.1

Some of the steps in this proof will be developed in problems at the end of the chapter. A proof of Theorem 8.1 is relatively simple after P8.2 and P8.3 make us realize that

$$\int_0^\infty \ln|S(j\omega)|d\omega = \frac{1}{2}\int_{-\infty}^\infty \ln S(j\omega)d\omega$$

and evaluate

$$\frac{1}{2}\int_{-\rho}^{\rho} \ln S(j\omega)d\omega = \frac{1}{2j}\int_\Gamma \ln S(s)ds + \frac{1}{2}\int_{-\frac{\pi}{2}}^{\frac{\pi}{2}} \rho e^{-j\theta} \ln S(\rho e^{-j\theta})d\theta,$$

where Γ is the contour in Fig. 7.14 in which the radius of the semi-circular segment, ρ, is made large. If L has no poles on the right-hand side of the complex plane then S has no zeros on the right-hand side of the complex plane. Because S is asymptotically stable it has no poles on the right-hand side either which means that the function $\ln S(s)$ is analytic on the right-hand side of the complex plane. From Cauchy's residue theorem (Theorem 3.1),

$$\int_\Gamma \ln S(s)ds = 0.$$

This simplification results in

$$\int_0^\infty \ln|S(j\omega)|d\omega = \frac{1}{2}\lim_{\rho \to \infty}\int_{-\frac{\pi}{2}}^{\frac{\pi}{2}} \rho e^{-j\theta} \ln S(\rho e^{-j\theta})d\theta. \tag{8.4}$$

The integral on the right-hand side converges if

$$\lim_{|s| \to \infty} s \ln S(s) = \lim_{|s| \to \infty} -sL(s) = -\kappa, \qquad |\kappa| < \infty. \tag{8.5}$$

You will prove the left-hand equality in (8.5) in P8.5. In this case,

$$\lim_{\rho \to \infty} \int_{-\frac{\pi}{2}}^{\frac{\pi}{2}} \rho e^{-j\theta} \ln S(\rho e^{-j\theta}) d\theta = -\pi \kappa,$$

resulting in

$$\int_0^\infty \ln|S(j\omega)| d\omega = \frac{1}{2} \lim_{\rho \to \infty} \int_{-\frac{\pi}{2}}^{\frac{\pi}{2}} \rho e^{-j\theta} \ln S(\rho e^{-j\theta}) d\theta = -\frac{\pi \kappa}{2},$$

which is a special case of (8.3) that holds only when L has no poles on the right-hand side of the complex plane.

If L has poles $p_i > 0$, $i = 1, \ldots, k$, on the right-hand side of the complex plane, define the auxiliary transfer-function:

$$\hat{S}(s) = S(s) \frac{s+p_1}{s-p_1} \frac{s+p_2}{s-p_2} \cdots \frac{s+p_k}{s-p_k},$$

in which every zero of S on the right-hand side of the complex plane, $p_i > 0$, is canceled and replaced by a zero on the left-hand side of the complex plane, $-p_i < 0$. In this way, both $\hat{S}(s)$ and $\ln \hat{S}(s)$ are analytic on the right-hand side of the complex plane. Moreover, after P8.7,

$$|\hat{S}(j\omega)| = |S(j\omega)| \prod_{i=1}^k \left| \frac{j\omega + p_i}{j\omega - p_i} \right| = |S(j\omega)|.$$

Repeating the same sequence of steps as was performed above but this time with $\hat{S}$ instead of S we obtain from (8.4)

$$\int_0^\infty \ln|S(j\omega)| d\omega = \frac{1}{2} \lim_{\rho \to \infty} \int_{-\frac{\pi}{2}}^{\frac{\pi}{2}} \rho e^{-j\theta} \ln \hat{S}(\rho e^{-j\theta}) d\theta. \quad (8.6)$$

In order to compute the integral on the right-hand side we use the fact that

$$s \ln \hat{S}(s) = s \ln S(s) + \sum_{i=1}^k s \ln \left(\frac{s+p_i}{s-p_i} \right), \quad \lim_{|s| \to \infty} s \ln \hat{S}(s) = -\kappa + \sum_{i=1}^k 2p_i,$$

after using (8.5) and P8.8. Evaluation of (8.6) leads to

$$\int_0^\infty \ln|S(j\omega)| d\omega = \frac{1}{2} \lim_{\rho \to \infty} \int_{-\frac{\pi}{2}}^{\frac{\pi}{2}} \rho e^{-j\theta} \ln \hat{S}(\rho e^{-j\theta}) d\theta = \pi \sum_{i=1}^k p_i - \frac{\pi \kappa}{2},$$

which is (8.3).

8.2 Robustness

Robustness to variations in the system properties and external disturbances is a much desired feature that continues to motivate research in the area of systems and control [DFT09; ZDG96]. Many ideas in this area stem from the frequency-response techniques you have studied in Chapter 7. For example, gain, phase, and stability margins have already been shown in Section 7.7 to relate to robustness to gain and phase disturbances in closed-loop. In many ways, robust analysis and controller design are closely related: in control design one seeks a system, the controller, that can stabilize a feedback loop; in robust analysis one seeks to prove that no perturbation, in the form of a system or disturbance signal, can de-stabilize a feedback loop. Most robust analysis methods are actually based on closed-loop stability analysis of the standard feedback diagram in Fig. 8.3. In this context, the feedback element Δ is the *uncertainty*, which is often used to represent nonlinearities and unknown or unmodeled components or sub-systems. The question of robust stability is posed after constraining Δ to live in a set with a representative property of interest.

For instance, recall the block-diagram in Fig. 6.4 which shows the connection of a feedback controller to the nonlinear model of the simple pendulum derived in Section 5.5. The only nonlinear block in this diagram is the nonlinear sine function. Upon renaming this block

$$\Delta(y) = \sin(y)$$

and rearranging the signal flow we arrive at the closed-loop diagram in Fig. 8.4, where the block Δ is referred to as an *uncertainty*. The particular sine function belongs to the set of functions bounded[5] by a linear function:

$$\mathbf{\Delta} = \{\Delta : \mathbb{R} \to \mathbb{R}, \quad |\Delta(y)| \le |y|\}. \tag{8.7}$$

Any function in $\mathbf{\Delta}$ must lie in the shaded area in Fig. 8.5, which also shows $\Delta(y) = \sin(y)$ as a solid curve.

If the closed-loop feedback connection in Fig. 8.4 is asymptotically stable for *all* $\Delta \in \mathbf{\Delta}$ we say that it is *robustly stable*. Of course, robust stability implies asymptotic stability in the case of a particular $\Delta(y)$, say $\Delta(y) = \sin(y) \in \mathbf{\Delta}$ for the pendulum.

8.3 Small Gain

Back to the block-diagram of Fig. 8.3, assume that G is an asymptotically stable linear time-invariant system model with H_∞ norm less than one, that is,

$$\|G\|_\infty = \sup_{\omega \in \mathbb{R}} |G(j\omega)| < 1. \tag{8.8}$$

This inequality appears frequently in the robust control literature, where it is known as a *small-gain* condition [DFT09; ZDG96]. After constructing a suitable[6] state-space representation for the transfer-function G, as in Section 5.2, we calculate the 2-norm of

---

[5] Because $-1 \le \sin(y)/y \le 1$.
[6] One that is minimal, that is observable and controllable. See Section 5.3 and [Kai80].

8.3 Small Gain

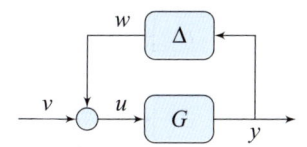

Figure 8.3 Closed-loop feedback configuration for robust analysis.

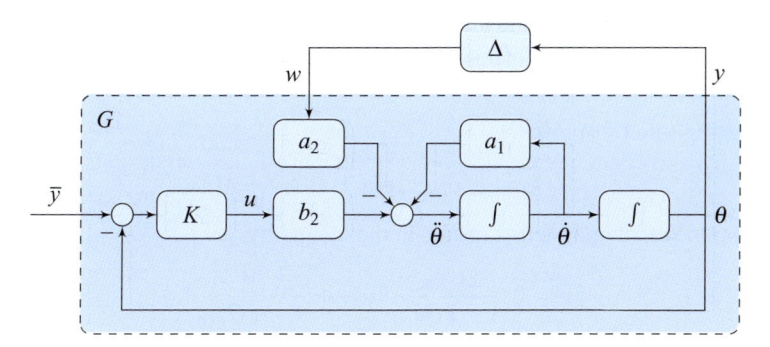

Figure 8.4 Feedback configuration of the controlled simple pendulum for robust analysis; $\Delta(y) = \sin(y)$; $\Delta \in \mathbf{\Delta}$.

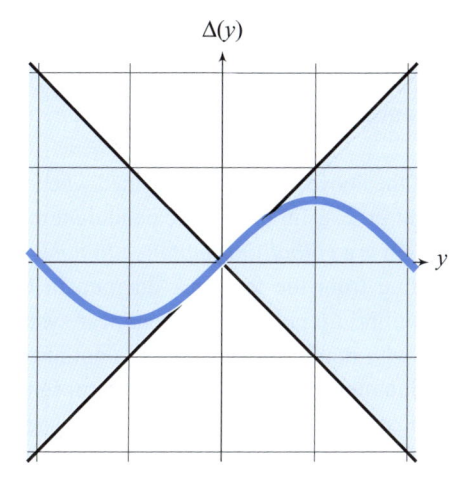

Figure 8.5 Functions in $\Delta \in \mathbf{\Delta}$ satisfy $|\Delta(y)| \leq |y|$ and must lie on the shaded area (sector); solid curve is $\sin(y) \in \mathbf{\Delta}$.

the output:

$$\|y\|_2 = \|G(w + v) + f(x_0)\|_2$$
$$\leq \|G\|_\infty \|w + v\|_2 + \|f(x_0)\|_2 \leq \|G\|_\infty (\|w\|_2 + \|v\|_2) + M\|x_0\|_2. \quad (8.9)$$

The inequalities come from applying the triangle inequality (3.47), (3.48) and (5.8). The signal $f(x_0)$ is used here as was done in Section 5.2 to represent the response to a possibly nonzero initial condition, and the constant M is related to the eigenvalues of the *observability Gramian* (see (5.8)). It might be a good idea to go back now to review the material in Sections 3.9 and 5.2 before continuing. If Δ belongs to $\mathbf{\Delta}$ then

$$\|w\|_2 = \|\Delta(y)\|_2 \leq \|y\|_2.$$

Substituting into (8.9) gives

$$\|y\|_2 \leq \|G\|_\infty (\|y\|_2 + \|v\|_2) + M\|x_0\|_2,$$

which upon rearranging terms results in the inequality

$$\|y\|_2 \leq \frac{\|G\|_\infty}{1 - \|G\|_\infty} \|v\|_2 + \frac{M}{1 - \|G\|_\infty} \|x_0\|_2. \quad (8.10)$$

Because $\|G\|_\infty < 1$, boundedness of the 2-norm of the input, v, and of the initial condition, x_0, implies boundedness of the 2-norm of the output, y. The 2-norm of the signal w is also bounded because $\Delta \in \mathbf{\Delta}$. Moreover, it is possible[7] to show that boundedness in the 2-norm sense actually implies that

$$\lim_{t \to \infty} y(t) = 0,$$

in other words, that all signals will eventually converge to zero. Convergence to zero from *any* bounded initial condition is guaranteed in the presence of *any* input v with bounded 2-norm.

An important particular instance of the above analysis is obtained when the input, v, is zero and $\Delta(y)$ is a nonlinear element in $\mathbf{\Delta}$ as in the simple pendulum model. In this case, it is possible to prove that the origin is a *globally asymptotically stable* equilibrium point for the nonlinear system resulting from the feedback diagram in Fig. 8.3.[8] In other words, the origin is the *only* equilibrium point of the nonlinear system and it is asymptotically stable. Unfortunately, with nonlinear systems, it is not always possible to go from asymptotic stability or input–output stability in the 2-norm sense to input–output stability in the ∞-norm sense, that is, BIBO stability.[9] In the special case of the diagram in Fig. 8.3 this is possible but requires technical devices which are beyond the scope of this book.[10] We will present an *ad hoc* methodology for the control of the simple pendulum in Section 8.4.

---

[7] This result is surprisingly hard to prove under the general framework we are working within. One path is to construct a state-space realization that is controllable and observable and from there infer that all signals are bounded in the 2-norm. That is, *both* $x(t)$ and $\dot{x}(t)$ have bounded 2-norms. Only then can one use an argument like P3.38 to establish that $\lim_{t \to \infty} x(t) = 0$ and therefore $\lim_{t \to \infty} y(t) = 0$.

[8] See [Kha96, Section 10.2].

[9] Some nonlinear systems are asymptotically stable but not BIBO stable: $\dot{x} = -x + 2xu$ is asymptotically stable at the origin when $u = 0$ but unstable when $u = 1$ [Son08].

[10] The linear system G provides a Lyapunov function that can be used as in Theorem 6.1 of [Kha96].

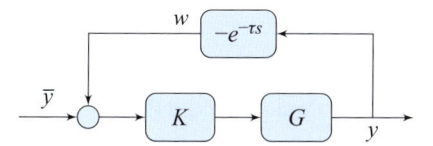

Figure 8.6 Closed-loop feedback configuration for analysis of closed-loop systems with feedback delay.

Speaking of the simple pendulum, note that its open-loop model has several stable equilibria, namely $\theta = 2\kappa\pi$, $\kappa \in \mathbb{Z}$, $\dot{\theta} = 0$. However, in closed-loop with a linear controller, the origin becomes the *only* stable equilibrium. This brings up a curious practical situation: if the sensor that measures the angle θ is not *reset* before closed-loop operation, that is, if its initial reading is not in the interval $(-\pi, \pi)$, then the controller will wind up or wind down the pendulum[11] until the sensor reading is zero!

The setup in Fig. 8.3 can also be used to prove stability for different classes of *uncertainty*. One example is when Δ is an asymptotically stable linear time-invariant system. In this case, stability of the loop in Fig. 8.3 can be assessed in terms of the loop transfer-function $L = G\Delta$. Recalling the frequency-domain definition of the H_∞ norm in Section 3.9,

$$\|L\|_\infty = \sup_{\omega \in \mathbb{R}} |L(j\omega)|$$
$$= \sup_{\omega \in \mathbb{R}} |G(j\omega)||\Delta(j\omega)| \leq \sup_{\omega \in \mathbb{R}} |G(j\omega)| \sup_{\chi \in \mathbb{R}} |\Delta(j\chi)| \leq \|G\|_\infty \|\Delta\|_\infty.$$

If G and Δ are asymptotically stable, and $\|G\|_\infty < 1$ and $\|\Delta\|_\infty \leq 1$, then L is asymptotically stable and $\|L\|_\infty < 1$. This means that the Nyquist plot of L is entirely contained[12] inside the unit circle, hence no encirclements of the point -1 can occur. That is, the linear closed-loop connection in Fig. 8.3 is guaranteed to be asymptotically stable for all asymptotically stable transfer-functions $\Delta(s)$ such that $\|\Delta\|_\infty \leq 1$.

A popular application of such results is to the study of the stability of closed-loop systems with delays in the loop.[13] Consider the closed-loop feedback control connection of the system, G, and controller, K:

$$y(t) = G(u(t)), \qquad u(t) = K(\bar{y}(t) - y(t - \tau)),$$

in which the feedback control at time t is based on the delayed signal $y(t - \tau)$, i.e. the signal produced $\tau \geq 0$ seconds earlier, a situation which is common in practice. When all blocks in the loop are linear time-invariant systems it is possible to write the loop in terms of the transforms to obtain the feedback diagram in Fig. 8.6. Compare this diagram with the standard diagram for robustness analysis in Fig. 8.3. Application of the small-gain condition to linear systems with feedback delays then follows from the

---

[11] One can only hope that this pendulum is not a rocket.
[12] We will have more to say about such circles in Section 8.5.
[13] See [GKC03] for many more connections between delay systems and robustness.

correspondence

$$\Delta(s) \leftarrow -e^{-s\tau}, \qquad G(s) \leftarrow G(s)K(s), \qquad V(s) \leftarrow \bar{Y}(s).$$

Therefore, if both G and K are asymptotically stable and $\|GK\|_\infty < 1$ then the closed-loop connection will be asymptotically stable in the presence of any constant delay $\tau \geq 0$ because

$$\|\Delta\|_\infty = \|e^{-\tau s}\|_\infty = \sup_{\omega \in \mathbb{R}} |e^{-j\tau\omega}| = 1.$$

As a simple example, consider the feedback connection of

$$G(s) = \frac{1}{s+1}, \qquad K(s) = K \geq 0,$$

which is asymptotically stable in the presence of any constant delay $\tau \geq 0$ if

$$\|KG\|_\infty = K \sup_{\omega \in \mathbb{R}} |G(j\omega)| = K|G(0)| = K < 1.$$

Conditions for stability in the presence of delay obtained in this way are often conservative [GKC03]. Even in this very simple example, we already know that the feedback connection is asymptotically stable in the presence of delays for all $K \leq 1$, which can be shown as an application of the Nyquist criterion. Indeed, the sensitivity transfer-function of the above feedback connection when $\tau = 1$ and $K = 1$ is

$$S(s) = \frac{1}{1 + e^{-s}GK} = \frac{s+1}{s+1+e^{-s}}.$$

That is the same as (7.12), which we have already shown to be asymptotically stable in Section 7.5.

8.4 Control of the Simple Pendulum – Part III

We will now apply robustness tools to the design of a controller for the simple pendulum. Instead of working with the linearized models at the equilibrium points $\theta = 0$ and $\theta = \pi$, we shall work directly with the closed-loop configuration for robust analysis from Fig. 8.4 and treat the nonlinearity $\Delta(y) = \sin(y)$ as a member of the sector shown in Fig. 8.5 as discussed earlier in Section 8.2.

Consider for now that the reference input $\bar{y}$ is zero. We will treat the case $\bar{y} \neq 0$ later. A transfer-function, $G(s)$, suitable for robust analysis of the closed-loop control of the simple pendulum, is obtained from the input w to the output y in Fig. 8.4:

$$G(s) = \frac{Y(s)}{W(s)} = \frac{-a_2 F(s)}{1 + b_2 K(s)F(s)}, \qquad F(s) = \frac{1}{s^2 + a_1 s}, \qquad (8.11)$$

where $K(s)$ is the transfer-function of the controller. We set the parameters as in (6.20) and (6.21):

$$a_1 = \frac{b}{J_r} = 0, \qquad a_2 = \frac{mgr}{J_r} \approx 49 \text{ s}^{-2}, \qquad b_2 = \frac{1}{J_r} \approx 66.7 \text{ kg}^{-1}\text{ m}^{-2}, \qquad (8.12)$$

8.4 Control of the Simple Pendulum – Part III

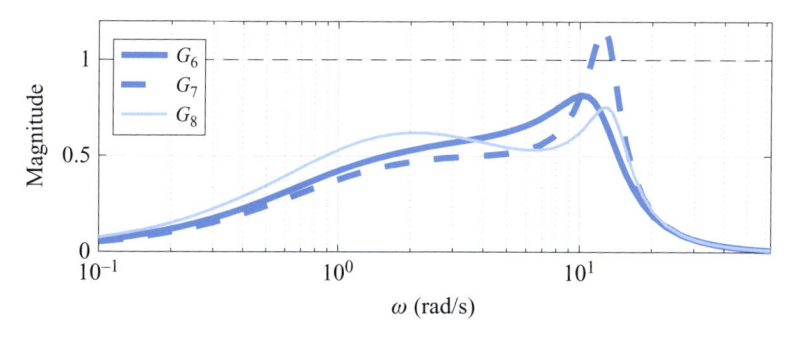

Figure 8.7 Magnitude of the frequency response of the transfer-function G, from (8.11), plotted on a linear scale for robust analysis of the simple pendulum in closed-loop with the controllers C_6 (thick solid), from (6.25), C_7 (thick dashed), from (7.20), and C_8 (thin solid), from (8.13); the H_∞ norm is the peak value.

and substitute $K(s)$ for the transfer-function of the controllers C_6, from (6.25), and C_7, from (7.20), to compute the corresponding transfer-functions G_6 and G_7 using (8.11). You will verify in P8.9 that G_6 and G_7 are both asymptotically stable. Moreover, the magnitude of their frequency response, plotted using a linear scale in Fig. 8.7, is such that

$$\|G_6\|_\infty \approx 0.82, \qquad \|G_7\|_\infty \approx 1.17.$$

Because $\|G_6\|_\infty < 1$ we conclude after using the small-gain argument discussed in Section 8.3 that when $\bar{y} = 0$ the origin is a globally asymptotically stable equilibrium point of the *nonlinear model* of the simple pendulum in closed-loop with the controller C_6, from (6.25). Unfortunately it is not possible to reach the same conclusion in closed-loop with the controller C_7, from (7.20), because $\|G_7\|_\infty \not< 1$. Note that the actual physical system may still be globally asymptotically stable with C_7; we were just not able to show it with our current combination of model and analysis tools. Indeed, we will succeed in proving that C_7 is also a suitable (robust) controller later using the more refined circle criterion to be introduced in Section 8.5. Before that, let us first try to adjust C_7 to make the closed-loop more robust.

The idea is to make changes in the zeros and poles of C_7 so as to lower the peaking in G_7 seen in Fig. 8.7. Compare Fig. 8.2 with Fig. 8.7 and note that, in the case of the simple pendulum, the peaking of the magnitude of the transfer-function G, from (8.11), is similar to the peaking of the magnitude of the sensitivity transfer-function, S. Therefore we hope that a reasoning similar to that employed in Section 8.1 to explain the peaking of the magnitude of the sensitivity can also explain the peaking of the magnitude of the transfer-function G. We expect that the peaking of $|S(j\omega)|$ will be reduced if we allow $|S(j\omega)|$ to be higher at lower frequencies. This can be accomplished by reducing the gain of the loop transfer-function, $L = G_\pi C$, at low frequencies, $|\omega| < 7$. One strategy is to stick with the general controller structure from (7.18), where $z_1 = 1$ is kept at the same location as in C_7 and the second zero, z_2, is shifted from 5 to 3 while preserving the same level of gain in the region $\omega > 7$ by enforcing the constraint $|C_8(j7)| = 2$.

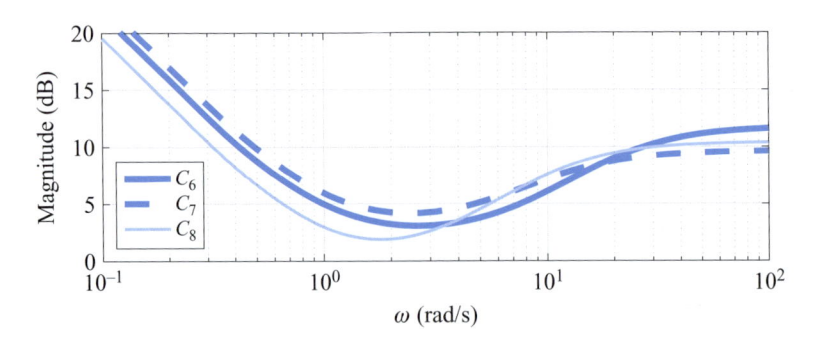

Figure 8.8 Magnitude of the frequency response of the controllers C_6 (thick solid), from (6.25), C_7 (thick dashed), from (7.20), and C_8 (thin solid), from (8.13).

Because the zero z_2 occurs earlier, the overall magnitude at low frequencies, $|\omega| < 7$, is reduced. We finally move the pole p_1 slightly from $p_1 = 11$ to $p_1 = 10.5$ to limit the gain at high frequencies. This results in the controller

$$C_8(s) \approx 3.28 \frac{(s+1)(s+3)}{s(s+10.5)}. \tag{8.13}$$

The features discussed above can be visualized in the Bode plots of the three controllers shown in Fig. 8.8. The impact in the closed-loop performance can be evaluated in the

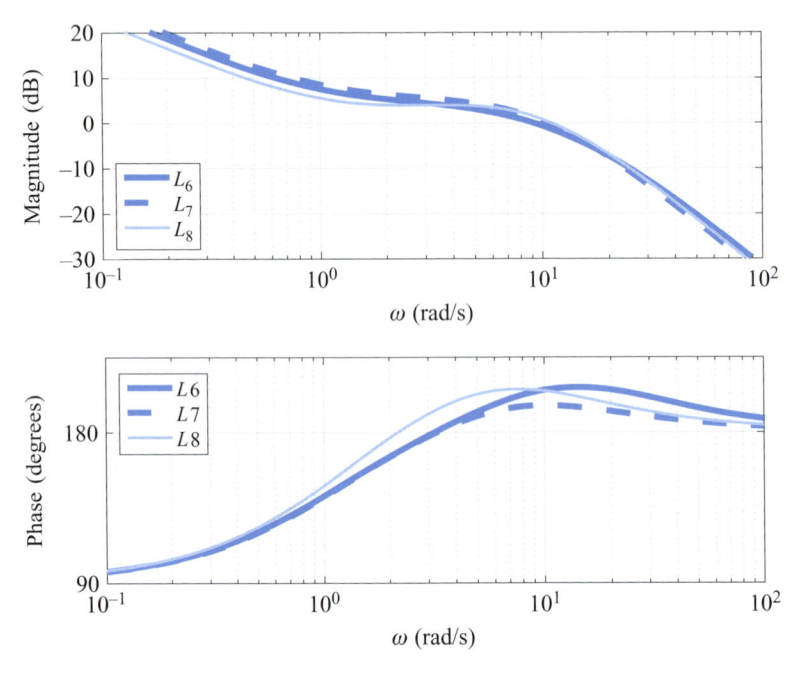

Figure 8.9 Bode plots of the loop transfer-functions $L_6 = G_\pi C_6$ (thick solid), $L_7 = G_\pi C_7$ (thick dashed), and $L_8 = G_\pi C_8$ (thin solid) calculated for the simple pendulum model linearized around $\theta = \pi$, G_π, for the controllers C_6, from (6.25), C_7, from (7.20), and C_8, from (8.13).

8.4 Control of the Simple Pendulum – Part III

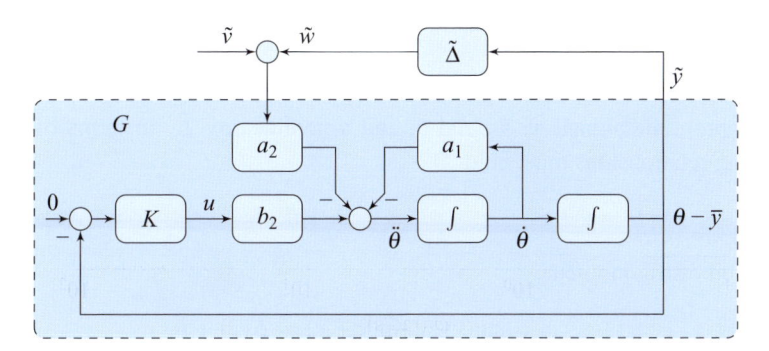

Figure 8.10 Modified feedback configuration of the controlled simple pendulum from Fig. 8.4 for robust tracking analysis; constant reference signal $\bar{y}$ has been incorporated into the state of the last integrator; $\tilde{\Delta}(\tilde{y}) = \cos(\bar{y})\sin(\tilde{y}) + \sin(\bar{y})(\cos(\tilde{y}) - 1)$, $\tilde{\Delta} \in \Delta$; $\tilde{v} = \sin(\bar{y})$.

Bode plots of the loop transfer-functions in Fig. 8.9. Note that C_8 accomplishes the goal of reducing the loop gain at frequencies below $\omega < 7$ while preserving the same level of loop gain as L_7 at frequencies higher than $\omega > 7$. More importantly, notice how the phase of L_8 is pushed further away from $180°$ around $\omega = 10$, where $|L| \approx 1$. Looking back at Fig. 8.7 we see how the peak of the magnitude of the corresponding transfer-function for robust analysis, G_8, has been significantly reduced with C_8. Indeed,

$$\|G_8\|_\infty \approx 0.76 < 1,$$

which is even lower than the value achieved by G_6. Of course, improved robustness has been achieved at the expense of some degradation of performance at lower frequencies, namely around 1 Hz. It is the control engineer's job to assess whether such trade-offs are possible in view of the intended application of the system in hand when confronted with control design decisions.

We shall now address the issue of tracking. As mentioned at the end of Section 8.3, because of the presence of nonlinearities in the closed loop, one needs to deploy heavy theoretical machinery to go from asymptotic stability to BIBO stability, which would allow one to extend the above analysis to a nonzero reference $\bar{y} \neq 0$. Even then, it is often the case that additional arguments need to be invoked to prove asymptotic tracking. This is the motivation for the discussion in the next paragraphs in which we consider the problem of asymptotic tracking of a constant reference input for the simple pendulum.

In general, as discussed in the end of Section 5.8, asymptotic tracking of a nonzero input requires a shift of the closed-loop equilibrium point. In response to a constant reference input, a useful strategy is to shift the state before analyzing stability. After that, one can prove asymptotic tracking *locally*, for instance, by using the linearization technique discussed in Section 5.4. In the next paragraphs we take a different route that will lead to a *global* result. We do so by manipulating signals in the block-diagram of Fig. 8.4 so as to incorporate the nonzero reference $\bar{y}$ into one of the states. One possible model is shown in the block-diagram in Fig. 8.10, where the signals $\tilde{y}$, $\tilde{w}$, and $\tilde{v}$, and the block $\tilde{\Delta}$ will be detailed next. Note how the constant nonzero reference signal has been replaced by a zero signal and incorporated into the state of the last integrator. Of course one could also reach the same conclusions by manipulating state-space formulas (see P8.10).

In order for the diagrams in Fig. 8.4 and Fig. 8.10 to be equivalent we need to determine appropriate signals $\tilde{y}$, $\tilde{w}$, and $\tilde{v}$, and a nonlinearity $\tilde{\Delta}$. In terms of signals the following relationships must hold:

$$y = \theta = \tilde{y} + \bar{y}, \qquad w = \tilde{w} + \tilde{v}, \qquad \tilde{w} = \tilde{\Delta}(\tilde{y}).$$

After a bit of trigonometry

$$w = \sin(y) = \sin(\tilde{y} + \bar{y}) = \tilde{\Delta}(\tilde{y}) + \tilde{v},$$

where

$$\tilde{\Delta}(\tilde{y}) = \cos(\bar{y})\sin(\tilde{y}) + \sin(\bar{y})(\cos(\tilde{y}) - 1), \qquad \tilde{v} = \sin(\bar{y}).$$

As you will show in P8.11, in this case, $\tilde{\Delta}(\tilde{y})$ is also in $\boldsymbol{\Delta}$, which is remarkable! Note that the transfer-function of the linear part of the system in Fig. 8.10, G, is the same as the one in Fig. 8.4, which is given by (8.11).

With the setup from Fig. 8.10 in mind we repeat the calculations as in Section 8.3:

$$\|\tilde{y}\|_2 = \|G(\tilde{w} + \tilde{v}) + f(x_0)\|_2 = \|G\tilde{w} + G\tilde{v} + f(x_0)\|_2$$
$$\leq \|G\|_\infty \|\tilde{w}\|_2 + \|G\tilde{v}\|_2 + M\|x_0\|_2 \leq \|G\|_\infty \|\tilde{y}\|_2 + \|G\tilde{v}\|_2 + M\|x_0\|_2,$$

in which the signal $f(x_0)$ represents the response to a possible nonzero initial condition as seen in Section 5.2, and rearrange to obtain

$$\|\tilde{y}\|_2 \leq \frac{\|G\tilde{v}\|_2 + M\|x_0\|_2}{1 - \|G\|_\infty}.$$

Note that we chose to keep $G\tilde{v}$ together since $\tilde{v}(t) = \bar{v} = \sin(\bar{y})$, $t \geq 0$, is a constant signal, therefore the signal norm $\|\tilde{v}\|_2$ is unbounded. On the other hand, as we will see soon, $\|G\tilde{v}\|_2$ is bounded. Indeed, because $\tilde{v}(t) = \bar{v}$, $t \geq 0$, is constant,

$$\mathscr{L}\{G\tilde{v}\} = G(s)\frac{\bar{v}}{s}.$$

Furthermore, in the case of (8.11), because $G(s)$ is strictly proper and asymptotically stable and $K(s)$ has a pole at the origin it follows that $G(s)$ has a zero at the origin. Therefore the transfer-function $G(s)/s$ is strictly proper and asymptotically stable, hence $\|G\tilde{v}\|_2$ is bounded. Like before, in Section 8.3, because G is asymptotically stable, $\|G\|_\infty < 1$, and $\|G\tilde{v}\|_2$ and $\|x_0\|_2$ are bounded, hence $\|\tilde{y}\|_2$ is bounded and

$$\lim_{t \to \infty} \tilde{y}(t) = \lim_{t \to \infty} (G\tilde{v})(t) = \lim_{t \to \infty} f(t, x_0) = 0,$$

which proves asymptotic tracking.

A final touch is the evaluation of stability with respect to various levels of damping. In Section 6.5 we stated that one might expect that controllers designed under the assumption of no damping, $b = 0$, should also perform well in the presence of natural damping, $b > 0$. Robust analysis with respect to more than one parameter is often a complicated

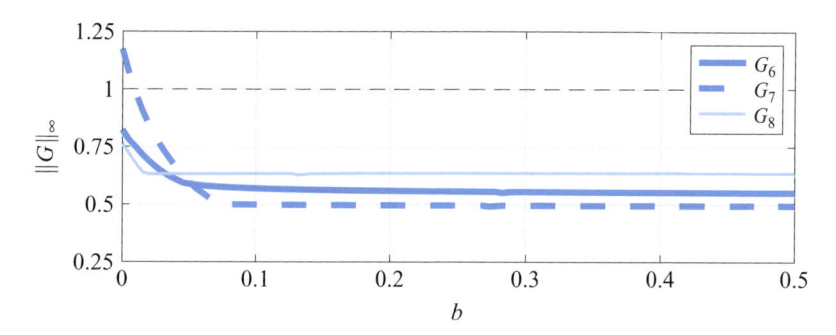

Figure 8.11 H_∞ norm of the transfer-function, G, from (8.11), for robust analysis of the simple pendulum in closed-loop with the controllers C_6 (thick solid), from (6.25), C_7 (thick dashed), from (7.20), and C_8 (thin solid), from (8.13), as a function of the damping b.

task.[14] What we do here is simply evaluate the H_∞ norm of the transfer-function G, from (8.11), for various values of $b > 0$. The result is shown in Fig. 8.11, from which we can see that as b grows the H_∞ norm of G decreases so that the closed-loop nonlinear connection in Figs. 8.4 and 8.10 with controllers C_6 and C_8 will remain asymptotically stable and achieve asymptotic tracking for any[15] constant level of damping $b > 0$. Note that for $b > 0.01$ even the norm of G_7, calculated with the controller C_7, from (7.20), is below one. Interestingly, for $b > 0.05$ the transfer-function G_7 has the lowest H_∞ norm, suggesting that C_7 is the most robust setup when high enough levels of damping are present.

As an illustration of the behavior of the closed-loop pendulum, Fig. 8.12 shows a nonlinear simulation of the simple pendulum in closed-loop with the controller C_6, from (6.25), displaying asymptotic stability and asymptotic tracking behavior for various values of the damping, b, and a reference input that swings the pendulum from 135° to 45° to 135°.

8.5 Circle Criterion

It is possible to generalize the small-gain robustness analysis of Section 8.3 to the slightly more general class of uncertainties

$$\Delta_{ab}(a,b) = \{\Delta : \mathbb{R} \to \mathbb{R}, \quad |\Delta(y) + ay| \leq b|y|, \quad b > 0\}. \tag{8.14}$$

The class $\Delta = \Delta_{ab}(0,1)$, from (8.7), is a particular case. Any function in $\Delta_{ab}(a,b)$ must lie in the shaded area in Fig. 8.13. For example, we could *tighten* the sector around the sine nonlinearity by considering the sector shown in Fig. 8.13 instead of Δ. It is sometimes easier to work with an alternative description of the uncertainty set $\Delta_{ab}(a,b)$

---

[14] *Structured uncertainty* is the terminology used in the literature [DFT09].
[15] In this simple example it is possible to show analytically that the H_∞ norm of G remains at the levels shown in Fig. 8.8 for any value of $b > 0.5$.

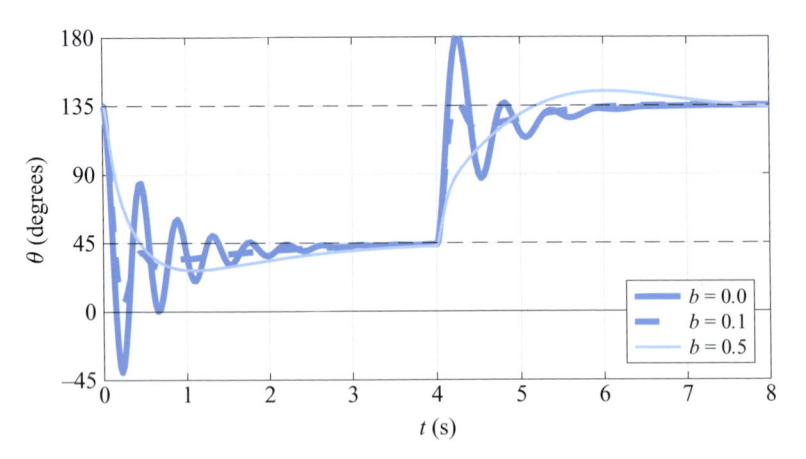

Figure 8.12 Nonlinear simulation of the simple pendulum (see Section 5.5) in closed-loop with the controller C_6, from (6.25), for various values of viscous damping coefficient, b. The reference $\bar{y}$ is set to 45° during the first 4 s and them moved to 135° in the last 4 s; note how the intrinsic damping of the pendulum helps to attenuate oscillations.

in terms of the parameters

$$(\alpha, \beta) = (a - b, a + b). \tag{8.15}$$

Conversely

$$(a, b) = (\tfrac{1}{2}(\alpha + \beta), \tfrac{1}{2}(\beta - \alpha)). \tag{8.16}$$

Relations (8.15) and (8.16) establish a one-to-one correspondence between a pair (a, b), $b > 0$, and a pair (α, β), $\beta > \alpha$. Moreover, as you will show in P8.12,

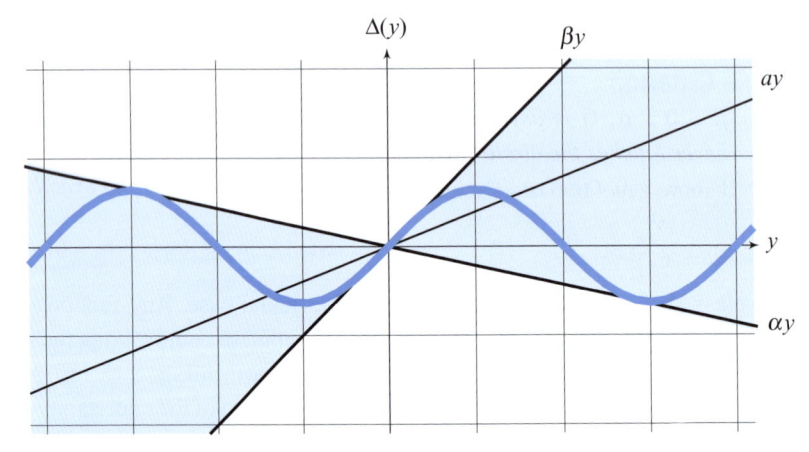

Figure 8.13 Functions in $\Delta \in \boldsymbol{\Delta}_{ab}(a, b) = \boldsymbol{\Delta}_{\alpha\beta}(\alpha, \beta)$ satisfy $|\Delta(y) - ay| \leq b|y|$ or, equivalently, $(\Delta(y) - \alpha y)(\Delta(y) - \beta y) \leq 0$, and must lie on the shaded area (sector). The solid curve is $\sin(y) \in \boldsymbol{\Delta}_{ab}(0.39, 0.61) = \boldsymbol{\Delta}_{\alpha\beta}(-0.22, 1) \subset \boldsymbol{\Delta}$.

8.5 Circle Criterion

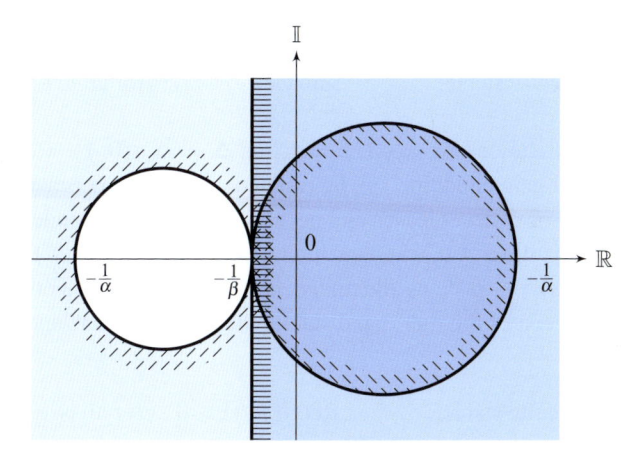

Figure 8.14 Circles $C(\alpha, \beta)$, for $\beta > \alpha$ and $\beta > 0$; the case $\alpha = 0$ degenerates into a half-plane; hashes indicate regions that map into the unit circle.

$\mathbf{\Delta}_{ab}(a, b) = \mathbf{\Delta}_{\alpha\beta}(\alpha, \beta)$, where

$$\mathbf{\Delta}_{\alpha\beta}(\alpha, \beta) = \{\Delta : \mathbb{R} \to \mathbb{R}, \quad (\Delta(y) - \alpha y)(\Delta(y) - \beta y) \le 0, \quad \beta > \alpha\}. \quad (8.17)$$

We use either $\mathbf{\Delta}_{ab}$ or $\mathbf{\Delta}_{\alpha\beta}$ interchangeably whenever convenient. For simplicity we also assume that $\beta > 0$.[16] A robust stability condition for the generalized uncertainty $\mathbf{\Delta}_{\alpha\beta}(\alpha, \beta)$ can be stated in geometric terms. For $\beta > \alpha$, $\beta > 0$, let $C(\alpha, \beta)$ be the circle centered on the real axis that intersects the real axis at the points $-\alpha^{-1}$ and $-\beta^{-1} < 0$, as illustrated in Fig. 8.14. The unit circle centered at the origin is $C(-1, 1)$. We need to consider two cases: $\beta > 0 > \alpha$, and $\beta > \alpha > 0$. The associated stability condition is stated next.

THEOREM 8.2 (Circle criterion) *Consider the diagram in Fig. 8.3. The origin of a state-space realization of the connection of the linear time-invariant system G with any $\Delta \in \mathbf{\Delta}_{\alpha\beta}(\alpha, \beta)$ is globally asymptotically stable if one of the following conditions holds:*

(a) $\beta > 0 > \alpha$, *G is asymptotically stable and the polar plot of G never leaves and never touches the circle $C(\alpha, \beta)$;*
(b) $\beta > \alpha > 0$, *the Nyquist plot of G encircles the circle $C(\alpha, \beta)$ exactly m times in the counter-clockwise direction, where m is the number of poles of G on the right-hand side of the complex plane, but never enters and never touches the circle $C(\alpha, \beta)$.*

Both conditions can be easily checked graphically by analyzing the behavior of the polar plot or Nyquist plot of G with respect to the circle $C(\alpha, \beta)$; this is the reason why this criterion is referred to as the *circle criterion*. Note that in condition (b) the transfer-function G need not be stable and one might need to indent poles on the imaginary axis, hence the need for a Nyquist rather than a simpler polar plot.

---
[16] If a negative β is needed simply *flip* the sign ahead of the uncertainty, i.e. use $-\Delta(y)$.

A special case of Theorem 8.2 is the small-gain condition from (8.8)

$$G \text{ is asymptotically stable,} \qquad \sup_{\omega \in \mathbb{R}} |G(j\omega)| < 1 \quad \text{for all } \omega \in \mathbb{R},$$

which corresponds to the choice of $\alpha = -1, \beta = 1$, and $\Delta = \Delta_{ab}(0, 1) = \Delta_{\alpha\beta}(-1, 1)$. Geometrically, the small gain condition can be restated as follows: G is asymptotically stable and the polar plot of G never leaves and never touches the unit circle $C(-1, 1)$.

Another special case is $\lim \alpha \uparrow 0$, when the interior of the circle $C(\alpha, \beta)$ degenerates into the half-plane $\text{Re}(s) > -\beta^{-1}$, as illustrated in Fig. 8.14. When both $\alpha \uparrow 0$ and $\beta \downarrow 0$, this half-plane becomes the right-hand side of the complex plane and asymptotic stability follows if

$$G \text{ is asymptotically stable,} \qquad \text{Re}(G(j\omega)) > 0 \quad \text{for all } \omega \in \mathbb{R}.$$

An asymptotically stable transfer-function G satisfying the above condition is known as strictly[17] *positive-real*. Because the frequency response of a strictly positive-real transfer-function has positive real part, its polar plot never ventures onto the left-hand side of the complex plane, and therefore never encircles the point -1 in closed-loop with any constant feedback gain $K \geq 0$. The corresponding sector description is

$$\lim_{\alpha \uparrow 0, \beta \downarrow 0} \Delta_{\alpha\beta}(\alpha, \beta) = \{\Delta : \mathbb{R} \to \mathbb{R}, \quad y\Delta(y) \geq 0\},$$

which is the entire first and third quadrants. Positive-realness implies that the phase of G is bounded, that is,

$$|\angle G(j\omega)| < \frac{\pi}{2}, \qquad \text{for all } \omega \in \mathbb{R}.$$

As with the small-gain condition, positive-realness also guarantees stability in closed-loop with any transfer-function $\Delta(s)$ which is asymptotically stable and positive-real. Without getting into details, if $G(s)$ is strictly positive-real and $\Delta(s)$ is positive-real then

$$|\angle G(j\omega)| < \frac{\pi}{2}, \qquad |\angle \Delta(j\omega)| \leq \frac{\pi}{2} \quad \text{for all } \omega \in \mathbb{R}$$

and therefore

$$|\angle L(j\omega)| = |\angle G(j\omega) + \angle \Delta(j\omega)| < \pi$$

and the Nyquist plot of the loop transfer-function, L, never crosses the negative real axis.

As an illustration of the use of the circle criterion, consider once again the control of the simple pendulum. As shown in Fig. 8.13, it is possible to find a tighter description for the nonlinearity $\Delta(y) = \sin(y)$ in terms of the sector uncertainty $\Delta_{ab}(0.39, 0.61) = \Delta_{\alpha\beta}(-0.22, 1)$ (see also P8.13). The polar plot of the transfer-function G, from (8.11),

---

[17] With the condition $\lim_{\omega \to \infty} \text{Re}(G(j\omega)) > 0$ relaxed to $\lim_{\omega \to \infty} \omega^2 \text{Re}(G(j\omega)) > 0$ in the case of strictly proper transfer-functions [Kha96, Section 10.1].

8.5 Circle Criterion

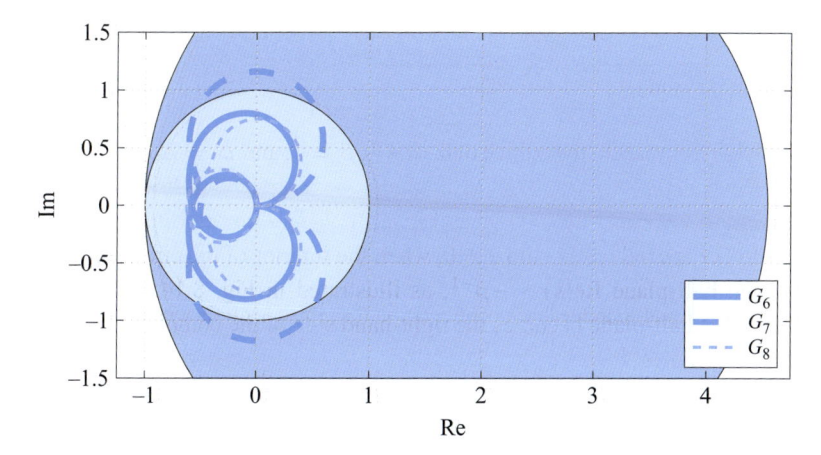

Figure 8.15 Polar plots of the transfer-function G, from (8.11), calculated for the simple pendulum model with the controllers C_6 (thick solid), from (6.25), C_7 (thick dashed), from (7.20), and C_8 (thin dashed), from (8.13). All controllers produce closed-loop responses which are contained in the circle $C(-0.22, 1)$; the closed-loop response with the controller G_7 is the only one that is not also contained in the circle $C(-1, 1)$.

calculated for the simple pendulum model with the controllers C_6, from (6.25), C_7, from (7.20), and C_8, from (8.13), is shown in Fig. 8.15 superimposed on the unit circle $C(-1, 1)$, from the small-gain condition, and the circle $C(-0.22, 1)$, from the circle criterion with the tighter uncertainty description. All controllers designed so far can now be proven to asymptotically stabilize the simple pendulum because the polar plots of G_6, G_7, and G_8 all lie inside the circle $C(-0.22, 1)$. Recall that the model with the controller C_7 previously failed the small-gain test, which can be visualized in Fig. 8.15 as the polar plot of G_7 leaving the unit circle.

Proof of the Circle Criterion

We start by noticing that $\Delta \in \mathbf{\Delta}_{ab}(a, b)$ if and only if

$$\tilde{\Delta}(y) = b^{-1}(\Delta(y) + ay) \in \mathbf{\Delta} = \mathbf{\Delta}_{\alpha\beta}(-1, 1).$$

By expressing Δ in terms of $\tilde{\Delta}$ we obtain

$$\Delta(y) = b\,\tilde{\Delta}(y) - ay.$$

In other words, the feedback connection of $\Delta \in \mathbf{\Delta}_{ab}(a, b)$ with G in Fig. 8.3 is equivalent to the feedback connection of $\tilde{\Delta} \in \mathbf{\Delta}_{\alpha\beta}(-1, 1)$ with $\tilde{G}$ in Fig. 8.16. From the diagram in Fig. 8.16,

$$\tilde{G} = \frac{bG}{1 + aG}, \qquad \tilde{w} = b^{-1}w, \qquad \tilde{v} = b^{-1}v.$$

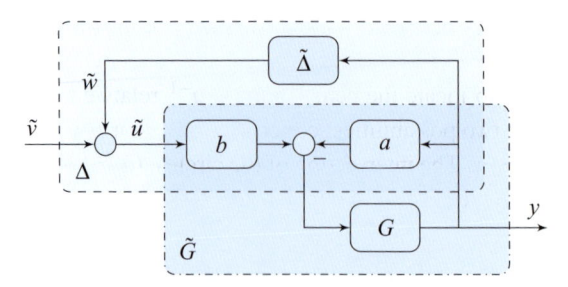

Figure 8.16 Closed-loop feedback configuration for robust analysis with the circle criterion.

Because $b > 0$, $\tilde{v}$ is bounded if and only if v is bounded. Using the small-gain analysis of Section 8.3, the closed-loop connection in Fig. 8.16 is robustly stable if

$$\frac{bG}{1+aG} \text{ is asymptotically stable,} \qquad \left\|\frac{bG}{1+aG}\right\|_\infty < 1. \tag{8.18}$$

The next paragraphs are dedicated to translating these two conditions in terms of the open-loop transfer-function G, instead of $\tilde{G} = bG/(1+aG)$.

The case $a = 0$ is the simplest, and reduces to G asymptotically stable and $\|G\|_\infty < b^{-1}$, as expected. When $a \neq 0$ the analysis is more complicated, and we must have recourse to properties of mappings in the complex plane, which you will have the opportunity to patch together in problems P8.14 and P8.15. Here we present only the summary of the relevant facts. Let

$$U(s_0, \rho) = \{s \in \mathbb{C}: \quad |s - s_0| = \rho > 0\}$$

be a circle centered at s_0 with radius $\rho > 0$ and define the one-to-one mapping

$$T: \mathbb{C} \to \mathbb{C}, \quad T(z) = \frac{bz}{1+az} = \frac{\frac{1}{2}(\beta - \alpha)z}{1 + \frac{1}{2}(\alpha + \beta)z}, \tag{8.19}$$

which is a particular case of the bilinear transformation [Neh52, Section V.2]. The second condition in (8.18) can be restated as follows: is the image of $T(G(j\omega))$ in the interior of the unit circle? This question can be answered graphically: the answer is affirmative if and only if the polar plot of $G(j\omega)$ lies in the interior of the image of the unit circle under the inverse mapping T^{-1}. This is made easy by the fact that the bilinear transformation maps a circle into another circle [Neh52, Section V.2] and the image of the unit circle under the inverse mapping T^{-1} is shown in P8.16 to be the circle $U(s_0, \rho)$ where

$$s_0 = -\tfrac{1}{2}\left(\alpha^{-1} + \beta^{-1}\right), \qquad \rho = \tfrac{1}{2}\left|\alpha^{-1} - \beta^{-1}\right|, \tag{8.20}$$

that is, the circle $C(\alpha, \beta)$ we introduced earlier. Some examples were shown in Fig. 8.14 for different choices of parameters α and $\beta > 0$.

For robust stability we must locate not only the image of the unit circle but also the image of its interior. A summary of the conclusions (see [Neh52, Section V.2] for details) is as follows: if the pole of the mapping T, i.e. the point $-a^{-1} = -2(\alpha + \beta)^{-1}$, is inside the circle $C(\alpha, \beta)$ then T maps the interior of the circle $C(\alpha, \beta)$ into the interior of the unit circle. Otherwise, if the point $-a^{-1}$ is outside the circle $C(\alpha, \beta)$ then T maps the exterior of the circle $C(\alpha, \beta)$ into the interior of the unit circle. With that information all that is left is to locate the center point $-a^{-1}$ relative to $C(\alpha, \beta)$. With $\beta > 0, \beta > \alpha \neq 0$, there are two possibilities: either $\beta > 0 > \alpha$ or $\beta > \alpha > 0$.

Consider first that $\beta > 0 > \alpha$. The intersection of the circle $C(\alpha, \beta)$ with the real axis is the segment $[-\beta^{-1}, -\alpha^{-1}]$ (see Fig. 8.14). If $\beta > |\alpha| > 0$ then

$$\beta > \frac{\alpha + \beta}{2} = a > 0 \quad \Longrightarrow \quad -a^{-1} < -\beta^{-1},$$

and if $|\alpha| > \beta > 0$ then

$$\alpha < \frac{\alpha + \beta}{2} = a < 0 \quad \Longrightarrow \quad -a^{-1} > -\alpha^{-1},$$

so that the point $-a^{-1}$ is always outside $C(\alpha, \beta)$. Consequently, the interior of the circle $C(\alpha, \beta)$ is mapped by T into the interior of the unit circle $C(0, 1)$. Recalling that

$$\left\| \frac{bG}{1 + aG} \right\|_\infty = \sup_{\omega \in \mathbb{R}} \left| \frac{bG(j\omega)}{1 + aG(\omega)} \right| = \sup_{\omega \in \mathbb{R}} |T(G(j\omega))|$$

we conclude that

$$\left\| \frac{bG}{1 + aG} \right\|_\infty = \sup_{\omega \in \mathbb{R}} |T(G(j\omega))| < 1$$

if and only if $G(j\omega) \in \text{int}\, C(\alpha, \beta)$ for all $\omega \in \mathbb{R}$. If the polar plot of $G(j\omega)$ never leaves the circle $C(\alpha, \beta)$ it also never encircles $-a^{-1}$. According to the Nyquist stability criterion, $1 + aG(s)$ will be asymptotically stable if and only if G does not have any poles on the right-hand side of the complex plane, that is, if G is itself asymptotically stable. This is condition (a) in Theorem 8.2.

Consider now the case $\beta > \alpha > 0$. Because

$$\beta > \frac{\alpha + \beta}{2} = a > \alpha > 0 \quad \Longrightarrow \quad -\beta^{-1} > -a^{-1} > -\alpha^{-1},$$

the point $-a^{-1}$ is always inside $C(\alpha, \beta)$. Consequently, the exterior of the circle $C(\alpha, \beta)$ is mapped by T into the interior of the unit circle $C(0, 1)$. As before,

$$\left\| \frac{bG}{1 + aG} \right\|_\infty = \sup_{\omega \in \mathbb{R}} |T(G(j\omega))| < 1$$

if and only if $\nexists \omega$ such that $G(j\omega) \in C(\alpha, \beta) \cup \text{int}\, C(\alpha, \beta)$. This time the polar plot of $G(j\omega)$ must never enter the circle $C(\alpha, \beta)$. It is, however, possible that $G(j\omega)$ encircles $C(\alpha, \beta)$. In this case, each encirclement of $C(\alpha, \beta)$ corresponds also to an encirclement of the point $-a^{-1}$. According to the Nyquist stability criterion, $1 + aG(s)$ will be asymptotically stable if and only if the number of net counter-clockwise encirclements of $G(j\omega)$ around the point $-a^{-1}$ is equal to the number of poles of G on the right-hand side of the complex plane. This is condition (b) in Theorem 8.2.

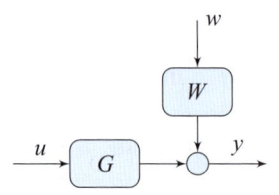

Figure 8.17 System with control input, u, and measurable disturbance, w.

8.6 Feedforward Control and Filtering

Continuing with our discussion on performance, consider now the system with a control input, u, and a disturbance input, w, shown in Fig. 8.17. The key assumption in this section is that the disturbance, w, is either known ahead of time or can be measured *online*.

One control architecture that can take advantage of this knowledge of the disturbance is the one in Fig. 8.18. In this figure the controller, C, makes use of three inputs: the measured output, y, the reference input, $\bar{y}$, and the measured disturbance, w. Internally, it is composed of a feedback term, the block K, and a *feedforward* term, the block F. The feedforward block, F, makes no use of the output, y, and in this sense it is essentially open-loop control. Any instabilities will have to be dealt with by the feedback block, K.

With the setup of Fig. 8.18 in mind, we calculate

$$e = S\bar{y} - (W - FG)Sw, \tag{8.21}$$

where $S = (1 + GK)^{-1}$ is the standard sensitivity transfer-function. Clearly if one can set

$$W = FG$$

then the tracking error, e, becomes completely decoupled from the disturbance, which would be a remarkable feat. Unfortunately it is not always possible to achieve this. Indeed, one would need to implement a feedforward controller:

$$F = G^{-1}W,$$

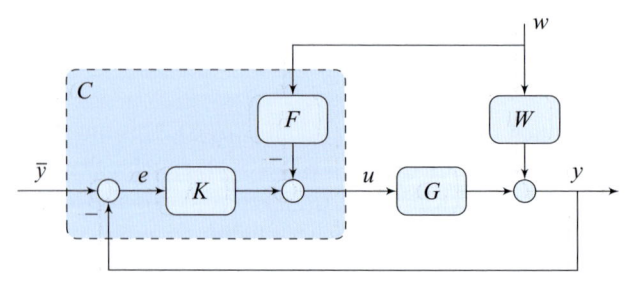

Figure 8.18 Closed-loop feedback configuration for tracking and measured disturbance, w, with feedback, K, and feedforward, F, control blocks.

which might not be possible, depending on G and W. For instance, if W is static, i.e. of zeroth order, and G is rational and strictly proper, then F would not be proper, and therefore would not be realizable (see Chapter 5). Even when the *filter F* is realizable, the need to invert the system, G, should bring to mind the many difficulties with open-loop control discussed in various parts of this book, such as the nefarious effects of cancellations of non-minimum-phase poles and zeros addressed in Section 4.7. For instance, if

$$W(s) = \frac{1}{s+1}, \qquad G(s) = \frac{s-1}{(s+2)(s+3)}, \tag{8.22}$$

then

$$F(s) = G^{-1}(s)W(s) = \frac{(s+2)(s+3)}{(s+1)(s-1)} \tag{8.23}$$

is unstable and therefore the connection in Fig. 8.18 will not be internally stable even though $W - FG = 0$. Indeed, because W and F appear in *series* with the rest of the components of the closed-loop, one should expect trouble whenever W or F are not asymptotically stable. Note that instability of the system, G, is not a problem as long as it is stabilized by the feedback controller, K.

A particular case of interest is that of input disturbances such as in Fig. 1.13 or Fig. 4.11 with $v = 0$. These diagrams become equivalent to Fig. 8.18 if we set $W = G$. In this case the choice $F = 1$ achieves perfect disturbance rejection, since $W = G = FG$. This means that if we know ahead[18] of time that a known disturbance will affect the control input of a system we can simply cancel it by applying an opposite input. For example, in the case of the car cruise control subject to a change of road slope discussed in Section 4.5, if one could *anticipate* the slope, and compute and apply the exact force necessary to counteract the additional component of gravitational force due to the slope, then one would achieve "perfect" disturbance rejection. Of course this is easier said than done since, in practice, accurately estimating an upcoming slope might require sophisticated resources, such as for instance fast vision-processing capabilities. Moreover, other forces and uncertainties, e.g. friction and drag forces, uncertainties on the total mass as well as the mass distribution of the car, etc. are guaranteed to torment the optimistic control engineer. Nevertheless, even if perfect rejection is not achieved, the solution proposed in Fig. 8.18 is certainly much better than simply open-loop control and can be better than simply closed-loop feedback. In fact, the combination of feedforward and feedback controllers can deliver enhanced performance, as the feedforward control attempts to invert the system model, while providing robustness to uncertainties through feedback.

A similar configuration that is widely deployed in tracking systems is depicted in Fig. 8.19. In this case the input w is more an auxiliary reference input than a disturbance. Analysis of the closed-loop connection provides

$$e = (W - FG)Sw,$$

---

[18] Or at least just in time.

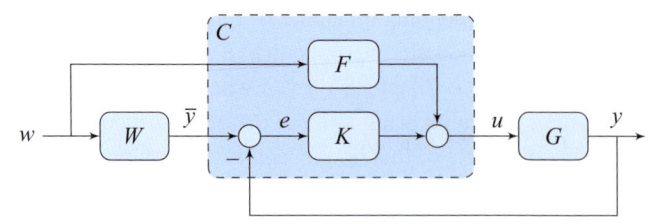

Figure 8.19 Closed-loop feedback configuration for tracking with reference input, w, feedback, K, and feedforward, F, control blocks.

which is very similar to the expression obtained for the configuration in Fig. 8.18. For this reason much of the former discussion applies verbatim to the diagram in Fig. 8.19. The case $W = G$ is of special importance here as well, not only because perfect tracking is possible with $F = 1$, but also because of its interpretation: when $W = G$, the auxiliary reference input, w, can be thought of as the control input required to drive the system, G, to a desired output, $\bar{y}$. The feedforward controller, $F = 1$, makes sure that this reference input, w, is applied to the actual system while the feedback controller, K, corrects any mistakes in the achieved trajectory, y. This is the scheme behind virtually all control systems in which open-loop control inputs are used. For a concrete example, w might be the calculated thrust input necessary to take a rocket from the surface of the earth to the surface of the moon along a desired trajectory, $\bar{y}$. The actual thrust input, u, is produced by the rocket in closed loop with a feedback controller that makes sure the desired trajectory is being closely tracked. What other way could we have made it to the moon?

When perfect input rejection is not possible we have to settle for more modest goals. For instance, we could seek to achieve asymptotic disturbance rejection. The setup is similar to the one in Chapter 4, as for instance in Lemma 4.1 and Lemma 4.2. Assuming that $\bar{y}(t) = \bar{y}\cos(\omega t + \phi)$, $t \geq 0$, if $S(j\omega) = 0$, or equivalently if G or K have a pole at $s = j\omega$, then the first term in (8.21) will approach zero asymptotically. Likewise, if $T(j\omega) = 0$, where $T = (W - FG)S$, then the second term in (8.21) will also approach zero asymptotically when $w(t) = \bar{w}\cos(\omega t + \psi)$, $t \geq 0$. Because the zeros of S are the poles of K and G (see Chapter 4), when K has a pole at $s = j\omega$ then S and GS will have at least one zero at $s = j\omega$ if no pole–zero cancellations occur. Likewise, if we assume[19] that W and F are asymptotically stable, then such zeros cannot be canceled by the stable poles of W or F and we conclude that $T(j\omega) = 0$ if K has a pole at $s = j\omega$. In other words, if K is chosen so that the closed-loop system achieves asymptotic tracking *and* disturbance rejection as in Lemma 4.2 then feedforward will not impact this property.

That feedforward will not destroy what feedback worked hard to achieve is good news. However, when K does not have a pole at $s = j\omega$, it may still be possible to achieve asymptotic disturbance rejection using feedforward. Indeed, all that is needed

---

[19] This is a necessity if internal stability is required.

is that a stable $F(s)$ be chosen so that

$$F(j\omega) = \lim_{s \to j\omega} G(s)^{-1}W(s),$$

and hence that $T(j\omega) = 0$. Note that this requirement is much weaker than asking that F be equal to $G^{-1}W$. Indeed, when $\omega = 0$, that is, in the case of step inputs, this can be done with a static feedforward gain $F = G(0)^{-1}W(0)$. A complete feedforward controller for the case $\omega > 0$ will be worked out in P8.21 and P8.22. Compare this with (8.23), where F is forced to be unstable. Finally, it may be possible that G has a pole at $s = j\omega$, in which case the closed-loop system already achieves asymptotic tracking and a choice of $F(j\omega) = 0$ or even $F(s) = 0$ will suffice to achieve asymptotic disturbance rejection.

Alternatively, one might seek to design an asymptotically stable F such that $(W - FG)S$ is *small*. For example, one could use the frequency-domain norms introduced in Section 3.9, e.g. the H_2 or the H_∞ norm, and attempt to solve an optimization problem of the form

$$\min_F \{\|(W - FG)S\| : F \text{ is asymptotically stable and realizable}\}. \tag{8.24}$$

When F is rational the realizability requirement is satisfied if F is proper. If we choose the H_2 norm then solving (8.24) is relatively easy. For the transfer-functions G and W in (8.22) and $S = (s+1)^{-1}$ you will show in P8.27–P8.29 that

$$\|(W - FG)S\|_2^2 = \|Z_+\|_2^2 + \|Z_- - F\tilde{Y}\|_2^2, \tag{8.25}$$

where

$$\tilde{Y}(s) = \frac{1}{(s+2)(s+3)}, \quad Z_+(s) = \frac{1}{2}\frac{1}{s-1}, \quad Z_-(s) = -\frac{1}{2}\frac{1}{s+1}, \tag{8.26}$$

from which the optimal feedforward filter is

$$F(s) = \tilde{Y}(s)^{-1}Z_-(s) = -\frac{1}{2}\frac{(s+2)(s+3)}{s+1}.$$

Note that this filter is not proper, and therefore not realizable. It is possible to formally show that for any $k \geq 1$ the proper filter $\tilde{F}(s) = F(s)/(\tau s + 1)^k$ has a performance which is as close as desired to that of $F(s)$ if τ is made small enough. See [DFT09, Chapter 10] for details. The argument here is the same as the one used in Section 6.3, where an extra pole far enough on the left-hand side of the complex plane was introduced to make a PID controller proper and the same practical considerations apply. This problem is often much more complicated if the H_∞ norm is used instead of the H_2 norm [DFT09, Chapter 9]. The solution in the case of the above example is, however, still simple enough to be computed in P8.30 and P8.31, where it is shown that

$$F(s) = -\frac{1}{4}\frac{(s+2)(s+3)^2}{s+1} \tag{8.27}$$

is the optimal H_∞ solution to (8.24). Note that this feedforward filter is even more aggressive at higher frequencies than the optimal H_2 filter.

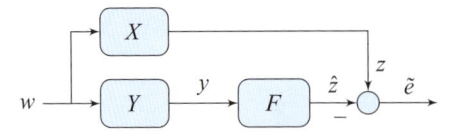

Figure 8.20 Standard configuration for filtering; F is the filter that is to be designed to render the filtered error, $\tilde{e} = z - \hat{z}$, as small as possible.

Broadly speaking, the optimization problem (8.24) is an *optimal filtering* problem. Filtering and estimation are widely studied problems in signals and systems as well as in the mathematics, statistics, and even finance literature. Optimal solutions can be computed for various norms and setups, including stochastic processes. We recommend the excellent treatises [AM05; KSH00] for further reading on this subject. The problem of optimal filtering or optimal estimation is often formulated with the help of the diagram in Fig. 8.20. Given the systems, X and Y, and some characterization of the input signal, w, the goal is to design an *estimator*, F, that produces an estimate, $\hat{z}$, that is as close as possible to the reference output, z. This is done by minimizing some measure of the estimation error, $\tilde{e}$, for example by solving the optimization problem

$$\min_F \{\|X - FY\| : \quad F \text{ is asymptotically stable and realizable}\}. \tag{8.28}$$

In some applications one can get away with an unstable estimator, for example when the estimator is used in closed-loop with a feedback controller. It may also be possible to relax some of the requirements on realizability if the filter is not to be used online. Contrast (8.28) with (8.24) to see that the optimal feedforward design problem (8.24) is in fact a particular case of the optimal filtering problem (8.28) where $X = WS$ and $Y = GS$.

It is worth noticing that the diagram in Fig. 8.20 shows its true versatility when the signals and systems are allowed to be multivariate. For instance, in the context of *filtering* we can let

$$w = \begin{pmatrix} r \\ v \end{pmatrix}, \qquad X = \begin{bmatrix} G & 0 \end{bmatrix}, \qquad Y = \begin{bmatrix} G & I \end{bmatrix}, \qquad z = r,$$

so that the reference output, z, is equal to an input *signal*, r, which is corrupted by *noise*, v, after traversing the *communication channel*, G. The goal is to produce a *filtered signal*, $\hat{z}$, which is as close as possible to the original signal, r.

Finally we close the discussion by highlighting some deeper connections between filtering and control. As we have seen so far, feedforward control could be cast and solved as a filtering problem. It may be somewhat surprising to realize that feedback control methods can also be used to solve filtering problems. For instance, the celebrated *Kalman filter* [AM05; KSH00] is deeply rooted in a duality between feedback control and filtering, with the state-space realization of the filter making use of feedback to compare the actual measurement with the filter predicted measurement. This relationship is made explicit in the diagram of Fig. 8.21, in which the filter, F, has been realized in the form of a feedback system. Indeed, as you will show in P8.32, when the

8.6 Feedforward Control and Filtering

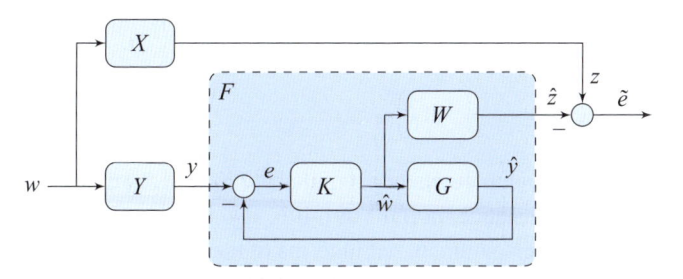

Figure 8.21 Filter estimates the input signal, $\hat{w}$, and the measurement, $\hat{y}$; feedback compares the estimate, $\hat{y}$, with the actual measurement, y.

models W and G match X and Y, that is when $W = X$ and $G = Y$, it follows that

$$\tilde{e} = WSw, \qquad S = \frac{1}{1+GK}, \qquad (8.29)$$

where S is the familiar sensitivity transfer-function studied in connection with closed-loop feedback systems. All methods used to design feedback controllers can therefore be used to design the "controller," K. For instance, a filter that can asymptotically track a step input signal, w, can be obtained by letting K have integral action. Note that the filter in Fig. 8.21 produces estimates of the measured output, $\hat{y}$, and the input signal, $\hat{w}$. Indeed, this structure is that of an *observer*, which is at the heart of MIMO output feedback controllers and the Kalman filter (see P8.33 and P8.34).

There are many more examples in the systems and control literature where the problems of filtering, estimation, and control are connected. Perhaps one of the most fundamental problems is that of parametrizing all stabilizing controllers, a procedure that converts problems in feedback control into the design of filters, at the heart of many developments in robust, H_2 and H_∞ control and estimation theory [DFT09]. The simplest case, that of a stable system, can be visualized easily in the block-diagram of Fig. 8.22, where the system block, G, has been duplicated and *absorbed* into the filter block, Q. Because G is asymptotically stable, the connection is internally stable if and only if Q is also asymptotically stable. Moreover, for any stabilizing controller, K, the filter, Q, is given by

$$Q = \frac{K}{1+KG},$$

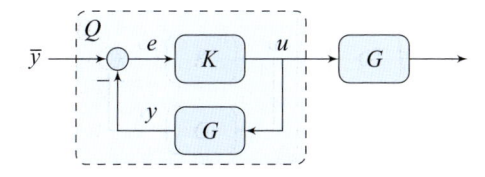

Figure 8.22 Parametrization of all stabilizing controllers: when G is asymptotically stable all stabilizing controllers are parametrized by $K = Q(1 - GQ)^{-1}$, with Q asymptotically stable.

which must be stable. Conversely, given an asymptotically stable filter, Q, the connection of Q and G is internally stable, therefore the controller

$$K = \frac{Q}{1 - QG}$$

must be stabilizing. The complete theory requires the introduction of many new ideas, which can be found, for instance, in [DFT09]. It is also remarkable that once the feedback controller, K, has been parametrized in terms of the auxiliary filter, Q, all transfer-functions in the feedback loop, say those in Fig. 4.18, become linear[20] functions of Q. This enables the formulation of an incredibly diverse set of optimization problems as convex problems. See for instance [BB91]. Unfortunately those problems happen to be infinite-dimensional[21] and the existing numerical algorithms may not be immediately practical since they tend to produce solutions with very large dimensions even when the input data is made of low-order systems.

Problems

8.1 The *maximum modulus principle* [BC14, Section 59] states that if a complex function f is analytic in an open set D then $|f(s)| \geq |f(z)|$ for any $z \in D$. Show that if f has a maximizer in the interior of D then f is a constant.

8.2 Show that

$$\ln S(s) = \ln|S(s)| + j(2\kappa\pi + \angle S(s)),$$

for all $\kappa \in \mathbb{Z}$.   *Hint: If $s = \rho e^{j\theta}$ then $\ln s = \ln \rho + j\theta$.*

8.3 Show that if $S^*(j\omega) = S(-j\omega)$ and $\rho \geq 0$ then

$$\int_{-\rho}^{\rho} \ln S(j\omega) d\omega = 2 \int_{0}^{\rho} \ln|S(j\omega)| d\omega,$$

Hint: Use P8.2.

8.4 Show that if (8.2) holds then

$$\lim_{|s| \to \infty} sL(s)^k = 0$$

for all $k \geq 2$.   *Hint: $sL(s)^k = s^{1-k} (sL(s))^k$.*

8.5 Show that for all $|L(s)| < 1$

$$s \ln \left(\frac{1}{1 + L(s)} \right) = -sL(s) + sO(L(s)^2),$$

---
[20] Technically affine.
[21] This is because the space of rational functions of a single complex variable is infinite-dimensional: a basis for the space of rational functions cannot be written using a finite number of rational functions, not even if the order is fixed. Contrast that with the space of polynomials of fixed degree, which can be written in terms of a finite number of monomials, e.g. $1, x, x^2, \ldots$.

and if (8.2) holds then
$$\lim_{|s|\to\infty} -s \ln S(s) = \lim_{|s|\to\infty} sL(s),$$
where $S(s) = (1 + L(s))^{-1}$. *Hint: Expand in power series* $\ln(1+x) = x - x^2/2 + \cdots$, $|x| < 1$ *and use P8.4.*

8.6 Explain why in P8.5 one does not have to worry about the assumption $|L(s)| < 1$ when evaluating $\lim |s| \to \infty$.

8.7 Show that
$$\left|\frac{j\omega + p}{j\omega - p}\right| = 1$$
for all $\omega \in \mathbb{R}$.

8.8 Show that
$$\frac{s+p}{s-p} = \left(1 - \frac{2p}{s+p}\right)^{-1}$$
and
$$\lim_{|s|\to\infty} s \ln\left(\frac{s+p}{s-p}\right) = \lim_{|s|\to\infty} \frac{2ps}{s+p} = 2p.$$

Hint: Use P8.5.

8.9 Show that G, from (8.11), is asymptotically stable when the parameters a_1, a_2, and b_2 are given by (8.12) and K is C_6, from (6.25), C_7, from (7.20), or C_8, from (8.13).

8.10 Construct state-space realizations for the block-diagrams in Fig. 8.4 and Fig. 8.10 and show how their states are related.

8.11 Let
$$\Delta(y) = \cos(\bar{y})\sin(y) + \sin(\bar{y})(\cos(y) - 1).$$
Show that $\Delta \in \boldsymbol{\Delta}$, from (8.7).

8.12 Assume $b > 0$. Show that $\boldsymbol{\Delta}_{ab}(a, b)$, from (8.14), and $\boldsymbol{\Delta}_{\alpha\beta}(\alpha, \beta)$, from (8.17), are such that $\boldsymbol{\Delta}_{ab}(a, b) = \boldsymbol{\Delta}_{\alpha\beta}(\alpha, \beta)$ if (8.15) and (8.16) hold. *Hint: Square both sides of the inequality in the definition of* $\boldsymbol{\Delta}_{ab}(a, b)$ *and show that the relations* (8.15) *and* (8.16) *are one-to-one.*

8.13 Show that
$$\Delta(y) = \sin(y) \in \boldsymbol{\Delta}_{ab}(0.39, 0.61) = \boldsymbol{\Delta}_{\alpha\beta}(-0.22, 1).$$

Hint: Calculate all tangents to the curve $\Delta(y) = \sin(y)$ *that pass through the origin and evaluate their coefficients.*

8.14 Show that the image of the imaginary axis $\mathbb{I} = \{j\omega : \omega \in \mathbb{R}\}$ under the bilinear mapping

$$T : \mathbb{C} \to \mathbb{C}, \quad T(s) = \frac{c-s}{c+s} \tag{8.30}$$

when $c \neq 0$ is real is $U(0, 1) \setminus \{-1\}$, where $U(0, 1)$ is the unit circle centered at the origin. Explain why the point -1 is not included in the mapping. *Hint: Evaluate the magnitude and the phase of $T(j\omega)$ for $\omega \in \mathbb{R}$.*

8.15 Show that the inverse of the mapping

$$T : \mathbb{C} \to \mathbb{C}, \quad T(z) = \frac{bz}{1+az} = \frac{\frac{1}{2}(\beta-\alpha)z}{1+\frac{1}{2}(\alpha+\beta)z},$$

is the mapping

$$T^{-1} : \mathbb{C} \to \mathbb{C}, \quad T^{-1}(z) = \frac{z}{b-az} = \frac{2z}{(\beta-\alpha)-(\alpha+\beta)z}.$$

8.16 Let T^{-1} be defined as in P8.15. Show that if $\beta > 0$ then

$$T^{-1}\left(\frac{1-s}{1+s}\right) = s_0 + \rho \frac{c-s}{c+s},$$

where

$$c = -\frac{\alpha}{\beta}, \quad s_0 = -\frac{1}{2}\left(\alpha^{-1} + \beta^{-1}\right), \quad \rho = \frac{1}{2}\left|\alpha^{-1} - \beta^{-1}\right|.$$

8.17 Let T^{-1} be defined as in P8.15. Show that if $\beta > 0$ the image of the unit circle $U(0, 1)$ under the mapping T^{-1} is the circle $U(s_0, \rho)$, where

$$s_0 = -\frac{1}{2}\left(\alpha^{-1} + \beta^{-1}\right), \quad \rho = \frac{1}{2}\left|\alpha^{-1} - \beta^{-1}\right|.$$

Hint: Use P8.14 and P8.16.

8.18 The controller in Fig. 8.23 is known as a *two degrees-of-freedom controller*. The transfer-function K is the feedback part of the controller and the transfer-function F is the *feedforward* part of the controller. Show that

$$y = (H + FD)\bar{y}, \quad H = \frac{GK}{1+GK}, \quad D = \frac{G}{1+GK}.$$

How does the choice of the feedforward term F affect closed-loop stability? Name one advantage and one disadvantage of this scheme when compared with the standard

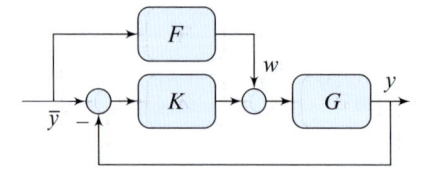

Figure 8.23 Diagram for P8.18.

diagram of Fig. 8.1 if you are free to pick any suitable feedforward, F, and feedback, K, transfer-functions.

8.19 With respect to P8.18 and the block diagram in Fig. 8.23 show that if
$$F = G^{-1}$$
then $y = \bar{y}$ regardless of K. Is it possible to implement the choice $F = G^{-1}$ (a) when G is not asymptotically stable, (b) when G has a zero on the right-hand side of the complex plane, or (c) when G has a delay in the form $G(s) = e^{-s\tau}\tilde{G}(s)$, $\tau > 0$? Compare with the open-loop solution discussed in Chapter 1.

8.20 Let
$$G(s) = \frac{s+1}{s-1}, \quad K(s) = K.$$
Select K and design F in the block-diagram in Fig. 8.23 so that $y = \bar{y}$. If $\bar{y}$ is a unit step, what is the corresponding signal w?

8.21 Show that if
$$F(j\omega) = \lim_{s \to j\omega} G(s)^{-1} W(s) = \alpha e^{j\theta}$$
is finite then it is always possible to calculate the gain K and the time-delay $\tau \geq 0$ of the feedforward controller
$$F(s) = Ke^{-\tau s}$$
so that $F(j\omega) = \alpha e^{j\theta}$.

8.22 Repeat P8.21 for a feedforward controller:
$$F(s) = \frac{\beta}{(s+\alpha)^k}, \quad k \geq 1.$$
Produce an example in which it is necessary to use a value of $k > 1$. Explain why k never has to be greater than 2.

8.23 Show that any rational transfer-function without poles on the imaginary axis can be factored as the product $G = U\tilde{G}$, where $\tilde{G}$ has all its poles and zeros on the left-hand side of the complex plane, that is, $\tilde{G}$ is *minimum phase*, and U is a product of factors of the form
$$U(s) = \pm \prod_i \frac{s + p_i}{s - p_i}. \tag{8.31}$$
Prove also that both $\tilde{G}$ and its inverse $\tilde{G}^{-1}$ are asymptotically stable.

8.24 Transfer-functions of the form (8.31) are known as *all-pass*. Show that $\|U\|_\infty = 1$ and that
$$\|UT\|_2 = \|T\|_2, \quad \|UT\|_\infty = \|T\|_\infty,$$
where T is any transfer-function that is asymptotically stable. *Hint: Use P8.7.*

8.25 Show that for any rational transfer-function, G, the poles of the transfer-function

$$\tilde{G}(s) = G(-s)$$

lie on the mirror image of the poles of G with respect to the imaginary axis.

8.26 Show that any rational transfer-function without poles on the imaginary axis can be factored as the sum $G = G_+ + G_-$, where G_- and $\tilde{G}_+$ are asymptotically stable. *Hint: Expand in partial fractions and use P8.25.*

8.27 Let $G = G_+ + G_-$ be as in P8.26. Assume that $\|G_+\|_2$ and $\|G_-\|_2$ are bounded. Show that

$$\int_{-\infty}^{\infty} G_+(j\omega)^* G_-(j\omega) d\omega = 0.$$

Use this result to prove that

$$\|G\|_2^2 = \|G_+\|_2^2 + \|G_-\|_2^2,$$

which is analogous to Pythagoras' theorem. *Hint: Use the fact that $G_+(j\omega)^* = \tilde{G}_+(j\omega)$, convert the integral to a contour integral and use Theorem 3.1.*

8.28 Let X and Y be asymptotically stable. Factor $Y = U\tilde{Y}$ as in P8.23 and show that

$$\|X - FY\|_2 = \|Z - F\tilde{Y}\|_2,$$

where $Z = U^{-1}X$. Explain why Z is in general not asymptotically stable. Use P8.26 and P8.27 to factor $Z = Z_+ + Z_-$ and prove that

$$\|X - FY\|_2^2 = \|Z_+\|_2^2 + \|Z_- - F\tilde{Y}\|_2^2$$

if F is asymptotically stable.

8.29 Let $X = WS$ and $Y = GS$ and use P8.28 to calculate (8.25) and (8.26) when W and G are as in (8.22) and $S = (s+1)^{-1}$.

8.30 Assume that X and Y are asymptotically stable and that the only zero of Y on the right-hand side of the complex plane is $s_0 > 0$. Use the maximum modulus principle (see P8.1) to show that

$$\|X - FY\|_\infty \geq |X(s_0)|$$

when F is asymptotically stable. Furthermore, prove that

$$F = Y^{-1}(X - X(s_0))$$

is asymptotically stable and is the optimal solution to problem (8.28). *Hint: Use the right-hand side of the complex plane as the open set D then show that $\|X - FY\|_\infty = |X(s_0)|$ when F is as above.*

8.31 Let $X = WS$ and $Y = GS$ and use P8.30 to verify the value of F given in (8.27) when W and G are as in (8.22) and $S = (s+1)^{-1}$.

8.32 Show that

$$\tilde{e} = \left(X - \frac{WKY}{1+KG}\right)w$$

in the diagram of Fig. 8.21 and verify (8.29) when $W = X$ and $G = Y$.

8.33 Consider the diagram in Fig. 8.21. Let X and Y be represented by the state-space equations

$$\dot{x} = Ax + Bw,$$
$$y = Cx + Dw,$$
$$z = Ex.$$

When $W = X$ and $G = Y$ the relationship among signals $\hat{w}, \hat{y},$ and $\hat{z}$ can be represented by the state-space equations

$$\dot{\hat{x}} = A\hat{x} + B\hat{w},$$
$$\hat{y} = C\hat{x} + D\hat{w},$$
$$\hat{z} = E\hat{x},$$

where the matrices (A, B, C, D, E) are equal to the ones in the representation of X and Y. Use the remaining signals in the diagram to show that when K is a scalar, i.e. $K(s) = K$, then

$$\dot{\hat{e}} = (A + BKC)\hat{e} + Bw,$$
$$\tilde{e} = E\hat{e},$$
(8.32)

where $\hat{e}$ is the *state estimation error* vector,

$$\hat{e} = x - \hat{x}.$$
(8.33)

What conditions are required in order for the system described by the state-space equations (8.32) to be asymptotically stable? Is it necessary that $W = X$ and $G = Y$ be asymptotically stable as well?

8.34 Consider the same setup as in P8.33, but instead of making $W = X$ and $G = Y$, let the relationship among signals $\hat{w}, \hat{y},$ and $\hat{z}$ be described by the state-space equations

$$\dot{\hat{x}} = A\hat{x} + F(y - \hat{y}),$$
$$\hat{y} = C\hat{x} + D\hat{w},$$
$$\hat{z} = E\hat{x},$$

where the matrices (A, B, C, D, E) are equal to the ones in the representation of X and Y and the vector F has compatible dimensions. This filter is known as a *state-observer*. Construct a block-diagram representing the connection of the system and the above filter and show that

$$\dot{\hat{e}} = (A + FC)\hat{e} + Bw,$$
$$\tilde{e} = E\hat{e},$$
(8.34)

where $\hat{e}$ is the state estimation error vector given in (8.33). What conditions are required in order for the system described by the state-space equations (8.32) to be asymptotically stable? Is it necessary that X and Y be asymptotically stable as well? Compare your answer with that for P8.33.

8.35 In P7.12, you designed an I controller for the simplified model of a rotating machine driven by a belt without slip as in Fig. 2.18(a). Using the data and the controller you designed in P7.12, plot the Bode diagram of the closed-loop sensitivity transfer-function and comment on the relationship of this plot, the closed-loop poles, and the stability margins you calculated in P7.12.

8.36 Repeat P8.35 for the PI controller you designed in P7.13. Compare your answer with that for P8.35.

8.37 In P7.15, you designed a dynamic feedback controller for the simplified model of the elevator in Fig. 2.18(b). Using the data and the controller you designed in P7.15, plot the Bode diagram of the closed-loop sensitivity transfer-function and comment on the relationship of this plot, the closed-loop poles, and the stability margins you calculated in P7.15.

8.38 In P7.17, you designed a dynamic feedback controller for the simplified model of the mass–spring–damper system in Fig. 2.19(b). Using the data and the controller you designed in P7.17, plot the Bode diagram of the closed-loop sensitivity transfer-function and comment on the relationship of this plot, the closed-loop poles, and the stability margins you calculated in P7.17.

8.39 In P7.18, you designed a dynamic feedback controller for the simplified model of the mass–spring–damper system in Fig. 2.20(b). Using the data and the controller you designed in P7.18, plot the Bode diagram of the closed-loop sensitivity transfer-function and comment on the relationship of this plot, the closed-loop poles, and the stability margins you calculated in P7.18.

8.40 Repeat problem P8.39 for the controller you designed in P7.19. Compare your answer with that for P8.39.

8.41 In P7.29, you designed a dynamic velocity feedback controller for the simplified model of the rotor of the DC motor in Fig. 2.24. Using the data and the controller you designed in P7.29, plot the Bode diagram of the closed-loop sensitivity transfer-function and comment on the relationship of this plot, the closed-loop poles, and the stability margins you calculated in P7.29.

8.42 Repeat P8.41 for the position controller you designed in P7.30.

8.43 Repeat P8.41 for the torque controller you designed in P7.31.

8.44 In P7.32, you designed a dynamic feedback controller to control the temperature of the water in a water heater. Using the data and the controller you designed in P7.32, plot the Bode diagram of the closed-loop sensitivity transfer-function and comment on the relationship of this plot, the closed-loop poles, and the stability margins you calculated in P7.32.

8.45 Repeat P8.44 for the controller you designed in P7.33. Compare your answer with that for P8.44.

8.46 Repeat P8.44 for the controller you designed in P7.34. Compare your answer with those for P8.44 and P8.45.

8.47 In P7.35, you designed a dynamic feedback controller for the simplified model of a satellite orbiting earth as in Fig. 5.18. Using the data and the controller you designed in P7.35, plot the Bode diagram of the closed-loop sensitivity transfer-function and comment on the relationship of this plot, the closed-loop poles, and the stability margins you calculated in P7.35.

References

[AM05] Brian D. O. Anderson and John B. Moore. *Optimal Filtering*. New York, NY: Dover Publications, 2005.

[ÅM08] Karl J. Åström and Richard M. Murray. *Feedback Systems: An Introduction for Scientists and Engineers*. v2.10d. Princeton, MA: Princeton University Press, 2008.

[BB91] Stephen P. Boyd and Craig H. Barratt. *Linear Controller Design: Limits of Performance*. Englewood Cliffs, NJ: Prentice-Hall, 1991.

[BC14] James Brown and Ruel Churchill. *Complex Variables and Applications*. Ninth Edition. New York, NY: McGraw-Hill, 2014.

[BD12] William E. Boyce and Richard C. DiPrima. *Elementary Differential Equations*. Tenth Edition. New York, NY: Wiley, 2012.

[Bro+07] Bernard Brogliato, Rogelio Lozano, Bernhard Maschke, and Olav Egeland. *Dissipative Systems Analysis and Control: Theory and Applications*. Second Edition. London: Springer, 2007.

[Bro15] Roger Brockett. *Finite Dimensional Linear Systems*. Classics in Applied Mathematics. Philadelphia, PA: SIAM: Society for Industrial and Applied Mathematics, 2015.

[Che98] Chi-Tsong Chen. *Linear System Theory and Design*. Third Edition, Oxford Series in Electrical and Computer Engineering. Oxford: Oxford University Press, 1998.

[Chu72] Ruel Churchill. *Operational Mathematics*. Third Edition. New York, NY: McGraw-Hill, 1972.

[DB10] Richard C. Dorf and Robert H. Bishop. *Modern Control Systems*. Twelfth Edition: Menlo Park, CA: Pearson, 2010.

[DFT09] John C. Doyle, Bruce A. Francis, and Allen R. Tannenbaum. *Feedback Control Theory*. New York, NY: Dover Publications, 2009.

[Fey86] Richard P. Feynman. "Appendix F – Personal Observations on Reliability of Shuttle." In *Report of the Presidential Commission on the Space Shuttle Challenger Accident*. NASA, 1986. URL: http://history.nasa.gov/rogersrep/genindex.htm.

[FPE14] Gene F. Franklin, J. David Powell, and Abbas Emami-Naeini. *Feedback Control of Dynamic Systems*. Seventh Edition. Menlo Park, CA: Pearson, 2014.

[GGS01] Graham C. Goodwin, Stefan F. Graebe, and Mario E. Salgado. *Control System Design*. Upper Saddle River, NJ: Prentice-Hall, 2001.

[GKC03] Keqin Gu, Vladimir Kharitonov, and Ji Chen. *Stability of Time-Delay Systems*. Boston, MA: Birkhäuser, 2003.

[GO03] Bernard R. Gelbaum and John M. H. Olmsted. *Counterexamples in Analysis*. New York, NY: Dover Publications, 2003.

[Gro14] Samantha Grossman. *1 in 4 Americans Apparently Unaware the Earth Orbits the Sun* 2014. URL: http://ti.me/NVEICn.

[Jaz08] Reza N. Jazar. *Vehicle Dynamics: Theory and Applications*. New York, NY: Springer, 2008.

[Jen14] Sharmaine Jennings. *Galileo's Work on Projectile Motion*. 2014. URL: http://galileo.rice.edu/lib/student_work/experiment95/paraintr.html.

[Kai80] Thomas Kailath. *Linear Systems*. Englewood Cliffs, NJ: Prentice-Hall, 1980.

[KF75] Andrei N. Kolmogorov and Sergei V. Fomin. *Introductory Real Analysis*. New York, NY: Dover Publications, 1975.

[Kha96] Hassan K. Khalil. *Nonlinear Systems*. Second Edition. Englewood Cliffs, NJ: Prentice-Hall, 1996.

[KSH00] Thomas Kailath, Ali H. Sayed, and Babak Hassibi. *Linear Estimation*. Upper Saddle River, NJ: Prentice-Hall, 2000.

[Lat04] Bhagawandas P. Lathi. *Linear Systems and Signals*. Second Edition. Oxford: Oxford University Press, 2004.

[LeP10] Wilbur R. LePage. *Complex Variables & Laplace Transform for Engineers*. New York, NY: Dover Publications, 2010.

[Lju99] Lennart Ljung. *System Identification: Theory for the User*. Second Edition. Englewood Cliffs, NJ: Prentice-Hall, 1999.

[LMT07] Kent H. Lundberg, Haynes R. Miller, and David L. Trumper. "Initial Conditions, Generalized Functions, and the Laplace Transform: Troubles at the Origin." In *IEEE Control Systems Magazine* **27**(1) (2007), pp. 22–35.

[Neh52] Zeev Nehari. *Conformal Mapping*. New York, NY: McGraw-Hill, 1952.

[Olm61] John M. H. Olmsted. *Advanced Calculus*. New York, NY: Appleton–Century–Crofts, 1961.

[PSL96] John J. Paserba, Juan J. Sanchez-Gasca, and Einar V. Larsen. "Control of Power Transmission." In *The Control Handbook*. Edited by William S. Levine. Boca Raton, FL: CRC Press, 1996, pp. 1483–1495.

[PW34] Raymond E. A. C. Paley and Norbert Wiener. *Fourier Transforms in the Complex Domain*. New York, NY: American Mathematical Society, 1934.

[Son08] Eduardo D. Sontag. "Input to State Stability: Basic Concepts and Results." In *Nonlinear and Optimal Control Theory*. Edited by P. Nistri and G. Stefani. Lecture Notes in Mathematics. New York, NY: Springer, 2008, pp. 163–220.

[Ste+03] Garry M. Steil, K. Rebrin, R. Janowski, C. Darwin, and M. F. Saad. "Modeling β-Cell Insulin Secretion – Implications for Closed-Loop Glucose Homeostasis." In *Diabetes Technology & Therapeutics* **5**(6) (2003), pp. 953–964.

[Tal07] Nassim Nicholas Taleb. *The Black Swan: The Impact of the Highly Improbable*. New York, NY: Random House, 2007.

[Vid81] Mathukumalli Vidyasagar. *Input–Output Analysis of Large-Scale Interconnected Systems*. Lecture Notes in Control and Information Sciences. New York, NY: Springer, 1981.

[Vid93] Mathukumalli Vidyasagar. *Nonlinear Systems Analysis*. Englewood Cliffs, NJ: Prentice-Hall, 1993.

[Wae91] Bartel L. van der Waerden. *Algebra*. Vol. I. New York, NY: Springer, 1991.

[WW06] Orville Wright and Wilbur Wright. "Flying Machine." US Patent No. 821.383. 1906. URL: http://airandspace.si.edu/exhibitions/wright-brothers/online/images/fly/1903_07_pdf_pat821393.pdf.

[ZDG96] Kemin Zhou, John C. Doyle, and Keith Glover. *Robust and Optimal Control*. Englewood Cliffs, NJ: Prentice-Hall, 1996.

Index

Abel–Ruffini theorem, 63
amplifier, 132
analog computer, 133
analytic function, 48, 57, 62, 221, 231, 260
   Cauchy–Riemann conditions, 57
   entire, 58
   meromorphic, 218
approximation, 4, 56, 141, 168, 203, 216
argument principle, 218, 223, 226, 229
   for negatively oriented contour, 220

ballcock valve, 34
bandwidth, 242, 258, 259
BIBO, *see* bounded-input–bounded-output
block-diagram, 1, 17
   all stabilizing controllers, 283
   closed-loop control, 8, 91, 255
   closed-loop with input disturbance, 12, 31
   closed-loop with input disturbance and measurement noise, 105, 138
   closed-loop with reference, input disturbance, measurement noise, and feedback filter, 113
   closed-loop with time-delay, 265
   differential equation, 20
   feedforward control, 278, 279
   filter as feedback system, 282
   filtering, 282
   first-order differential equation, 127
   integrator, 19
   linear control of nonlinear systems, 153
   open-loop control, 7
   open-loop with input disturbance, 12
   pendulum, 144
      proportional control, 170
      robust control, 262
      robust tracking, 269
   phase disturbance, 237
   proportional–derivative control, 172
   proportional–integral control, 99
   proportional–integral–derivative control, 174
   robust analysis, 262
   second-order differential equation, 127
   series connection, 91
   system with input disturbance, 11
   system with measured disturbance, 278
Bode diagram, 185, 201, 216, 232, 236, 255
   asymptotes, 201, 203, 206, 232
   first-order model, 203
   magnitude, 201, 202
   phase, 201, 202
   poles and zeros at the origin, 208
   rational transfer-function, 202
   second-order model, 205, 207
Bode's sensitivity integral, 258
bounded signals, 69, 75, 93, 114, 137, 264

capacitor, 52, 126, 132
   as differentiator, 130
   as integrator, 133
   impulse, 131
   losses, 131
car steering, 148
   equilibrium trajectory, 149
   linearized model, 150
   non-minimum-phase model, 216
   nonlinear model, 149
Cauchy's residue theorem, 57, 59, 221, 260
causality, 54, 62
   Paley–Wiener, 62
characteristic equation, 21, 175, 228
   closed-loop, 176
   first-order, 94, 98
   rational function, 63
   second-order, 98, 100
   state-space, 136
   transcendental, 226
charge, 126
circle criterion, *see* robustness
closed-loop control, 8, 17
   internal stability, 115
communication channel, 282
complementary sensitivity, 11, 92, 111
continous-time model, 17

Index

contour, 58, 62, 218
  for assessing stability, 223
  for stability analysis, 260
  indentation, 226, 231, 243
  inverse Laplace transform, 60
  orientation, 58, 62, 219, 220
controllability, 139–141, 143, 262, 264
convolution, 50, 53, 55, 56, 69
cruise control
  closed-loop frequency response, 101, 108
  closed-loop sensitivity, 111
  differential equation, 19
  disturbance rejection
    constant slope, 31
    rolling hills, 109
  impulse response, 56
  integral control, 96, 98, 107, 169
    root-locus, 179
  internal stability, 116
  nonlinear model, 28
  proportional control, 24, 92, 98, 107
    root-locus, 178
    step response, 94
  proportional–integral control, 99, 108, 116, 169, 171
    nonlinear response, 102
    pole–zero cancellation, 116
    root-locus, 182
    step response, 101
  ramp response, 66, 73
    steady-state, 67
  saturation, 3
  sensitivity, 11, 25
  static model, 6
  step response, 21, 64, 73
  transfer-function, 56, 92
curve fitting, 3, 6, 23
  least-squares, 3, 4

damper, 153
damping, 35, 150, 167, 174, 189, 239, 270
  overdamped, 167
  ratio, *see* second-order model, 257
  underdamped, 167, 191
dB, *see* decibel
DC gain, 106, 202
DC motor, 104, 144
dead-zone, 104
decibel, 186, 201, 203
derivative control, 165
determinant, 136
differential equations
  homogeneous solution, 21
  initial conditions, 21
  linear, 22, 24, 56
  nonlinear, 28, 126

  ordinary, 17, 19, 22, 24, 29, 56
  particular solution, 20
  Runge–Kutta, 29
differentiator, 130, 171
  realization, 130
digital, 6, 17
discrete-time model, 17
  aliasing, 18
  sampling, 17, 127
distribution, 51
disturbance, 11, 17, 105, 151, 174, 261, 278
  input, 30
  measured, 278
  rejection, 30, 35, 280, 281
    asymptotic, 35
dynamic model, 52

eigenfunction, 75
eigenvalue, 136, 142
eigenvector, 136
equilibrium, 142, 148, 150
  closed-loop, 152
  point, 141
  trajectory, 142, 149
error signal, 7, 9, 24, 189, 255
  with measurement noise, 105
estimation, 140
estimator, 282
examples
  car steering, *see* car steering
  cruise control, *see* cruise control
  inverted pendulum, *see* pendulum in a cart
  pendulum, *see* pendulum
  pendulum in a cart, *see* pendulum in a cart
  toilet water tank, *see* toilet water tank
experiments, 1, 6, 18, 19, 22, 23, 28, 47, 56, 74, 154

feedback, 5, 9, 20
  controller, 7
  in state-space, 138
feedforward, 255, 278–280, 282
filter, 113, 152, 279, 281
  Kalman, 282, 283
filtering, 255, 282
first-order model
  Bode diagram, 202
  characteristic equation, 21
  rise-time, 22
  time-constant, 22
Fourier series, 54
Fourier transform, 47, 51, 54, 61, 62
frequency domain, 47, 49, 92
frequency response, 73, 74, 94, 201, 255
  magnitude, 74
  phase, 74

Index

gain, 8, 24, 26, 97

harmonic oscillator, 167
Hurwitz, 136, 137, 139, 142, 145, 153

I controller, *see* integral control
impulse response, 54, 69
impulse signal, 51, 52, 77, 173
   in capacitors, 131
   sifting property, 77
initial conditions
   state, 135
input saturation, 102, 153
input signal, 1, 5, 17
integral action, 33
integral control, 110, 151
   nonlinear model, 153
   transfer-function, 96
   wind-up, 103, 104
integrator, 19, 95, 96, 110, 126, 153, 171, 189, 242, 269
   gain, 97
   in the controller, 35
   in the system, 34, 35
   initial conditions, 128
   realization, 133
   trapezoidal rule, 127
inverted pendulum, *see* pendulum in a cart

Kalman filter, *see* filter
kinematic model, 150

lag control, 210
Laplace transform, 17, 25, 47, 51, 91
   Bromwich integral, 61, 62
   convergence, 48, 63, 70, 92
      boundedness, 70
   exponential growth, 47, 48
   inverse, 49, 51, 57, 60, 71, 166
      causality, 62
      from residues, 60
      rational function, 63
      unicity, 61
   table of pairs, 48
   table of properties, 50
Laurent series, 58, 62
lead control, 185, 208
   transfer-function, 185
Lebesgue integral, 75
linear model, 4, 47
linearity, 50, 52, 128
linearized model, 9, 126, 141, 150, 151
   time-varying, 142
loop transfer-function, 176, 185, 189, 227, 234, 236, 238, 243–245, 255, 258, 259, 265, 267, 274
Lyapunov, 141, 145

linearized model stability, 141
Lyapunov function, 145, 264

margin
   gain, 235, 236, 243
   phase, 235, 237, 243
   stability, 238, 257
mass, 18, 126, 142, 143, 146, 150, 279
mass matrix, 147
matrix exponential, 137
maximum modulus principle, 260
measurement, 7, 140, 184, 278
measurement noise, 105, 111
meromorphic, *see* analytic function
MIMO, *see* multiple-input–multiple-output
minimum-phase model, 214, 240
model
   dynamic, 17, 18
   static, 17, 19
modulator, 53
moment of inertia, 126, 143, 144, 146, 150
multiple-input–multiple-output, 135, 138, 140, 283

neighborhood, 48, 57, 141
Newton's laws, 18, 30, 140, 142, 143
noise signal, 282
   measurement, 256
nominal model, 5, 6
non-minimum-phase model, 205, 214, 258, 279
   step response, 214
   time-delay, 216
nonlinear model, 4, 9, 28, 102, 141
   integral control, 153
   linear control, 150
nonlinearity, 154, 262, 266, 270, 271, 274
norms of signals and systems, 75
   1-norm, 76
   2-norm, 77, 262
   H_2 norm, 77, 281, 283
   H_∞ norm, 77, 238, 262, 265, 271, 281, 283
   ∞-norm, 76
   p-norms, 75
Nyquist diagram, 201, 229, 232, 235–238, 255, 265, 273, 274
Nyquist stability criterion, 226–229, 236, 266, 277

observability, 139–141, 143, 262, 264
   Gramian, 137, 264
open-loop control, 6
   internal stability, 114
optimization, 281, 282
oscillations, 96, 99, 104, 108–110, 148, 166, 191, 257
output signal, 1, 5, 17
overshoot, 103, 104, 117, 167

Index

P controller, *see* proportional control
Padé approximation, *see* time-delay
Parseval's theorem, 77
PD controller, *see* proportional–derivative control
pendulum, 143, 183, 265
   block-diagram, 144
   Bode diagram, 239, 242
   data, 183
   direct design with root-locus, 191
   equilibrium, 144
   frequency-domain control, 238
   lead control, 185
      frequency response, 191
      root-locus, 185, 187
      with integral action, 189, 191
   linearized model, 145
   loop transfer-function, 240
   margins, 241, 245
   nonlinear model, 144
      simulation, 271
   Nyquist diagram, 240, 241, 243
   proportional control, 151, 169, 170
   proportional–derivative control, 172, 183
      root-locus, 183
   proportional–integral control, 184
      root-locus, 184
   robust control, 266
      circle criterion, 274
   robust tracking, 269
   sensitivity, 257
   transfer-function, 145
pendulum in a cart, 146
   equilibrium, 147
   inverted, 147
   linearized model, 147
   nonlinear model, 146
   vector second-order model, 146
physical realization, 5, 20, 112, 133
PI controller, *see* proportional–integral control
PID controller, *see* proportional–integral–derivative control
piecewise
   continuous, 47, 49, 222
   smooth, 47, 51, 52, 222
Plancherel's theorem, 77
polar plot, 217, 223, 273, 277
pole–zero cancellation, 56, 95, 97, 100, 106, 108, 111, 115, 178, 182, 187, 234, 279, 280
   in open-loop, 112
   in state-space, 139
poles
   complex-conjugate, 68, 69, 115, 165
   definition, 58
   dominant, 171, 191
   multiplicity, 63, 66, 181

   rational function, 63
   unstable, 113
positive-real, *see* transfer-function
proportional control, 8
   transfer-function, 24
proportional–derivative control, 172
   differential equation, 173
proportional–integral control, 99, 110
   implementation with circuits, 133
   transfer-function, 99
proportional–integral–derivative control, 165, 174, 189, 281
   transfer-function, 175
pulse-width modulation, 105
PWM, *see* pulse-width modulation

ramp signal, 52, 66, 97
rational function, 57, 63, 93
   partial fractions, 65, 67, 71, 221
   poles, *see* poles
   proper, 57, 65, 138, 175
      realization, 130
   strictly proper, 57, 63, 64, 138, 178, 180, 234
   zeros, *see* zeros
realizability, 281, 282
reference signal, 6, 17, 93, 255
   sinusoidal function, 94
   step function, 94
regulation, 32, 107, 189
regulator, 152, 153
relative degree, 258, 259
residue, 49, 57, 59, 62, 71, 166, 221
   rational function, 63
resistor, 132
rise-time, 22, 26, 98, 100, 101, 103
robust control, 262, 266, 283
robustness, 154, 170, 236–238, 243, 255, 258, 261, 265, 266, 279
   circle criterion, 267, 273
   small gain, 262, 271
root-locus, 165, 175, 227, 234, 236, 237, 259
   asymptotes, 180, 185, 187, 190
   real roots, 179
   rules, 176, 177

saturation, 103, 104, 154
   input, 26, 28, 153
   nonlinearity, 28
second-order model, 165
   Bode diagram, 205
   characteristic equation, 165
   damping ratio, 165, 167, 206
   differential equation, 127
   discriminant, 99, 100
   natural frequency, 165, 206, 207
      damped, 167

peak time, 167
percentage overshoot, 168
poles, 165
realization, 129
resonance frequency, 207
rise-time, 168
settling-time, 168
simple pendulum, 143
state-space realization, 134
step response, 167
time-constant, 168
vector form, 146, 147
sensitivity, 11, 97, 110, 238, 255, 278, 283
  peaking, 258, 267
  poles, 97
  rational, 97
  transfer-function, 93
  zeros, 97, 106
sensor, 24, 265
simple pendulum, *see* pendulum
simulation, 26, 102, 126, 133, 271
single-input–single-output, 135, 138, 176, 208
singularity, 57, 58, 62
  isolated, 58
sinusoidal signal, 73
  amplitude, 73, 130
  phase, 73
SISO, *see* single-input–single-output
small gain, *see* robustness
stability, 69, 93, 112, 201, 218, 223
  asymptotic, 69–71, 93, 114, 137, 141, 224
  bounded-input–bounded-output, 69, 75, 264, 269
  closed-loop, 229, 235
  global, 264
  internal, 114, 136, 139, 283
  stabilizing controllers, 283
  unstable, 74, 99, 148, 151
state-space model, 126, 269
  characteristic equation, 136
  controllability, *see* controllability
  differential equations, 134
  feedback connection, 138
  impulse response, 136
  initial conditions, 135, 137
  matrix exponential, 137
  minimal realization, 139, 140, 262, 273
  observability, *see* observability
  poles, 136
  realization, 282
  state vector, 135, 141
  strictly proper, 136

  transfer-function, 135
steady-state response, 71–73
step response
  experiment, 23
step signal, 50–52, 111, 281, 283
stepper motor, 104
summer, 128
superposition, 129, 166
system, 1
system identification, 24, 56

Taylor series, 11, 57, 141
time domain, 47, 49, 92
time-constant, 19, 22, 26, 28, 98, 100, 101, 103
time-delay, 216, 237, 265, 266
  Padé approximation, 216
  stability, 226
time-invariant model, 47, 53, 91, 92, 150, 201, 223
time-varying model, 53, 142, 150
  impulse response, 55
  stability, 142
toilet water tank, 33, 126
  differential equation, 33
  disturbance rejection, 110
  integrator, 33
  proportional control, 34, 95
  transfer-function, 95
tracking, 6, 8, 93, 153, 184, 189, 255
  asymptotic, 35, 94, 95, 97, 98, 100, 107, 256, 269, 280
  asymptotic with input-disturbance rejection, 106–108
  feedforward, 279
tracking error, 25, 93, 105, 278
  steady-state, 94
transfer-function, 8, 47, 91
  from differential equation, 55
  from impulse response, 55
  from input disturbance to output, 106
  from reference to control, 100
  from reference to error, *see* sensitivity
  from reference to output, *see* complementary sensitivity
  from state-space, 135
  initial conditions, 55, 56
  order, 57
  poles, *see* poles
  positive-real, 274
  rational, *see* rational function
  zero initial conditions, 56, 74
  zeros, *see* zeros
transient, 97
transient response, 67, 71, 180
triangle inequality, 76, 256, 264

uncertainty, 154, 262, 265, 271, 274, 279
   structured, 271

washout filter, 152
water heater, 1

waterbed effect, 258
Wright brothers, 9, 12

zeros
   complex-conjugate, 115
   rational function, 63